Experientia Supplementum

Volume 105

For further volumes:
http://www.springer.com/series/4822

Christopher P. Toseland • Natalia Fili
Editors

Fluorescent Methods for Molecular Motors

Springer

Editors
Christopher P. Toseland
Chromosome Organisation and Dynamics
Max-Planck Institute of Biochemistry
Martinsried
Germany

Natalia Fili
Department of Cellular Physiology
Ludwig-Maximilians-Universität München
Munich
Germany

ISSN 1023-294X
ISBN 978-3-0348-0855-2 ISBN 978-3-0348-0856-9 (eBook)
DOI 10.1007/978-3-0348-0856-9
Springer Basel Heidelberg Dordrecht New York London

Library of Congress Control Number: 2014947103

Printed on acid-free paper

Springer is part of Springer Science+Business Media (www.springer.com)

For Thalia. . .You remind us of how much we can learn every day.

Preface

The term "Fluorescence" was initially coined by George Gabriel Stokes in 1852, but the phenomenon had already been recorded for over 300 years. Our knowledge of fluorescence expanded through the pioneering work of Perrin, Jablonski and Förster and Weber. This work coincided with the synthesis of organic dyes—the fluorophores. Our ways to manipulate fluorophores have grown to suit our needs and our ways to detect them have developed more than could be imagined a few years ago. Almost all laboratories utilize fluorescence, whether it is microscopy or as a reagent. . .we could consider that we now live in the "Age of Fluorescence".

Fluorescence and motor proteins have a long interconnected relationship. Indeed, the progress in fluorescence methods has directly advanced our knowledge of these molecular machines.

Written by experts, this book focuses on the application of fluorescence to study motor proteins. It is intended for a large community of biochemists, biophysicists and cell biologists who study a diverse collection of motor proteins. It can be used by researchers to gain an insight into their first experiments, or by experienced researchers who are looking to expand their research to new areas. Within each chapter you will find valuable advice for executing the experiments, along with detailed background knowledge in order to develop your own experiments.

We thank all the authors for their hard work making this book a reality.

Martinsried, Germany
Munich, Germany

Christopher P. Toseland
Natalia Fili

Contents

Chapter 1
Fluorescence and Labelling: How to Choose and What to Do

Natalia Fili and Christopher P. Toseland

Abstract This chapter provides an overview of fluorescent labelling of different reactants related to the biochemistry of motor proteins. The fluorescent properties of different labels and the advantages and disadvantages of the labelling methods are discussed. This will allow for a careful selection of fluorescent proteins for different applications relating to motor proteins.

Keywords Fluorescence • Single molecule • Motor proteins • Fluorophores • Fluorescent proteins • Quantum dots

Abbreviations

Deac	Diethylaminocoumarin
edaATP	2′(3′)-*O*-[*N*-(2-aminoethyl)-carbamoyl]ATP
FP	Fluorescent protein
FRET	Förster (or fluorescence) resonance energy transfer
Mant	2′(3′)-*O*-(*N*-methylanthraniloyl)
MDCC	7-diethylamino-3-((((2-maleimidyl)ethyl)amino)carbonyl)coumarin
pdaATP	2′(3′)-*O*-[*N*-(3-aminopropyl)carbamoyl]ATP
QDots	Quantum dots
SNR	Signal to noise
sptPALM	Single particle tracking photoactivated localisation microscopy

N. Fili
Department of Cellular Physiology, Ludwig-Maximilians-Universität München, Schillerstrasse. 44, 80336 Munich, Germany
e-mail: natali.fili@med.lmu.de; natali.fili@gmail.com

C.P. Toseland (✉)
Chromosome Organisation and Dynamics, Max-Planck Institute of Biochemistry, Martinsried 82152, Germany
e-mail: toseland@biochem.mpg.de

C.P. Toseland and N. Fili (eds.), *Fluorescent Methods for Molecular Motors*, Experientia Supplementum 105, DOI 10.1007/978-3-0348-0856-9_1,

TIRF	Total internal reflection microscopy
TMR	Tetramethylrhodamine

1.1 Introduction

Fluorescence methods provide rich and varied information on motor proteins. Indeed, due to the activity of motor proteins, fluorescence can be used to investigate all aspects of their activity [1–7]. This book describes a range of fluorescence-based assays for motor proteins which allow the detailed dissection of their molecular mechanisms. The assays can be performed in real time, in solution, with high time resolution and sensitivity down to single molecules. This is exemplified by the chapters of this book which present assays suitable to investigate biochemical cycles, interactions, conformation changes and tracking of motor proteins in vitro and in vivo.

Fluorescence provides a sensitive and rapid readout with high signals. Because emission is in the nanosecond range (Fig. 1.1), compared to many molecular transitions which are in the microsecond to millisecond range, fluorescence does not limit the experimental time resolution. However, detection can be a greater limiting factor, restricting measurements from milliseconds.

A fluorophore reports upon the molecule it is attached to and its environment. In the simplest case, this allows the motor protein, or its substrates, to be visualised or detected. More advanced assays can measure protein–track interactions, conformation changes and entire reaction cycles. This is achieved by measuring fluorescence intensity, anisotropy or fluorescence/Förster resonance energy transfer (FRET). Such measurements can be performed in an ensemble or single-molecule format. Briefly, ensemble measurements offer a greater flexibility with the least requirements in terms of fluorescence properties, whereas single-molecule measurements provide greater sensitivity allowing the detection of rare, or short-lived, states which can be masked by the ensemble averaging [8]. However, there are strict requirements on the selection of the fluorophore.

Obviously, there are potential disadvantages of using fluorescence. This includes difficulty in interpreting the fluorescent signal. Aberrant fluorescent signals can be problematic, especially if it is on the timescale of the biological process. Likewise, a premature loss of the signal can give incomplete description of even mask the biological processes under study. In addition, fluorescent signals cannot be quantified, as it is the case with absorbance measurements. Finally, and most importantly, extrinsic probes have to be added to the system. This is usually achieved through coupling a fluorophore to one of the reactants. However, this can dramatically perturb the system, and therefore, it is vitally important to select the most suitable label, as discussed here and in [9].

A wide range of approaches are currently available for fluorescent labelling depending upon the nature of the target molecule–protein, nucleotide or nucleic acid. Specific labels are optimised for different measurements, based upon their

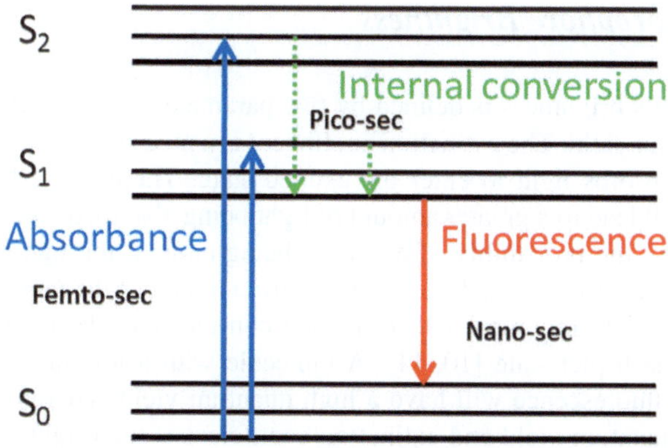

Fig. 1.1 Excitation and emission of a fluorophore. Absorption of light at a particular wavelength causes the fluorophore to move from a ground state (S_0) and become excited in a higher electronic energy level (either S_1 or S_2). Different vibrational levels exist in each electronic energy level. Following this excitation, the fluorophore relaxes to the lowest vibrational state of S_1 through internal conversion in ~10^{-10} s. Following internal conversion, the fluorophore remains in the S_1 energy state for a defined amount of time, termed the fluorescence lifetimes (10^{-9} s). The fluorophore returns to the ground state through release of a photon at a particular wavelength in the process of fluorescence

photophysical properties, size and ease of coupling. Therefore, before selecting a label, it is important to define the type of experiment to be performed. For instance, is it a single molecule or an ensemble measurement? Is the aim of the experiment to measure a conformation change, a FRET signal or visualise a protein? Each category has specific requirements for the label. However, there are some general features. Ideally, a fluorescent label should be small, bright and stable, without causing any perturbation to the biological system.

1.2 Fluorophore Properties

To be able to effectively select the most suitable label for a particular application, it is important to be able to compare the fluorescent candidates. Such comparison can be based several properties which are highlighted below. These properties are all related to the excitation and emission cycle of the fluorophore (Fig. 1.1) [10, 11]. A fluorophore is excited to high-energy singlet states (S_1 and/or S_2) when it absorbs a photon. From the excited state, the molecule must relax to the lowest energy level of S_1 through a non-radiative loss of energy. From this state, the fluorophore can release a photon as fluorescence to return to the ground state. The molecule can also cross from the singlet state to the triplet state (T_1) through intersystem crossing. Relaxation from T_1 back to the ground state occurs through nonfluorescent processes or phosphorescence.

1.2.1 Fluorophore Brightness

A fluorophore's brightness is defined by two parameters, its extinction coefficient and its quantum yield. The extinction coefficient is a measure of the ease in which a fluorophore absorbs light to enter the excited state. Therefore, a high extinction coefficient will lead to a greater amount of light being absorbed. Once a photon has been absorbed, the probability of a photon being emitted through fluorescence is related to the quantum yield. This is derived from a ratio of the rate of fluorescence over all rates of deactivation, which include non-radiative decay and intersystem crossing to the triplet state [10, 11]. A molecule with a low rate of deactivation compared to fluorescence will have a high quantum yield. Therefore, a molecule with a high quantum yield and extinction coefficient is the brightest. To compare the brightness of different fluorophores, the relative brightness of fluorophore is taken by multiplying the extinction coefficient with the quantum yield [12].

1.2.2 Fluorescent Lifetime

The fluorescent lifetime is the average time the molecule spends in the excited state. This is determined by the rates of fluorescence, internal conversion and intersystem crossing to the triplet state. This time is highly variable and dictates how long the molecule has to react with the environment. The lifetime is dependent upon the fluorophore's environment, and corresponding changes in lifetime correlate with intensity changes. Conformation changes or substrate binding can dramatically change the local environment of the fluorophore and therefore its lifetime. As will be described below, the lifetime is an important parameter for fluorescence polarisation/anisotropy measurements.

In single-molecule imaging, the greater the number of photons detected, the greater is the localisation accuracy of the single molecules. Therefore, the fluorescent lifetime is particularly important. A long fluorescence lifetime will reduce the number of excitation and emission cycles, which in turn reduces the overall number of photons detected within a given time frame. For this reason, fluorophores with short lifetimes are preferable (<5 ns) [13].

1.2.3 Stoke's Shift

During the relaxation in S_1, the loss of energy due to non-radiative decay, before the release of a photon, leads to an emission maximum at a lower energy (higher wavelength) than the excitation. The difference between the excitation and emission wavelength is the Stoke's shift, and the degree of shift is related to the molecular structure of the fluorophore [10, 11]. Cy3 and Cy3B are good examples, whereby molecules with a greater internal flexibility (Cy3) lose a high amount of energy to non-radiative decay before release of a photon. Therefore, the Stoke's shift of Cy3 is larger than that of Cy3B, 14 nm and 11 nm, respectively [14–16]. Should the fluorophore become restricted within a protein binding site, then the internal flexibility will be reduced, causing a change in the Stoke's shift.

While the degree of Stoke's shift determines the relative position of the excitation and emission peaks, the shape of the peaks is important to consider. Overall, fluorescent proteins (FP) have broad excitation and emission peaks. Organic fluorophores can have very narrow peaks, but this is dependent upon the specific fluorophore. And quantum dots have very broad excitation peaks but narrow emission peaks. Narrow peaks can be advantageous for selecting emission with minimal bleed-through from the excitation. Furthermore, narrow-defined emission peaks allow multiple colours to be used and selected for individually.

1.2.4 Photostability

Fluorophores will go through many cycles of excitation and emission, but all will eventually be irreversibly deactivated or photobleached. During these cycles, approximately 1:1,000 excitations lead to a transition from the singlet state (S_1/S_2) to the triplet state (T_1) [13]. This state is long lived (millisecond), and the fluorophore will not emit again until it reaches the ground state and is re-excited. These "dark" periods give rise to the phenomena of fluorophore blinking. It is during the long-lived triplet state that the fluorophore can react with oxygen and free radicals leading to irreversible deactivation. Established methods exist to remove oxygen from solution using scavenger enzymes such as glucose oxidase and catalase [17]. Triplet state quenchers, such as Trolox, and redox systems, such as methyl viologen and ascorbic acid, can further reduce the photobleaching rate by quenching the highly reactive state [13, 18, 19].

Fluorophore longevity may range from desired to required depending upon the experiment being performed. In the latter case, the fluorophore has to be chosen carefully to ensure low blinking and bleaching rates. Also, optimisation of the scavenger system and triplet state quencher has to be performed specifically for the fluorophore chosen.

Overall, fluorophore stability is the major requirement for tracking motor proteins. Therefore, the fluorophore must have a low bleaching rate; ideally this would be slower than the detachment rate of the protein from its track. In terms of stability, enhanced fluorescent proteins such as eGFP are brighter, but they can frequently be observed only for 10 s before they are photobleached. Quantum dots (QDots) are the most stable forms of label, since under most circumstances they do not bleach. Organic fluorophores provide intermediates between fluorescent proteins and QDots.

A signature feature of QDots is photoblinking, the random alternation of their fluorescent signal between a bright and a dark state. This temporal loss of trajectory data complicates single-molecule tracking and analysis (for further details, see Chaps. 7 and 10). The past decade, various strategies to synthesise non-blinking QDots have been reported, each with its downsides [20–22]. Alternatively, blinking can be suppressed by the use of thiols [23], which is not compatible with in vivo studies. Intensity fluctuations also occur with fluorescent proteins and organic fluorophores to various extends. This can be overcome with using suitably optimised oxygen scavengers and triplet state quenchers [13].

1.3 Types of Measurement

Fluorescence experiments are typically divided into two classes: those conducted on a large number of molecules, leading to ensemble-averaged fluorescence intensities, or those on the single-molecule level, which reveal intensity distributions over a large number of individually observed fluorophores. Ensemble fluorescence experiments are performed with spectrofluorimeters which can be coupled to stopped-flow devices for time-resolved measurements (Chap. 3). Single-molecule experiments are performed using fluorescence microscopes with laser excitation. Most frequently, this is in the form of total internal reflection microscopy for surface immobilised molecules. Alternatively, confocal microscopy can be used for freely diffusing molecules in solution (Chap. 7).

Among the measurements described below, some may directly utilise the fluorescence of the fluorophore, like, for instance, in the case of measuring intensity changes, visualising and tracking a molecule. Other methods, such as FRET and anisotropy/polarisation, require specialised approaches, analysis and interpretation.

1.3.1 Fluorescence Intensity

To visualise or detect substrate binding, it can be advantageous to have an intensity increase. In bulk measurements, this allows bound and unbound species to be identified, whereas in single-molecule measurements, it allows isolation of protein-bound molecules from background fluorescence. Intensity changes are largely restricted to organic fluorophores. However, some fluorescent proteins do display a degree of environmental fluorescence changes.

Intensity changes relate to environmental sensitivity of the fluorescent probe. The quantum yield and extinction coefficient are frequently affected by environmental factors such as pH, solvents and viscosity. Furthermore, they are also affected by structural features of the fluorophore whereby rigid, planar molecules have a higher fluorescence [11]. Therefore, if a fluorophore becomes restricted in a binding site, an intensity change will be observed. If a fluorophore has a high quantum yield while free in solution, then it is unlikely to undergo large intensity increases. Intensity decreases may also be observed, but increases are always preferred, so that the measurement is not perturbed by photobleaching.

As binding sites are frequently more polar than the solvent and in these sites fluorophores can be restricted to planar conformations, positioning of fluorophores in binding sites can allow detection of conformation changes. This has been harnessed in the design of protein biosensors (Chap. 2). For the coumarin phosphate-binding protein, a conformation change upon binding phosphate is reported by change in the coumarin fluorescence [24]. The change in the solvent exposure is coupled with a structural confinement, resulting in a planar conformation of the fluorophore. The most prominent effect is an eightfold increase in coumarin's quantum yield. Overall, this leads to a large fluorescent increase of up to 13-fold. In a similar way, selecting the appropriate labelling position on a nucleotide or nucleic acid molecule can generate similar responses upon protein–ligand interaction.

1.3.2 Fluorescence/Förster Resonance Energy Transfer

FRET is a mechanism of energy transfer between donor and acceptor fluorophores [25]. Energy transfer occurs when two fluorophores with overlapping donor emission and acceptor excitation spectra are in close proximity (<10 nm). Upon exciting the donor molecule, there is non-radiative energy transfer to the acceptor. Therefore, fluorescence from the acceptor is only observed when the fluorophores are in close proximity. The efficiency of energy transfer is inversely proportional to the sixth power of the distance between donor and acceptor, meaning that FRET is very sensitive to small distance changes [Eq. (1.1)]. More detailed description of this approach is given in Chap. 8.

$$E = \left[1 + \left(\frac{R}{R_0}\right)^6\right]^{-1} \tag{1.1}$$

whereby R is the distance between donor and acceptor fluorophores and R_0 is the Förster distance. This is the distance where the efficiency of energy transfer is 50 %.

In general, one should keep in mind that it is possible that the Stoke's shift of one or both of the fluorophores can change during the conformation change. With FRET measurements, the degree of spectral overlap is directly related to the observed FRET signal; therefore, any changes in Stoke's shift will give false FRET changes. Likewise, fluorescent intensity changes attributed to environmental sensitivity may complicate the FRET signal, making it difficult to interpret the conformation change from additional local rearrangements in the protein.

Fluorescent proteins can be used for FRET measurements which are largely based upon protein–protein interactions. However, they can be used to probe conformation changes if there is a large change in structure [5, 26]. Organic fluorophores can be used to probe intra- and intermolecular conformation changes, as well as ligand binding. QDots cannot be used in FRET assays due to the broad excitation spectrum.

1.3.3 Fluorescence Polarisation/Anisotropy

The application of fluorescent polarisation/anisotropy is based on the fact that molecules will rotate more slowly in solution when bound to a large complex, due to hydrodynamic effects. A fluorescently labelled molecule will have a high fluorescence polarisation/anisotropy when bound to a macromolecule complex than when alone, as the fluorophore is less mobile (Fig. 1.2). Therefore, this method can be applied to measure complex formation (protein to protein, protein to DNA and protein to ATP). This is particularly useful because fluorescence intensity changes do not always occur and sometimes numerous fluorescent phases occur, which makes it difficult to distinguish which phase corresponds to binding or to conformation changes. Fluorescence polarisation/anisotropy circumvents this problem by directly reporting upon complex formation [27].

Fluorescence anisotropy and fluorescence polarisation are frequently used interchangeably. While the values do differ, they essentially report the same process. Anisotropy values are preferred for analysis because they are additive and, therefore, mathematically simpler to handle [27]. This allows free and protein-bound species to be easily calculated from an anisotropy measurement.

Fluorescence anisotropy uses polarised light to selectively excite fluorophores whose absorption transition dipole is parallel to that of the polarised excited light. This process is termed "photoselection". The range of dipoles selected is dependent upon the angle of the dipole in relation to the excitation light. The chance of a dipole absorbing a photon is proportional to the cosine squared of the angle between the excitation light and the dipole. More simply, the greater the angle, the lower the absorbance [10]. Following photoselection, fluorophores can tumble in solution during the excited state and emit light in an alternative plane to that of the excited light plane. When a fluorophore excited by polarised light is immobile during its fluorescence lifetime (typically 0.1–30 ns), the polarisation of the emitted light will also be unchanged. However, if the fluorophore is mobile during this time, the orientation of the molecules will be randomised, and the emitted light will be less polarised. As a rough guide, the molecular weight of the protein in Daltons divided by 1,000 gives an estimation of the rotation time in nanoseconds.

The fluorescence lifetime is important for these measurements. If the lifetime is too short, the molecule will not have the time to rotate, so the emitted light will always appear polarised. However, if the lifetime is too long, the emitted light will always be depolarised. For many biological systems, the fluorescence lifetime should be relatively short. However, if the fluorescently labelled species are very large, then a longer lifetime is required to allow anisotropy measurements. Due to these requirements, it is important to select a suitable fluorophore for these

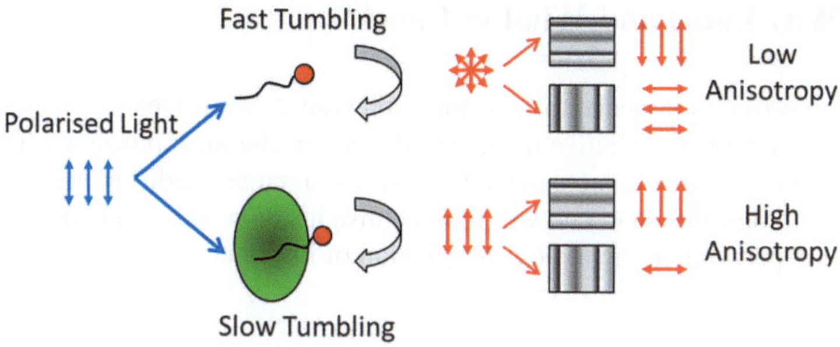

Fig. 1.2 Fluorescence anisotropy. A fluorescently labelled species is excited with vertically polarised light. Emission is measured through polarisers that are parallel and perpendicular to the plane of the excitation. Fluorophores in the excited state are able to tumble in solution within the fluorescence lifetime. The rate of tumbling is related to the size of the macromolecular complex. A large complex, depicted by the protein shown in *green*, has a high anisotropy because the polarised light is not dispersed through rotation

measurements. For instance, fluorescein is a suitable fluorophore for fluorescence anisotropy measurements, as its lifetime (4 ns) is long enough for many biological applications. In addition, linkers that would space the fluorophore away from the protein, ligand or DNA will increase flexibility of the fluorophore. It is useful to think of a fluorophore bound to a linker as a propeller spinning freely. This acts to decrease the rotation correlation time (described below), which will decrease the anisotropy of the bound species [27, 28].

One of the first equations to describe polarisation (P) was proposed by Perrin in 1926 [29]:

$$\frac{1}{P}-\frac{1}{3}=\left(\frac{1}{P_0}-\frac{1}{3}\right)\left(1+\frac{3\tau}{\rho}\right) \tag{1.2}$$

where P_0 is the limiting polarisation of the fluorophore, typically 0.5, ρ is the Debye rotational relaxation time which is dependent upon the volume and shape (molecular weight) of the rotating species, and τ is the fluorescence lifetime. Equation (1.2) is simplified for anisotropy (r) here:

$$\frac{r_0}{r}=\left(1+\frac{\tau}{\tau_c}\right) \tag{1.3}$$

where r_0 is the limiting anisotropy of the fluorophore in solution, typically 0.4, and τ_c is the rotational correlation time ($\rho = 3\tau_c$). Therefore, larger molecules or complexes have a slower rotation correlation time in relation to the fluorescence lifetime and thus a higher anisotropy.

1.4 Why Label and What to Label?

Labelling provides a means to perform fluorescence measurements, which themselves provide a very sensitive method to dissect biochemical processes. The three sections below discuss in more detail the reasons for targeting different components within a biological system and will provide insights on which component it is best to target depending on the biological question of interest.

1.4.1 *Protein Labels*

Labelling the protein is probably the most common application as it is the component most biological questions are focused on. It allows the protein to be visualised in vivo and in vitro for single-molecule measurements, and it also allows for conformations and binding events to be observed.

Proteins have intrinsic fluorescence due to residues such as tryptophan, but this obviously depends upon their content. Tryptophan fluorescence changes upon nucleotide binding have proven to be a useful tool for investigating the ATPase kinetics of myosins [30]. However, these changes are not always present. Furthermore, it may be better to measure the ligand and not the proteins response to the binding because these processes are not necessarily the same. Therefore, extrinsic labels provide far greater advantages for observing biological processes. These include more flexibility in choosing the location of the label and enhanced fluorescence properties suitable for a greater number of techniques [9].

Proteins provide the greatest flexibility when choosing a label, but they are also the most easily perturbed by fluorescent labelling. As will be discussed below (Sect. 1.5), proteins can be coupled to fluorescence proteins, QDots and organic fluorophores via various linkers or tags (Fig. 1.3) [9]. Through specific attachment schemes, labelling can be achieved with a high specificity at various locations. This provides a high degree of freedom for probing biological processes from entire cells to single molecules. However, these locations have to be carefully selected so as not to compromise the activity of the protein. Furthermore, larger labels such as fluorescent proteins and QDots can sterically interfere with the protein activity. Due to the vast possibilities of labels, it is important to determine the best label based upon the application and question. Organic fluorophores are relatively small labels but with good photophysical properties. The most commonly used organic dyes for tracking single motors are tetramethylrhodamine (TMR) [31], Cy3 [32] and Cy5 [33] (Chaps. 9 and 10). These labels are used in combination with fluorescently labelled actin or microtubule tracks (Chap. 5). Organic fluorophores are also ideal candidates for FRET measurements because of their small size and they can be positioned throughout the protein. This flexibility in the position of the labelling also allows conformation changes to be probed with environmentally sensitive fluorophores (Chaps. 2 and 8). Fluorescent proteins are ideal if an accurate

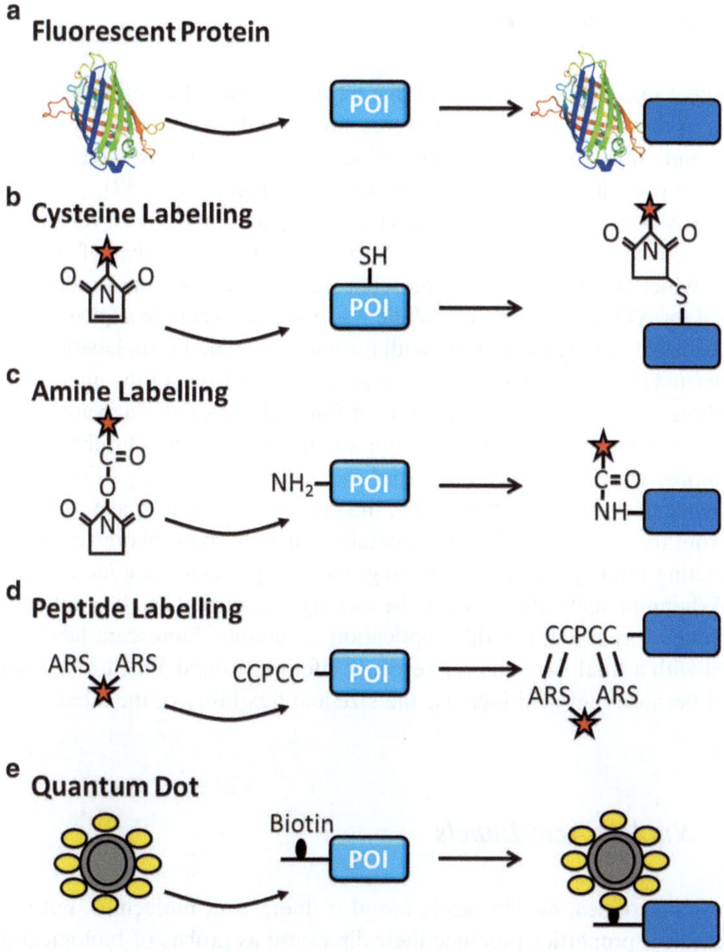

Fig. 1.3 Fluorescent labelling of proteins. (**a**) An example of fluorescent protein conjugation to the protein of interest (POI) (**b**) Cysteine labelling reaction with a maleimide-conjugated fluorophore. (**c**) Amine-reactive succinimidyl-ester conjugated fluorophore and N-terminal amine on POI. (**d**) Peptide labelling example of a self-labelling tetracysteine tag which binds to the bis-ARSenic fluorescein (FlAsH) probe. (**e**) Example of biotinylation protein conjugation with streptavidin-functionalised quantum dot. Adapted from [9]

labelling stoichiometry is important, and they provide the best method for visualising motors in vivo (Chap. 13). Finally, QDots are better for visualising single molecules, as long as their large size does not perturb the activity of the protein they are attached to. QDots can also be used as substitutes for molecular cargo, as described in Chap. 11

1.4.2 Nucleotide Labels

Fluorescent nucleotides are used to dissect the individual components steps of an ATPase cycle (Chap. 4). These steps are fundamental to the activity of the motor protein and, thus, represent crucial biochemical transitions which have to be deciphered to understand the motor protein's mechanism [34–37].

While ATP is a fluorescent molecule, the quantum yield is very low (10^{-4}) [38]. Moreover, the absorbance, at 260 nm, overlaps with that of tryptophan at 280 nm, which would mask the fluorescent signal. Therefore, the structure has to be modified for ATP to be observed. While it is possible to couple a QDot onto ATP, it would be highly likely to interfere with the catalysis. Therefore, labelling strategies are restricted to organic fluorophores. Due to the small size of the nucleotide, small fluorophores are preferred, but the type of fluorophore strictly depends on the type of measurement. Single-molecule measurements require bright, photostable fluorophores which are therefore larger molecules, such as Cy3 [39]. To prevent interference of the label with the ATP, linkers can be used to space the fluorophore away from the nucleotide. Environmentally sensitive fluorophores are very useful for detecting binding and distinguishing between protein-bound/non-bound molecules. Where intensity changes may be lacking, anisotropy can be applied to probe the complex formation. For this application, a suitable fluorescent label should be selected with a small size and fluorescence lifetime around 5 ns. Linkers should be avoided because they will increase the size and flexibility of the label.

1.4.3 Nucleic Acid Labels

As with nucleotides, nucleic acids are also fluorescent molecules, but their poor photophysical properties preclude their direct use as probes of biological activity. Observing DNA/RNA is important for visualising a motor protein's track, binding events, unwinding and conformation changes (Chaps. 6 and 8). Smaller organic fluorophores are preferred labels in order to prevent perturbations. Synthetic oligonucleotide substrates provide the greatest flexibility by allowing multiple forms of reaction chemistries which can be located throughout the strand. Therefore, multiple labels can be coupled as desired through the molecule to probe the biological event of choice.

Fluorophores can be selected for intensity changes and anisotropy measurements. Along with FRET between protein and DNA (Chap. 8), FRET between different labels on the DNA can also be performed to probe DNA conformations (Chaps. 6 and 12).

1.5 Types of Labels and Labelling Strategies

The sections below discuss several types of labels and labelling strategies. The section of organic fluorophores provides a detailed overview of labelling strategies not only for protein targets but also nucleotides and nucleic acids.

1.5.1 Fluorescent Proteins

Molecular motors can be genetically labelled through their N- or C-terminal fusion with genetically encoded fluorescent probes. Genetic labelling presents the unique advantage of absolute labelling specificity. FPs also ensures a defined labelling stoichiometry, which is crucial in single-molecule measurements. Recent decades have witnessed a spectacular increase in the number of bright FPs of various colours, which enables simultaneous visualisation of the motor with its interacting network and partners [40, 41]. Therefore, FPs offer a flexible and easy labelling option, suitable for both in vitro and in vivo experiments.

FPs present disadvantages which one should take into account when designing experiments and in particular single-molecule assays. Compared to organic dyes, FPs are limited by their low brightness. For instance, the relative brightness of eGFP is threefold lower than Cy3B. Another downside of FPs is their low photostability and high photobleaching rate. This is crucial especially in single-molecule imaging, where maximising the signal-to-noise ratio (SNR) often requires imaging at high illumination intensities. In in vitro assays, photostability can be improved by the presence of oxygen scavenger systems (Chap. 7). However, this is not feasible for in vivo imaging. Another disadvantage of FPs is their relatively large size (typically 2–5 nm) [42], which could interfere with the function and properties of the motor of interest. Also, genetic labelling has the risk of generating truncated products rather than the intact labelled motor, due to degradation or partial translation by the expression system. In addition, FPs have variable maturation rates, and some have the tendency to oligomerise [40], which could compromise the function of the labelled motor and generate artefacts in the detection of single molecules and their dynamics. Few commonly used FPs, such eGFP and eYFP, tend to weakly dimerise, and when used, appropriate controls should be made [40]. Alternatively, monomeric FPs or monomeric FPs mutants, such as the A206K eGFP mutant, should be preferred. FPs are also discussed in detail in Chap. 13 and should always be addressed with control experiments before proceeding to single-molecule measurements.

Traditionally, GFP and its brighter enhanced version eGFP have been used for imaging single molecules both in vitro and in vivo [43–46]. However, due to its low quantum efficiency, blinking behaviour (Dickson RM et al. Nature 1997) and photobleaching rate, its suitability is limited, especially for studying long-lasting or highly dynamic events which require high imaging rate. One way to increase the brightness of the label is to use tandem monomeric fluorophores [47], although one should make sure that the considerably large size of the label does not interfere with the motor studied. Alternatively, monomeric FPs, which are more suitable for single-molecule imaging, are currently available in whole range of the visible spectrum [40, 41], such as red mRuby [48], the orange mKO [49] and the yellow–green eYFP. However, there is always a trade-off between the fluorophore's brightness and photostability, and therefore, the suitability depends on the application.

1.5.2 Organic Fluorophores

Synthetic organic fluorophores have been a popular labelling option given their significantly smaller size and their considerably higher brightness and photostability. However, labelling is based on the chemical coupling of the dye to the target, which can be limited by the efficiency of the reaction, its specificity and the resulting stoichiometry. The use of organic fluorophores is not restricted to proteins; therefore, nucleotide and nucleic acid labelling is also highlighted.

1.5.2.1 Protein Labelling

The fluorophore can be either coupled directly on the protein, on natural or genetically introduced cysteine and amine groups, or be conjugated to genetically introduced tags which could be a full enzyme (such as HaloTag, SNAP-tag and CLIP-tag), peptides (such as tetraCys, HexaHis and polyAsp) or unnatural amino acids (UAAs) (Fig. 1.3). Each labelling strategy has its advantages and disadvantages and should be selected depending on the application [9]. However, the use of chemical conjugation inside cells requires the extra step of incorporating the organic label or the labelled protein into the cells by microinjection, electroporation or endocytosis.

Cysteines provide the greatest flexibility for choosing the location of the labelling due to their low frequency. Maleimide-fluorophore conjugates are highly specific for the thiol group of cysteine, and the reactions are rapid (10 min–2 h) in typical buffers with moderate pH and temperatures. Another conjugate, iodoacetamide, is frequently used to target thiols due to their tolerance to reducing agents, such as dithiothreitol [50], which are both used before labelling cysteines to ensure that the residues are in their reduced form. Site-directed mutagenesis can be used to introduce specific sites for labelling or remove additional cysteines.

Amine-reactive conjugates, such as *N*-succinimidyl-esters or isothiocyanates, can be used to label lysine residues or N-terminal amines. Lysine residues are very frequent in proteins, and therefore, labelling is likely to result in binding many fluorophores. However, it is possible to specifically target the N-terminal α-amino group due to a lower p*K*a (p*K*a 7) compared to the amino groups of lysine (p*K*a 10–11). This facilitates successful labelling at a specific but limited location [51, 52].

For proteins sensitive to internal labelling or mutagenesis, many different types of peptide labelling strategies are available. There are two main approaches: (a) label a synthetic peptide, through thiol or amine-reactive conjugates as describe above, which is then attached to the target protein through ligation, and (b) genetically engineer peptide tags onto the target protein which are then reacted with the corresponding fluorophore conjugates. Examples of the latter case include short, self-labelling peptides, for instance, the tetracysteine peptide [53] which reacts with corresponding fluorescent bis-ARSenic fluorescein (FlAsH) probes.

Finally, protein tags such as Halo [54], SNAP [55] and CLIP [56] are self-labelling enzymes that covalently link to fluorescently labelled substrates. These tags are highly specific, and many commercial kits are available, making the process relatively easy.

In general, chemical conjugation requires rigorous control of the labelling. Given that the reactions are unlikely to be 100 % complete, the label protein has to be purified from the free dye and the unlabeled protein, in order to minimise the fluorescent background and to not compromise the experimental conditions. Since labelling is likely to change the electrostatic properties of the protein, ion-exchange chromatography can usually be applied. Also, the low-molecular-weight free dyes can easily be removed by size-exclusion chromatography. In addition, indirect labelling methods through peptides containing additional purification tags can provide further means of purifying the final construct.

Moreover, the labelling stoichiometry, which is crucial for single-molecule measurements, has to also be thoroughly optimised. In the case of in vivo studies, chemical conjugation is limited by the potentially high toxicity of the label, the nonspecificity of the reaction due to the affinity of the label for endogenous components and the high background due to incomplete labelling [57]. A recently developed strategy to minimise the fluorescent background generated by incomplete labelling inside cells is to use fluorescent labels which get activated only upon conjugation with the target protein [58].

1.5.2.2 Nucleotide Labelling

Unlike your protein of interest, several fluorescent nucleotides can be purchased. This can provide useful alternatives to the chemical synthesis route which can require specialised equipment. Nevertheless, it is important for one to consider wisely the type of label which is optimal for the biological system under study, especially when such a large variety exists. Overall, purity is a major concern, since contaminating ADP can significantly impede the motor protein. Therefore, the purity should always be assessed by high resolution HPLC.

The ATP structure can be modified at several locations. Firstly, purine ring modifications can be made. However, these interactions may disrupt the protein–

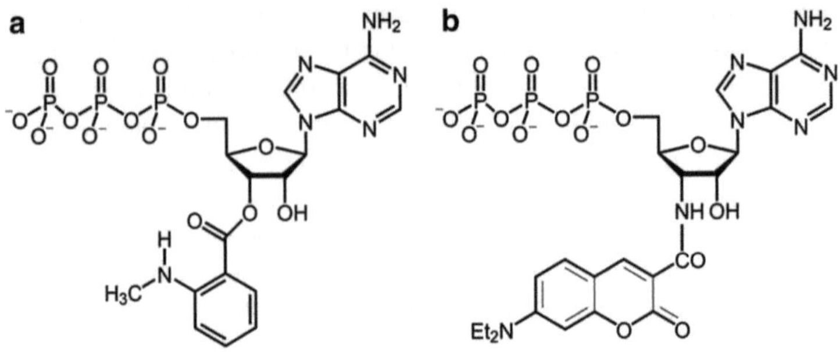

Fig. 1.4 ATP Labelling. Two examples of ribose-modified ATPs: (**a**) 2′(3′)mantATP and (**b**) deac-aminoATP

nucleotide interactions due to selectivity of the base [38]. The phosphates can also be modified, as done with γ-AMNS-ATP which was used in studies on *Escherichia coli* RNA polymerase [59]. However, these modifications frequently disrupt the cleavage step, as access to the γ-phosphate is blocked. Therefore, ribose modifications are frequently targeted. An example of such a modification is *N*-methylanthraniloyl-ATP (mantATP) (Fig. 1.4a) [60]. This nucleotide is made from an acyl link to the 2′ or 3′ hydroxyl by mixing ATP with methylisatoic anhydride. Performing this reaction is very simple as it occurs in water, at 37 °C, for 1 h, at pH between 9 and 9.6. The small mant fluorophore often leads to minimal perturbations.

However, ribose modifications can lead to problems, as the modification is through one of two hydroxyl groups (2′ or 3′). This leads to the formation of isomeric forms of the nucleotide, whereby one isomer may not be catalysed. It is possible to isolate either isomer at low pH, but the assays will also have to be carried out at low pH. This is because the isomers will quickly equilibrate ($t½=7$ min) at pH 7 [38]. It is also possible to derivatise the 2′ or 3′ hydroxyl group, so that only one remains for labelling. This has been successfully done with 3′-mant-2′-deoxy-ATP. Furthermore, using an ATP derivative with one hydroxyl group switch for an amine (3′-amino-2′-deoxy-ATP) allows for selective labelling with amine-reactive compounds.

There is a diverse collection of fluorescent labels which range from small aromatic rings (typically UV excitation range) through to large multi-ring labels (through the visible to near-red range). In the latter case, the label can be larger than the parent nucleotide. Longer chemical linkers to use can be different lengths of diamino-*n*-alkanes, such as 2′(3′)-*O*-[*N*-(2-aminoethyl)-carbamoyl]ATP (edaATP [61]) and 2′(3′)-*O*-[*N*-(3-aminopropyl)carbamoyl]ATP (pdaATP [62]). In this case, the fluorophore will be positioned further away from the protein, and therefore, it should not interfere significantly with the nucleotide association and catalysis. However, moving the fluorophore further away from the protein may also reduce any changes on its fluorescent properties upon binding or conformation changes. Ideally, it is best to choose the smaller label possible. Therefore, mantATP and deac-aminoATP analogues (Fig. 1.4) are very useful for in solution assays. For single-molecule methods, a larger and more stable fluorophore, within the visible spectrum, is required, such as Cy3. For further details on selecting a fluorophore and producing fluorescent nucleotides, please see [38, 39].

Purification of ATP (or ADP) is performed in the same manner for labelled and unlabelled nucleotides. Therefore, purifying a fluorescent ATP also separates ATP from ADP. Ion-exchange chromatography is applied, whereby the negatively charged ATP will bind to diethylaminoethyl (DEAE) cellulose. Volatile buffers (triethylammonium bicarbonate) are used to enable easy isolation of the purified nucleotide in a powder form. Assessment of the purity is performed via HPLC, equipped with a strong anion exchange column running an isocratic gradient of ammonium phosphate pH4 [38, 39].

1.5.2.3 Nucleic Acid Labelling

Fluorophores are introduced into oligonucleotides during chemical synthesis. Therefore, most labs will purchase ready-to-use fluorescent substrates. The fluorescent moiety can be added at the 3′ or 5′ end or at any internal location through thymidine analogues. Labelling is frequently performed through amine-reactive compounds, such as succinimidyl-esters. But, thiol, hydroxyl and phosphate reactive compounds can also be used. All of these reactive groups are also available with spacer groups. Due to this large flexibility, it is possible to choose the most suitable fluorophore for a specific application, as discussed above for nucleotides.

Purchased oligonucleotides are limited to around 200 bases. Longer fragments can be made by ordering fluorescently labelled PCR primers, and in combination with ligation and hybridisation reactions, numerous substrates can be generated.

1.5.3 Quantum Dots

In recent years, QDots have been an increasingly popular alternative probe especially for single-molecule measurements. This is due to their significantly advantageous optical properties over both FPs and organic dyes. These properties are defined by their inorganic semiconductor nanocrystal core and shell, in a size-dependent manner. Biocompatibility, surface passivation and functionalisation of the nanoparticle are achieved through an organic coat which combines amphiphilic polymers and peptides for solubility, poly(ethylene glycol) (PEG) for minimising nonspecific interaction and reactive biomolecules such as antibodies, DNA or streptavidin for conjugation to the molecule of interest [63]. Currently QDots are commercially available in various colours and with various functionalisation options, which provides high flexibility for multiparametric single-molecule experiments.

QDots are characterised by a large Stokes' shift, with a wide absorbance profile and a narrow emission spectrum with a size-dependent emission maximum. Key optical advantages of QDots are the exceptionally high quantum yields and extinction coefficients which are tenfold higher that the best organic fluorophores [64, 65]. In addition, their resistance to photobleaching is significantly higher than any conventional fluorescent probe. Together these properties allow the imaging of single QDot-tagged molecules for long periods of time, with higher SNR and a spatial resolution of only a few nanometres.

However, QDots present limitations that should be taken into account when designing single-molecule experiments. First, the organic coat which renders to QDots their labelling specificity and biocompatibility is bulky and results to a final size of 20–40 nm. Such large size could interfere with the properties of the motor, its interaction with partners and cargo and its dynamics, especially in the crowded intracellular environment. Recently, many approaches have been developed to reduce their size [63]. Another downside of QDots is their polyvalency, i.e. their ability to bind multiple target motors. This is due to their coating with multiple reactive molecules such as antibodies or streptavidin, which in turn have multiple binding sites for the target protein. Polyvalency is particularly problematic for single-molecule studies which require well-defined stoichiometry of fluorophore labelling. Stoichiometry can be controlled by controlling the molar ratio of the QDot versus the target molecule [66, 67]. However, this approach does not ensure a defined stoichiometry and can provide only an approximation. To tackle this, some groups have successfully engineered monofunctional QDots [68, 69]; however, these QDots are not commercially available. How to control the number of motors bound to QDot is discussed in Chap. 11. Another complication in the use of QDots, in particular for in vivo studies, is their delivery in the cytoplasm. This can be achieved by nonspecific endocytosis, lipid encapsulation, osmotic burst of pinocytotic vesicles, microinjection or electroporation [63]. Cytoplasmic delivery can be challenging, and it has to be selected and optimised depending on the application.

1.5.4 Fluorescent Probes for Single-Molecule Super-resolution Imaging

While these probes are FPs or organic fluorophores, there are specific requirements for these measurements, as discussed below. Super-resolution imaging relies on the ability of certain fluorophores to change their spectral properties upon irradiation with light of specific wavelength [70]. This change can occur in three different ways: (a) irreversible photoactivation from a dark state to bright emission upon exposure to UV or violet light, (b) irreversible photoconversion from one emission wavelength to another, and (c) photoswitching, which refers to the reversible switching between two emission wavelengths. Such properties can be found in both FPs and synthetic dyes. Unlike the ensemble-based super-resolution imaging, the single-molecule approach mainly relies on the fluorophore generating enough photons to allow its accurate localisation (Chap. 7). Therefore, the main requirement in single-molecule super-resolution is high brightness and not necessarily high photostability which can limit the time resolution. In addition, probes should show minimal light-independent interconversion and high photoswitching reliability. Also, as mentioned above, FPs should be monomeric and have quick and efficient chromophore maturation.

Among the most promising FP candidates for single-molecule super-solution imaging include (a) the dark-to-red photoactivatable PAmCherry [71] and the significantly brighter PATagFRP [72] which both have been used for single particle tracking photoactivated localisation microscopy (sptPALM) in living cells; (b) the red-to-green photoconvertible Dendra2 [73], mEos2 [74] and mKikGR [75]; and (c) the reversibly photoswitchable dark-to-green Dronpa [76, 77]. In addition, synthetic, reversibly photoswitchable probes of a broad range of colours are now available for super-resolution imaging by stochastic optical reconstruction microscopy (STORM), including Cy5 paired with various secondary activator dyes, such as Cy3 [78], Cy2 and Alexa Fluor 405, pairs of Cy7-Cy3 and Cy5.5-Cy3 [79] and Alexa405–Alexa647 [80].

1.5.5 Labelling of Tracks

A feature of motor proteins is that they use ATP to modulate the interactions with a track: filamentous actin, DNA, RNA or microtubules. Due to the importance of these tracks, this section highlights several approaches to visualise these components through direct and indirect labelling. In general, many labels will be coupled to the track, which allows observing the track for a long period of time before it is photobleached.

Filamentous actin can be stabilised by phalloidin which binds between two adjacent actin monomers. Phalloidin is a toxic heptapeptide from *Amanita phalloides* which acts to prevent actin depolymerisation [81]. Fluorescently labelled phalloidin is available in many colours, which allows simultaneous visualisation of a fluorescently labelled myosin and filamentous actin [6, 82]. Since this labelling option also acts to stabilise the filament, it is not suitable for measurements on dynamic actin filaments or in live cells. The Lifeact probe provides a useful alternative [83]. This is a short actin-binding peptide which can be coupled to organic fluorophores or fluorescent proteins. The Lifeact binding does not stabilise the filaments, and therefore, it allows observation of myosin upon dynamic filaments in live cells.

Microtubules are usually visualised by supplementing nonfluorescent tubulin with a fluorescent version, such as TAMRA (rhodamine)-labelled tubulin [84]. This scheme does not stabilise the filaments. Taxol is used for this, but the organic dye is only suitable for in vitro approaches. QDots can be coupled to microtubules but may interfere with motor translocation. GFP tubulin has also been used for visualising filaments in vivo.

There are numerous DNA-binding dyes available; YOYO, DAPI and the Hoechst range are very common [2, 85, 86]. Hoechst dyes bind to the minor groove without any major perturbations on the motor protein's translocation or DNA unwinding. But this is not a universal fact. Many different Hoechst dyes are available with various fluorescent properties; therefore, they can be selected depending on the application. In addition, they are applicable to both in vitro and in vivo approaches.

1.6 Conclusions

Fluorescence methods provide rich and varied information on motor proteins, revealing various aspects of their function and regulation which are difficult to observe by other approaches. This chapter has presented various methods for targeting fluorescent labels onto motor proteins and their tracks. However, there is ideal label or labelling strategy. New methods and labels are constantly being developed in order to find further ways to probe the mechanistic details of these complex biological systems.

References

1. Altman D, Goswami D, Hasson T, Spudich JA, Mayor S (2007) Precise positioning of myosin VI on endocytic vesicles in vivo. PLoS Biol 5(8):e210. doi:10.1371/journal.pbio.0050210
2. Bianco PR, Brewer LR, Corzett M, Balhorn R, Yeh Y, Kowalczykowski SC, Baskin RJ (2001) Processive translocation and DNA unwinding by individual RecBCD enzyme molecules. Nature 409(6818):374–378. doi:10.1038/35053131
3. Chisty LT, Toseland CP, Fili N, Mashanov GI, Dillingham MS, Molloy JE, Webb MR (2013) Monomeric PcrA helicase processively unwinds plasmid lengths of DNA in the presence of the initiator protein RepD. Nucleic Acids Res 41(9):5010–5023. doi:10.1093/nar/gkt194
4. Fili N, Mashanov GI, Toseland CP, Batters C, Wallace MI, Yeeles JT, Dillingham MS, Webb MR, Molloy JE (2010) Visualizing helicases unwinding DNA at the single molecule level. Nucleic Acids Res 38(13):4448–4457. doi:10.1093/nar/gkq173
5. Miyawaki A, Llopis J, Heim R, McCaffery JM, Adams JA, Ikura M, Tsien RY (1997) Fluorescent indicators for Ca2+ based on green fluorescent proteins and calmodulin. Nature 388(6645):882–887. doi:10.1038/42264
6. Sakamoto T, Webb MR, Forgacs E, White HD, Sellers JR (2008) Direct observation of the mechanochemical coupling in myosin Va during processive movement. Nature 455(7209): 128–132. doi:10.1038/nature07188
7. Theissen B, Karow AR, Kohler J, Gubaev A, Klostermeier D (2008) Cooperative binding of ATP and RNA induces a closed conformation in a DEAD box RNA helicase. Proc Natl Acad Sci USA 105(2):548–553. doi:10.1073/pnas.0705488105

8. Joo C, Balci H, Ishitsuka Y, Buranachai C, Ha T (2008) Advances in single-molecule fluorescence methods for molecular biology. Annu Rev Biochem 77:51–76. doi:10.1146/annurev.biochem.77.070606.101543
9. Toseland CP (2013) Fluorescent labeling and modification of proteins. J Chem Biol 6(3):85–95. doi:10.1007/s12154-013-0094-5
10. Lakowicz JR (2006) Principles of fluorescence spectroscopy, 3rd edn. Springer, New York
11. Shanker N, Bane SL (2008) Basic aspects of absorption and fluorescence spectroscopy and resonance energy transfer methods. Methods Cell Biol 84:213–242. doi:10.1016/S0091-679X(07)84008-8
12. Waggoner A (1995) Covalent labeling of proteins and nucleic acids with fluorophores. Methods Enzymol 246:362–373
13. Ha T, Tinnefeld P (2012) Photophysics of fluorescent probes for single-molecule biophysics and super-resolution imaging. Annu Rev Phys Chem 63:595–617. doi:10.1146/annurev-physchem-032210-103340
14. LiCata VJ, Wowor AJ (2008) Applications of fluorescence anisotropy to the study of protein-DNA interactions. Methods Cell Biol 84:243–262. doi:10.1016/S0091-679X(07)84009-X
15. Kapanidis AN, Weiss S (2002) Fluorescent probes and bioconjugation chemistries for single-molecule fluorescence analysis of biomolecules. J Chem Phys 117(24):10953–10964
16. Mujumdar RB, Ernst LA, Mujumdar SR, Lewis CJ, Waggoner AS (1993) Cyanine dye labeling reagents: sulfoindocyanine succinimidyl esters. Bioconjug Chem 4(2):105–111
17. Harada Y, Sakurada K, Aoki T, Thomas DD, Yanagida T (1990) Mechanochemical coupling in actomyosin energy transduction studied by in vitro movement assay. J Mol Biol 216(1):49–68. doi:10.1016/S0022-2836(05)80060-9
18. Rasnik I, McKinney SA, Ha T (2006) Nonblinking and long-lasting single-molecule fluorescence imaging. Nat Methods 3(11):891–893. doi:10.1038/nmeth934
19. Vogelsang J, Kasper R, Steinhauer C, Person B, Heilemann M, Sauer M, Tinnefeld P (2008) A reducing and oxidizing system minimizes photobleaching and blinking of fluorescent dyes. Angew Chem Int Ed Engl 47(29):5465–5469. doi:10.1002/anie.200801518
20. Chen Y, Vela J, Htoon H, Casson JL, Werder DJ, Bussian DA, Klimov VI, Hollingsworth JA (2008) "Giant" multishell CdSe nanocrystal quantum dots with suppressed blinking. J Am Chem Soc 130(15):5026–5027. doi:10.1021/ja711379k
21. Mahler B, Spinicelli P, Buil S, Quelin X, Hermier JP, Dubertret B (2008) Towards non-blinking colloidal quantum dots. Nat Mater 7(8):659–664. doi:10.1038/nmat2222
22. Wang X, Ren X, Kahen K, Hahn MA, Rajeswaran M, Maccagnano-Zacher S, Silcox J, Cragg GE, Efros AL, Krauss TD (2009) Non-blinking semiconductor nanocrystals. Nature 459(7247):686–689. doi:10.1038/nature08072
23. Hohng S, Ha T (2004) Near-complete suppression of quantum dot blinking in ambient conditions. J Am Chem Soc 126(5):1324–1325. doi:10.1021/ja039686w
24. Brune M, Hunter JL, Howell SA, Martin SR, Hazlett TL, Corrie JE, Webb MR (1998) Mechanism of inorganic phosphate interaction with phosphate binding protein from Escherichia coli. Biochemistry 37(29):10370–10380. doi:10.1021/bi9804277
25. Ha T (2001) Single-molecule fluorescence resonance energy transfer. Methods 25:78–86. doi:10.1006/meth.2001.1217
26. Calleja V, Alcor D, Laguerre M, Park J, Vojnovic B, Hemmings BA, Downward J, Parker PJ, Larijani B (2007) Intramolecular and intermolecular interactions of protein kinase B define its activation in vivo. PLoS Biol 5(4):e95. doi:10.1371/journal.pbio.0050095
27. James NG, Jameson DM (2014) Steady-state fluorescence polarization/anisotropy for the study of protein interactions. Methods Mol Biol 1076:29–42. doi:10.1007/978-1-62703-649-8_2
28. Eccleston JF, Hunchinson JP, White HD (2001) Stopped-flow techniques. In: Harding SE, Chowdry BZ (eds) Protein ligand interactions: structure and spectroscopy. Oxford University Press, Oxford, pp 201–237

29. Perrin F (1926) Polarisation de la lumière de fluorescence. Vie moyenne des molécules dans l'état excité. Journal de Physique et Le Radium 7:390–401
30. Trentham DR, Eccleston JF, Bagshaw CR (1976) Kinetic analysis of ATPase mechanisms. Q Rev Biophys 9(2):217–281
31. Reck-Peterson SL, Yildiz A, Carter AP, Gennerich A, Zhang N, Vale RD (2006) Single-molecule analysis of dynein processivity and stepping behavior. Cell 126(2):335–348. doi:10.1016/j.cell.2006.05.046
32. Yildiz A, Forkey JN, McKinney SA, Ha T, Goldman YE, Selvin PR (2003) Myosin V walks hand-over-hand: single fluorophore imaging with 1.5-nm localization. Science 300(5628): 2061–2065. doi:10.1126/science.1084398
33. Okten Z, Churchman LS, Rock RS, Spudich JA (2004) Myosin VI walks hand-over-hand along actin. Nat Struct Mol Biol 11(9):884–887. doi:10.1038/nsmb815
34. Toseland CP, Martinez-Senac MM, Slatter AF, Webb MR (2009) The ATPase cycle of PcrA helicase and its coupling to translocation on DNA. J Mol Biol 392(4):1020–1032. doi:10.1016/j.jmb.2009.07.071
35. Toseland CP, Powell B, Webb MR (2012) ATPase cycle and DNA unwinding kinetics of RecG helicase. PLoS One 7(6):e38270. doi:10.1371/journal.pone.0038270
36. Toseland CP, Webb MR (2010) Fluorescence tools to measure helicase activity in real time. Methods 51(3):259–268. doi:10.1016/j.ymeth.2010.02.012
37. Toseland CP, Webb MR (2013) ATPase mechanism of the 5′-3′ DNA helicase, RecD2: evidence for a pre-hydrolysis conformation change. J Biol Chem 288(35):25183–25193. doi:10.1074/jbc.M113.484667
38. Jameson DM, Eccleston JF (1997) Fluorescent nucleotide analogs: synthesis and applications. Methods Enzymol 278:363–390
39. Toseland CP, Webb MR (2011) Fluorescent nucleoside triphosphates for single-molecule enzymology. Methods Mol Biol 778:161–174. doi:10.1007/978-1-61779-261-8_11
40. Shaner NC, Steinbach PA, Tsien RY (2005) A guide to choosing fluorescent proteins. Nat Methods 2(12):905–909. doi:10.1038/nmeth819
41. Wu B, Piatkevich KD, Lionnet T, Singer RH, Verkhusha VV (2011) Modern fluorescent proteins and imaging technologies to study gene expression, nuclear localization, and dynamics. Curr Opin Cell Biol 23(3):310–317. doi:10.1016/j.ceb.2010.12.004
42. Ormo M, Cubitt AB, Kallio K, Gross LA, Tsien RY, Remington SJ (1996) Crystal structure of the Aequorea victoria green fluorescent protein. Science 273(5280):1392–1395
43. Pierce DW, Hom-Booher N, Vale RD (1997) Imaging individual green fluorescent proteins. Nature 388(6640):338. doi:10.1038/41009
44. Telley IA, Bieling P, Surrey T (2009) Obstacles on the microtubule reduce the processivity of Kinesin-1 in a minimal in vitro system and in cell extract. Biophys J 96(8):3341–3353. doi:10.1016/j.bpj.2009.01.015
45. Zhang W, Jiang Y, Wang Q, Ma X, Xiao Z, Zuo W, Fang X, Chen YG (2009) Single-molecule imaging reveals transforming growth factor-beta-induced type II receptor dimerization. Proc Natl Acad Sci USA 106(37):15679–15683. doi:10.1073/pnas.0908279106
46. Ulbrich MH, Isacoff EY (2007) Subunit counting in membrane-bound proteins. Nat Methods 4(4):319–321. doi:10.1038/nmeth1024
47. Cai D, Verhey KJ, Meyhofer E (2007) Tracking single Kinesin molecules in the cytoplasm of mammalian cells. Biophys J 92(12):4137–4144. doi:10.1529/biophysj.106.100206
48. Kredel S, Oswald F, Nienhaus K, Deuschle K, Rocker C, Wolff M, Heilker R, Nienhaus GU, Wiedenmann J (2009) mRuby, a bright monomeric red fluorescent protein for labeling of subcellular structures. PLoS One 4(2):e4391. doi:10.1371/journal.pone.0004391
49. Tsutsui H, Karasawa S, Okamura Y, Miyawaki A (2008) Improving membrane voltage measurements using FRET with new fluorescent proteins. Nat Methods 5(8):683–685. doi:10.1038/nmeth.1235

50. Getz EB, Xiao M, Chakrabarty T, Cooke R, Selvin PR (1999) A comparison between the sulfhydryl reductants tris(2-carboxyethyl)phosphine and dithiothreitol for use in protein biochemistry. Anal Biochem 273(1):73–80. doi:10.1006/abio.1999.4203
51. Galletto R, Amitani I, Baskin RJ, Kowalczykowski SC (2006) Direct observation of individual RecA filaments assembling on single DNA molecules. Nature 443(7113):875–878. doi:10.1038/nature05197
52. Amitani I, Liu B, Dombrowski CC, Baskin RJ, Kowalczykowski SC (2010) Watching individual proteins acting on single molecules of DNA. Methods Enzymol 472:261–291. doi:10.1016/S0076-6879(10)72007-3
53. Griffin BA, Adams SR, Tsien RY (1998) Specific covalent labeling of recombinant protein molecules inside live cells. Science 281(5374):269–272
54. Los GV, Encell LP, McDougall MG, Hartzell DD, Karassina N, Zimprich C, Wood MG, Learish R, Ohana RF, Urh M, Simpson D, Mendez J, Zimmerman K, Otto P, Vidugiris G, Zhu J, Darzins A, Klaubert DH, Bulleit RF, Wood KV (2008) HaloTag: a novel protein labeling technology for cell imaging and protein analysis. ACS Chem Biol 3(6):373–382. doi:10.1021/cb800025k
55. Sun X, Zhang A, Baker B, Sun L, Howard A, Buswell J, Maurel D, Masharina A, Johnsson K, Noren CJ, Xu MQ, Correa IR Jr (2011) Development of SNAP-tag fluorogenic probes for wash-free fluorescence imaging. Chembiochem 12(14):2217–2226. doi:10.1002/cbic.201100173
56. Gautier A, Juillerat A, Heinis C, Correa IR Jr, Kindermann M, Beaufils F, Johnsson K (2008) An engineered protein tag for multiprotein labeling in living cells. Chem Biol 15(2):128–136. doi:10.1016/j.chembiol.2008.01.007
57. Xia T, Li N, Fang X (2013) Single-molecule fluorescence imaging in living cells. Annu Rev Phys Chem 64:459–480. doi:10.1146/annurev-physchem-040412-110127
58. Komatsu T, Johnsson K, Okuno H, Bito H, Inoue T, Nagano T, Urano Y (2011) Real-time measurements of protein dynamics using fluorescence activation-coupled protein labeling method. J Am Chem Soc 133(17):6745–6751. doi:10.1021/ja200225m
59. Yarbrough LR, Schlageck JG, Baughman M (1979) Synthesis and properties of fluorescent nucleotide substrates for DNA-dependent RNA polymerases. J Biol Chem 254(23):12069–12073
60. Hiratsuka T (1983) New ribose-modified fluorescent analogs of adenine and guanine nucleotides available as substrates for various enzymes. Biochim Biophys Acta 742(3):496–508
61. Cremo CR, Neuron JM, Yount RG (1990) Interaction of myosin subfragment 1 with fluorescent ribose-modified nucleotides. A comparison of vanadate trapping and SH1-SH2 cross-linking. Biochemistry 29(13):3309–3319
62. Webb MR, Reid GP, Munasinghe VR, Corrie JE (2004) A series of related nucleotide analogues that aids optimization of fluorescence signals in probing the mechanism of P-loop ATPases, such as actomyosin. Biochemistry 43(45):14463–14471. doi:10.1021/bi0486334
63. Pinaud F, Clarke S, Sittner A, Dahan M (2010) Probing cellular events, one quantum dot at a time. Nat Methods 7(4):275–285. doi:10.1038/nmeth.1444
64. Resch-Genger U, Grabolle M, Cavaliere-Jaricot S, Nitschke R, Nann T (2008) Quantum dots versus organic dyes as fluorescent labels. Nat Methods 5(9):763–775. doi:10.1038/nmeth.1248
65. Pons T, Mattoussi H (2009) Investigating biological processes at the single molecule level using luminescent quantum dots. Ann Biomed Eng 37(10):1934–1959. doi:10.1007/s10439-009-9715-0
66. Courty S, Bouzigues C, Luccardini C, Ehrensperger MV, Bonneau S, Dahan M (2006) Tracking individual proteins in living cells using single quantum dot imaging. Methods Enzymol 414:211–228. doi:10.1016/S0076-6879(06)14012-4
67. Courty S, Luccardini C, Bellaiche Y, Cappello G, Dahan M (2006) Tracking individual kinesin motors in living cells using single quantum-dot imaging. Nano Lett 6(7):1491–1495. doi:10.1021/nl060921t
68. Howarth M, Liu W, Puthenveetil S, Zheng Y, Marshall LF, Schmidt MM, Wittrup KD, Bawendi MG, Ting AY (2008) Monovalent, reduced-size quantum dots for imaging receptors on living cells. Nat Methods 5(5):397–399. doi:10.1038/nmeth.1206

69. You C, Wilmes S, Beutel O, Lochte S, Podoplelowa Y, Roder F, Richter C, Seine T, Schaible D, Uze G, Clarke S, Pinaud F, Dahan M, Piehler J (2010) Self-controlled monofunctionalization of quantum dots for multiplexed protein tracking in live cells. Angew Chem Int Ed Engl 49(24):4108–4112. doi:10.1002/anie.200907032
70. Patterson G, Davidson M, Manley S, Lippincott-Schwartz J (2010) Superresolution imaging using single-molecule localization. Annu Rev Phys Chem 61:345–367. doi:10.1146/annurev.physchem.012809.103444
71. Subach FV, Patterson GH, Manley S, Gillette JM, Lippincott-Schwartz J, Verkhusha VV (2009) Photoactivatable mCherry for high-resolution two-color fluorescence microscopy. Nat Methods 6(2):153–159. doi:10.1038/nmeth.1298
72. Subach FV, Patterson GH, Renz M, Lippincott-Schwartz J, Verkhusha VV (2010) Bright monomeric photoactivatable red fluorescent protein for two-color super-resolution sptPALM of live cells. J Am Chem Soc 132(18):6481–6491. doi:10.1021/ja100906g
73. Gurskaya NG, Verkhusha VV, Shcheglov AS, Staroverov DB, Chepurnykh TV, Fradkov AF, Lukyanov S, Lukyanov KA (2006) Engineering of a monomeric green-to-red photoactivatable fluorescent protein induced by blue light. Nat Biotechnol 24(4):461–465. doi:10.1038/nbt1191
74. McKinney SA, Murphy CS, Hazelwood KL, Davidson MW, Looger LL (2009) A bright and photostable photoconvertible fluorescent protein. Nat Methods 6(2):131–133. doi:10.1038/nmeth.1296
75. Habuchi S, Tsutsui H, Kochaniak AB, Miyawaki A, van Oijen AM (2008) mKikGR, a monomeric photoswitchable fluorescent protein. PLoS One 3(12):e3944. doi:10.1371/journal.pone.0003944
76. Ando R, Mizuno H, Miyawaki A (2004) Regulated fast nucleocytoplasmic shuttling observed by reversible protein highlighting. Science 306(5700):1370–1373. doi:10.1126/science.1102506
77. Habuchi S, Ando R, Dedecker P, Verheijen W, Mizuno H, Miyawaki A, Hofkens J (2005) Reversible single-molecule photoswitching in the GFP-like fluorescent protein Dronpa. Proc Natl Acad Sci USA 102(27):9511–9516. doi:10.1073/pnas.0500489102
78. Bates M, Blosser TR, Zhuang X (2005) Short-range spectroscopic ruler based on a single-molecule optical switch. Phys Rev Lett 94(10):108101
79. Bates M, Huang B, Dempsey GT, Zhuang X (2007) Multicolor super-resolution imaging with photo-switchable fluorescent probes. Science 317(5845):1749–1753. doi:10.1126/science.1146598
80. Jones SA, Shim SH, He J, Zhuang X (2011) Fast, three-dimensional super-resolution imaging of live cells. Nat Methods 8(6):499–508. doi:10.1038/nmeth.1605
81. Yanagida T, Nakase M, Nishiyama K, Oosawa F (1984) Direct observation of motion of single F-actin filaments in the presence of myosin. Nature 307(5946):58–60
82. Snyder GE, Sakamoto T, Hammer JA III, Sellers JR, Selvin PR (2004) Nanometer localization of single green fluorescent proteins: evidence that myosin V walks hand-over-hand via telemark configuration. Biophys J 87(3):1776–1783. doi:10.1529/biophysj.103.036897
83. Riedl J, Crevenna AH, Kessenbrock K, Yu JH, Neukirchen D, Bista M, Bradke F, Jenne D, Holak TA, Werb Z, Sixt M, Wedlich-Soldner R (2008) Lifeact: a versatile marker to visualize F-actin. Nat Methods 5(7):605–607. doi:10.1038/nmeth.1220
84. Gell C, Bormuth V, Brouhard GJ, Cohen DN, Diez S, Friel CT, Helenius J, Nitzsche B, Petzold H, Ribbe J, Schaffer E, Stear JH, Trushko A, Varga V, Widlund PO, Zanic M, Howard J (2010) Microtubule dynamics reconstituted in vitro and imaged by single-molecule fluorescence microscopy. Methods Cell Biol 95:221–245. doi:10.1016/S0091-679X(10)95013-9
85. Graneli A, Yeykal CC, Prasad TK, Greene EC (2006) Organized arrays of individual DNA molecules tethered to supported lipid bilayers. Langmuir 22(1):292–299. doi:10.1021/la051944a
86. Greene EC, Wind S, Fazio T, Gorman J, Visnapuu ML (2010) DNA curtains for high-throughput single-molecule optical imaging. Methods Enzymol 472:293–315. doi:10.1016/S0076-6879(10)72006-1

Chapter 2
Fluorescent Biosensors: Design and Application to Motor Proteins

Simone Kunzelmann*, Claudia Solscheid*, and Martin R. Webb

Abstract Reagentless biosensors are single molecular species that report the concentration of a specific target analyte, while having minimal impact on the system being studied. This chapter reviews such biosensors with emphasis on the ones that use fluorescence as readout and can be used for real-time assays of concentration changes with reasonably high time resolution and sensitivity. Reagentless biosensors can be designed with different types of recognition elements, particularly specific binding proteins and nucleic acids, including aptamers. Different ways are described in which a fluorescence signal can be used to report the target concentration. These include the use of single, environmentally sensitive fluorophores; FRET pairs, often used in genetically encoded biosensors; and pairs of identical fluorophores that undergo reversible stacking interactions to change fluorescence intensity. The applications of these biosensors in different types of real-time assays with motor proteins are described together with some specific examples. These encompass regulation and mechanism of motor proteins, using both steady-state assays and single-turnover measurements.

Keywords Biosensor • Fluorescence • Assay • ATPase • Helicase • Kinesin • Myosin

Abbreviations

DCC-SSB	(G26C)SSB adduct with *N*-[2-(iodoacetamido)ethyl]-7-diethylaminocoumarin-3-carboxamide
ELISA	Enzyme-linked immunosorbent assay
FISH	Fluorescence in situ hybridization

*Both authors contributed equally.

S. Kunzelmann • C. Solscheid • M.R. Webb (✉)
MRC National Institute for Medical Research, Mill Hill, London NW7 1AA, UK
e-mail: mwebb@nimr.mrc.ac.uk

C.P. Toseland and N. Fili (eds.), *Fluorescent Methods for Molecular Motors*, Experientia Supplementum 105, DOI 10.1007/978-3-0348-0856-9_2,

FP	Fluorescent protein
FRET	Förster (or fluorescence) resonance energy transfer
MDCC	7-diethylamino-3-((((2-maleimidyl)ethyl)amino)carbonyl)coumarin
MDCC-PBP	(A197C) Phosphate binding protein adduct with MDCC
PNA	Peptide nucleic acid
SELEX	Systematic evolution of ligands by exponential enrichment
TIRFM	Total internal reflection microscopy

2.1 Introduction

This chapter reviews different types of biosensors and how some of them are being applied to probe molecular motors. The biosensors field is wide, in part because of a somewhat diffuse definition of biosensor, which means different things to different people. This chapter is restricted to what are often called reagentless biosensors, essentially a single species in solution that reports the concentration of a target molecule. The term "reagentless" implies that no other reagent, apart from the biosensor, needs to be added to the system being studied. This is in contrast with, for example, a coupled enzyme assay, in which several extra components may need to be present. This chapter focuses on reagentless biosensors that can be applied to continuous, time-resolved measurements with fluorescence as the readout, as these have found widespread use in assays with motor proteins to measure either the movement, ATP hydrolysis, or such aspects as regulation.

There are two essential elements of the reagentless biosensors of the type described here. The example of a biosensor for inorganic phosphate (P_i) [1, 2] in Fig. 2.1 will be used to illustrate these concepts, as this is a simple entity and has been widely used in the motility field [3–5]. Firstly, a recognition element senses the target analyte with some degree of selectivity over similar molecules. The recognition element can be an enzyme that catalyses a reaction of the analyte, a macromolecule, such as protein or nucleic acid, or a small molecule that binds the analyte. In the case of the P_i biosensor, this element is a phosphate binding protein that binds P_i tightly and with a high degree of selectivity. The second element is a reporter, which produces a signal, in response to interaction of the analyte with the recognition element that can be measured without interference from the rest of the assay components. The signal can be electrons or light, for example. The P_i biosensor has a coumarin fluorophore covalently bound to a single cysteine on the protein surface. At this position, the label responds with a ~tenfold increase in fluorescence intensity when P_i binds. The signal is transduced by a protein conformation change on binding P_i, thereby changing the environment of the fluorophore and hence its fluorescence: in this case, there is no separate transducing element as the protein itself does this. In some biosensors, such an element needs to be included within the biosensor to couple the recognition and reporter elements.

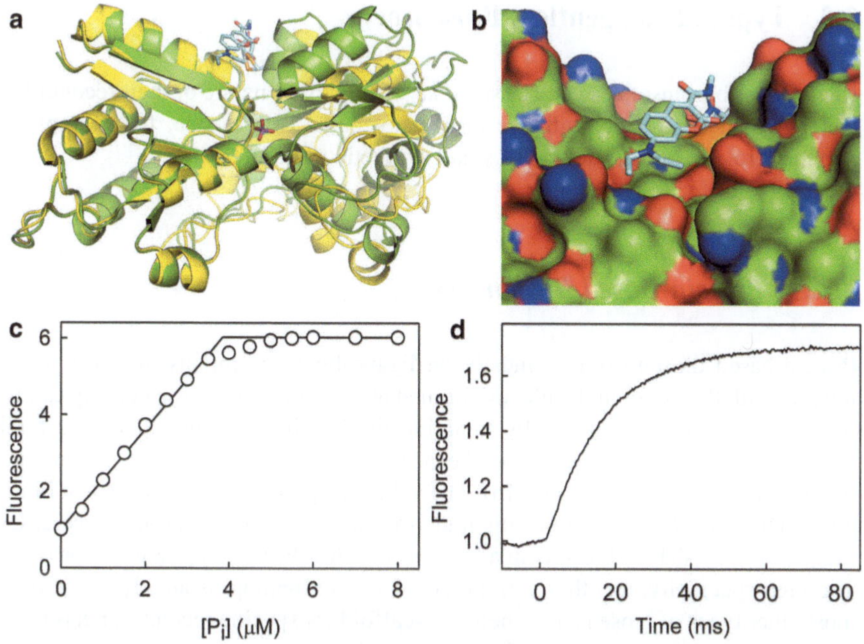

Fig. 2.1 P_i biosensor, MDCC-PBP. (**a**) Structure, showing P_i-bound form (*green/dark*) overlaid with unlabelled apo-PBP (*yellow/pale*). P_i is bound between the two domains. One domain (*left*) is overlaid to show movement of the other (*right*) on P_i binding [31, 77]. (**b**) Diethylaminocoumarin sits in a cleft between the two domains in the P_i-bound structure. In the apo form, it is likely that the coumarin position is flexible and pointing into the medium [31]. (**c**) Titration of P_i into 5 μM MDCC-PBP, to show region of linear response, capacity and calibration [2]. (**d**) Measurement of binding kinetics by mixing 0.25 μM P_i with 0.1 μM MDCC-PBP in a stopped-flow apparatus [2]

A major advantage of reagentless biosensors is that only a single component is added to the system being studied, so it causes only a minimal modification or interference. Essentially, reagentless biosensors can be "impactless" on a solution of proteins, etc. being assayed. For example, the P_i biosensor has been used to measure P_i produced by the ATP hydrolysis of myosins, kinesins, helicases, etc. In measuring such a product, the motility system itself is not directly altered: the biosensor interacts with P_i only after it dissociates from the motor protein. This is in contrast to some other fluorescence techniques, such as labelling components of the system itself (e.g. proteins or substrates), which potentially affects the studied reaction significantly. However, it should be pointed out that binding products of a reaction does limit the reversibility of that reaction, should that be an important feature: for example, P_i can inhibit muscle ATPase activity.

2.2 Types of Reagentless Biosensor

Reagentless biosensors can be classified according to the nature of their recognition element, which interacts with the target molecule. This can be, for example, a peptide, protein, nucleic acid, or a small synthetic molecule.

2.2.1 *Protein-Based Biosensors*

Protein-based biosensors rely mostly on ligand-binding proteins that selectively interact with the target molecule, as outlined above. If necessary, selectivity can be further enhanced or modified by mutating the binding site, based on structural information. This has, for example, been done on an ADP biosensor, based on a bacterial actin homologue, ParM [6]. The wild-type protein is an ATPase that binds both ADP and ATP, but introducing mutations in the active site strongly decreases the affinity for ATP and results in a biosensor with >400-fold preference for ADP. The high specificity, together with potentially tight binding, is an advantage over most other types of biosensors, where the scaffold for specific recognition has to be developed first. However, it largely restricts such biosensors to target molecules for which a suitable, specific binding protein is available. There are only a few examples of protein-based biosensors, where a new ligand specificity has been generated, for example, by computational design [7, 8].

A fluorescent reporter is coupled to the ligand-binding protein either by chemical labelling of the protein with small fluorescent probes or by genetic fusion to fluorescent proteins like GFP (green fluorescent protein) and its variants to produce genetically encoded biosensors. Chemical labelling at specific sites is often achieved via cysteine-reactive probes, mostly iodoacetamide or maleimide derivatives, which are commercially available for many fluorescent dyes. This enables the label to be positioned specifically on the surface. Small synthetic dyes generally give larger signal changes and may be more suitable for applications in vitro, whereas genetically encoded biosensors are the method of choice for measurements in cells or in vivo [9, 10]. There are also approaches combining the two strategies by genetic fusion with protein tags, such as SNAP-tag or CLIP-tag [11, 12], which can then be chemically modified with fluorescent dyes [13, 14], as outlined in Sect. 2.3. The chemical reaction is specific enough for labelling of a genetically encoded protein scaffold on a cell surface or even within cells, and thus, the functional biosensor can be generated in situ.

Another advantage of using a protein framework is that the normal specificity and activity of the protein may be harnessed as the transducing element. That is particularly so, when using a protein conformation change, linked to ligand binding, to change the environment of an attached fluorophore. This was outlined above for the P_i biosensor (Fig. 2.1). However, a number of different strategies have been used to transduce the binding event into a fluorescence change of the reporter in these types of biosensors, some of which are discussed in Sect. 2.3.

2.2.2 Nucleic Acid-Based Biosensors

Oligonucleotide-based sensors fall into two groups: firstly, hybridization probes that anneal to complementary sequences of DNA and RNA and, secondly, aptamers that recognize targets other than nucleic acids.

In most cases, hybridization probes are made up of synthetic nucleic acids, which are complementary to their target sequence and linked to a fluorescent reporter group [15, 16]. Many such probes are used in techniques requiring either the analyte or the hybridization probe to be surface bound, as is the case in single-molecule studies, microarrays and FISH (fluorescence in situ hybridization). Hybridization probes can also be applied in solution, for example in quantitative PCR or kinetic assays. The reporter element is formed by labels attached to the 3′- and/or 5′-ends of the probe(s), either as a single fluorophore or as a FRET (fluorescence resonance energy transfer) or fluorophore–quencher pair, depending on the strategy [15] (also see Sect. 2.3).

Aptamers form a versatile subclass of nucleic acid probes. In contrast to hybridization probes, aptamers are generated through "systematic evolution of ligands by exponential enrichment" (SELEX) specifically to bind small molecule ligands, metal ions, proteins, or even cell surfaces [17, 18]. The wide range of target molecules, which can be addressed by aptamer-based biosensors, is an advantage over the protein-based biosensors described above, where obtaining new ligand specificities usually requires a new recognition protein. Due to this, aptamers have found widespread applications in therapeutics and medical screens [19, 20], as well as in basic research [21]. While showing similar affinity and selectivity to antibody-based systems (see Sect. 2.2.3), single-stranded DNA or RNA aptamers have several advantages over the former class of biosensors. Their wide range of targets, small size, high thermostability, low immunogenicity, ease of synthesis and modification have led to an increase in their popularity [20–22].

Aptamers often undergo conformational changes when binding their target. Such structural changes can be exploited by strategically introducing chromophores to report on the structural alterations in different ways. One aptamer, making use of a conformational change to detect ATP with a >threefold selectivity over guanosine and pyrimidines, consists of a 26-bp DNA molecule, labelled with two pyrenyl groups [23]. In the absence of a ligand, the two pyrene rings are thought to interact with neighbouring bases on the single-stranded DNA (ssDNA), resulting in a low fluorescence quantum yield. Upon binding two ATPs, the DNA folds back on itself, becoming double stranded. The newly formed base interactions release the pyrenes, which become free to form a bis-pyrene excimer and so have an increased fluorescence [24]. While excimers and FRET pairs have been used in the development of a number of aptamers, the most common approach consists of a fluorophore and a quencher moiety attached to the 3′- and 5′-ends [22]. These "molecular beacons" are a large class of biosensors that are mostly based on nucleic acids but some also on peptides or PNA (peptide nucleic acid). They possess a typical stem–loop structure, which is disrupted by target association, resulting in the separation of the fluorophore–quencher pair [25]. A comprehensive description of the various types of aptamer sensors and their applications has been covered by several recent reviews [15, 26–28].

2.2.3 Antibody-Based Biosensors (Immunosensors)

Immunosensors use antibodies or antibody fragments as recognition elements to detect antigens with high specificity and high affinity. A wide range of reporters have been used, including those giving a fluorescence signal. Most antibody-based probes require other molecules as part of the biosensor and are therefore not reagentless. However, due to their importance in research, the principles of some immunosensors are outlined here.

Most immunoassays are heterogeneous with recognition taking place at the surface of a solid matrix. This is achieved either by direct binding of a fluorescently labelled antibody to an immobilized target or by secondary labelling, whereby a fluorescent molecule binds to an unlabeled, surface-bound antibody. Alternatively, there is a sandwich approach, in which an antigen–antibody complex, attached to a surface, is recognized by a labelled antibody for detection, as is used in the enzyme-linked immunosorbent assay, ELISA [29]. All these assays require the target or antibody to be separated from a homogeneous sample via interaction with the complementary component, fixed to a solid matrix. The accumulation of fluorescent molecules at a surface provides the readout. In a homogeneous format, target and sensor interact in solution, dispensing with a separation step. This interaction must result in a fluorescence change, and so these are similar in principle to other protein-based biosensors (Sect. 2.2.1). For example, two pools of antibodies raised against two different sites on an antigen can be labelled with compatible fluorescent dyes and used in a FRET-based manner [29]. To date the majority of immunosensors are used in surface-immobilized assays, which typically have low time resolution, rather than being suitable for time-resolved measurements [22].

2.2.4 Other Recognition Elements

Chemosensors, based on small synthetic molecules, form a large group of probes and include what are often simply called indicators [30]. Their design principles are largely very different from those of the biosensors described here. However, they often target biological molecules and are used similarly to biosensors, based on proteins or nucleic acids. An example, to measure calcium ions, is given in Sect. 2.5.

2.3 Design Principles: How Target Recognition Is Coupled to a Fluorescence Signal

A critical step in biosensor development is to couple target recognition to a fluorescence change of the reporter. There are a number of different designs, depending in part on the nature of the sensing element. Here, the main focus is on protein-based biosensors, but some design principles also apply to biosensors based on nucleic acid aptamers.

2.3.1 Single Fluorophore

Most simple designs rely on the fluorescence intensity change of a single, environment-sensitive fluorescent reporter that is attached to the recognition element. A conformation change in the recognition element upon target molecule binding is translated into a change of the local environment of the fluorophore (Fig. 2.2a). This can lead to a fluorescence intensity change by several mechanisms, such as changing the solvent accessibility and thus dynamic quenching, a change in polarity of the surrounding environment, or a change of specific interactions of the fluorophore with the protein. The last mechanism has been shown to be the main contribution for the phosphate biosensor, MDCC-PBP (Fig. 2.1). In this case, there is a specific mechanism by which the diethylaminocoumarin has high fluorescence in the phosphate-bound state, namely, by becoming coplanar due to an interaction with the protein [31]. A large number of other examples for this type of design, based on periplasmic binding proteins similar to PBP, exist, where ligand binding induces a large movement of two subdomains relative to each other leading to closure of the ligand-binding cleft [1, 32, 33].

Alternatively, a direct interaction of the fluorophore with the bound target molecule can cause a change in fluorescence intensity (Fig. 2.2b). An example is a biosensor for single-stranded DNA (DCC-SSB), based on the single-stranded DNA-binding protein (SSB). DNA binding causes an increase in the fluorescence of a single, environment-sensitive fluorophore (diethylaminocoumarin) in the absence of a recognized conformation change [34]. A third method of obtaining a fluorescence change is by ligand-induced modification of the biosensor, and an example is the biosensor that responds to ATP/ADP ratio [35]. A fluorophore-modified nucleoside-diphosphate kinase emits fluorescence depending on the degree of phosphorylation of the protein. Normally such phosphorylation is a transient part of the natural enzyme mechanism, but the process was harnessed so the biosensor responds to phosphorylation by ATP and dephosphorylation by ADP.

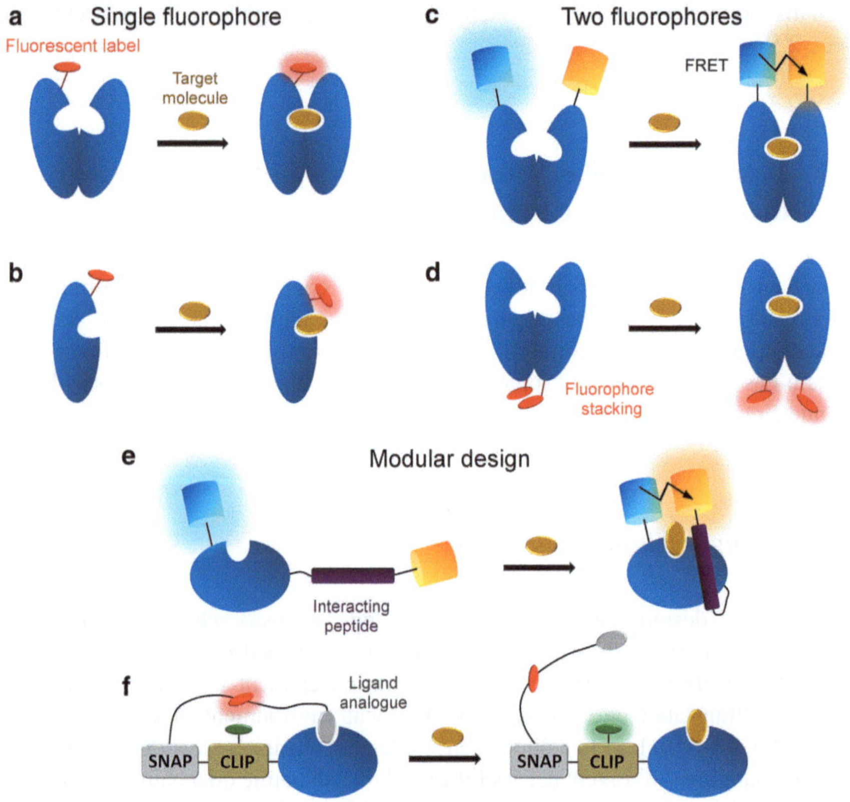

Fig. 2.2 Different designs of fluorescent reagentless biosensors. These design features mainly are applicable to protein-based biosensors. *Single-fluorophore biosensors*: conformation change (**a**) or interaction with target (**b**) alters the fluorophore environment. *Biosensors with two fluorophores*: readout is by FRET between two different fluorophores, shown here as fluorescent proteins (**c**), or by disruption of stacking between two identical fluorescent dyes (**d**). *Examples of modular designs*: part of the fused system interacts with the recognition element in either the target-bound (**e**) or target-free state (**f**), so transducing the signal when the target binds. In both types shown, the readout is by FRET. See Sect. 2.3 for details

In contrast to small synthetic dyes, in genetically encoded biosensors that use fluorescent proteins (FP), the chromophore is shielded inside the protein structure, which renders it less susceptible to environmental changes. The strategy for signal transduction is thus different. Several genetically encoded biosensors rely on the intensity change of a single FP, for example, biosensors reporting on the ATP/ADP ratio [36], Ca^{2+} [37, 38], Ins(1,3,4,5)P_4 [39] or glutamate [40]. By inserting FPs into a structural element that undergoes a conformation change upon ligand binding, this can alter the local structure or protonation state of the chromophore and thus the fluorescence intensity. Often circularly permuted versions of FPs are used, in which the N- and C-termini of the FP are in close proximity to the chromophore. In this way, when fused to the recognition protein, circularly permuted FPs are more likely to give a fluorescence change than the normal FP version. The dependence of fluorescence on the protonation state has also been exploited to use FPs as pH biosensors. Mutations around the chromophore have yielded variants that respond to changes in different pH ranges and with different spectral properties [41].

2.3.2 Two Fluorophores

A large number of biosensor designs are based on FRET (Fig. 2.2c). Here, a pair of fluorophores, specifically a FRET donor and acceptor, are fused to the recognition element. If a ligand-induced conformational change alters the distance or relative orientation of the FRET pair, this can result in a change in FRET. In comparison to probes based on fluorescence intensity of a single fluorophore, FRET provides a ratiometric signal, the ratio of acceptor to donor emission. This is an advantage for quantitative measurements in a complex environment, such as in cells, as the ratio should be independent of biosensor concentration. Due to the practical difficulty of specific labelling, a FRET-based design is not easily implemented with two different chemical dyes. In contrast, FRET-based mechanisms are very common among genetically encoded biosensors, such as for Ca^{2+} [42–44], cAMP [45, 46] and cGMP [47, 48], sugars [49–51], phosphate [52] and ATP [53].

In a similar way, a fluorophore–quencher pair of labels can be used, whereby the fluorescence depends on the proximity of the quencher to the fluorophore. Fluorescence quenching occurs by FRET or static quenching or, mostly, by a combination of both. Static quenching involves a physical interaction between fluorophore and quencher and thus requires the pair to be closer to each other than in a FRET pair. This principle is realized in "molecular beacons", as described in Sect. 2.2.2.

A more straightforward approach is to take advantage of stacked dimer formation with two identical fluorescent labels. Stacking can occur with various aromatic dyes, such as xanthenes, which show strongly quenched fluorescence in the stacked state (Fig. 2.2d). This requires labelling the protein with only one type of fluorophore, but at two positions. The strategy has successfully been applied to develop protein-based biosensors for phosphate and ADP/GDP [54–56]. These consist of a specific binding protein labelled with two tetramethylrhodamines, positioned so they can form a stacked dimer in the ligand-free state with a shift in absorbance wavelength from monomeric to dimeric state. A conformation change upon ligand binding causes disruption of the dimer and thereby a fluorescence increase. While the stacked rhodamine dimer forms in the ground state of the molecules, other dyes form excited state dimers (excimers). Excimers show a shift in emission wavelength and a signal increase can be observed upon target binding when monitoring the dimer emission wavelength. An example of the application of an excimer label is given by the bis-pyrene-labelled aptamer, described in Sect. 2.2.2.

2.3.3 Modular Designs

In all biosensor types described so far, target recognition and signal transduction are in a single protein or nucleic acid recognition element. However, another option is a modular design, in which additional transducing elements are introduced into the same molecule to create a functional signal transduction mechanism. This is realized, for example, with a peptide or protein domain that binds to the recognition element only in the ligand-bound form (Fig. 2.2e). Target molecule binding brings the recognition domain and the interacting domain in close proximity, which can be converted to a fluorescence signal, for example, by a FRET pair. This principle is the basis for a group of genetically encoded, Ca^{2+} biosensors (cameleons) [42, 57]. They consist of calmodulin (CaM) as the recognition element and a calmodulin-binding peptide derived from myosin light chain kinase (M13), which are fused to either BFP/GFP or CFP/YFP as a FRET pair. Binding of Ca^{2+} to CaM induces wrapping of CaM around the M13 peptide. This conformation change brings the FPs into closer proximity and increases FRET. Calmodulin has four Ca^{2+} binding sites with a large difference in affinity (>100-fold) between the binding sites in the N- and C-terminal domain. This enables the biosensor to report a very wide range of Ca^{2+} concentrations ($<10^{-7}$ to $>10^{-4}$ M), which could be further increased by mutations in CaM to create versions with different affinities [42]. An example of these Ca^{2+} biosensors being applied to a motility system is a study in live *Caenorhabditis elegans*, where they have been used to measure Ca^{2+} transients in pharyngeal muscle [58]. Since the first cameleons were described, several new versions of CaM-based biosensors have been developed with improved characteristics such as increased brightness and dynamic range, improved pH-stability, folding efficiency and photostability [59, 60].

Another type of modular biosensor design has been described using an intramolecular ligand analogue together with the recognition domain [14] (Fig. 2.2f). In these semisynthetic sensors, the ligand analogue is coupled to the recognition protein using a genetically encoded SNAP-tag. In the absence of a target, the ligand analogue occupies the binding site of the recognition element giving the biosensor a closed conformation. The target molecule competes off the intramolecular ligand leading to an open conformation. The signal readout is by FRET, whereby one fluorophore is introduced on the linker of the ligand analogue and a second fluorophore is either an FP or a dye coupled to another self-labelling tag (CLIP-tag). Several examples of these biosensors, including ones for glutamate and GABA, have been described [13, 61, 14, 62].

2.4 Other Considerations: Suitability to Particular Types of Assay

Most of the description of biosensor development so far has concentrated on the overall design process. Some of the basic properties of fluorescent biosensors, which are important in terms of their applicability, are discussed below.

2.4.1 *Fluorophore*

Important properties of the fluorophore are the wavelength and whether it is (or they are) suitable for the selected type of readout, such as environmental change of fluorophore or FRET. Interference with the fluorescence signal can be a problem, and this may depend on the type of assay intended. For example, are the wavelengths compatible with other components of the assay, or do other components absorb light and so may be damaged by the high intensity of exciting light? Generally, fluorophores with longer excitation and emission wavelengths are less prone to artefacts due to optical interference.

The sensitivity of the assay is in part dependent on the choice of fluorophores, the inherent signal change and the intensity of exciting light. An inherently large fluorescence change, usually severalfold, is desirable with a preference for a fluorescence increase rather than decrease. For a biosensor with a single fluorophore, a large change in fluorescence quantum yield is needed to report on a change in the environment of the fluorophore. Choice of FRET pairs depends on the overlap of the spectra, and the literature cited in this review gives a number of suitable pairs of fluorescent proteins. However, as pointed out already, labelling a protein with two different chemical dyes can be a challenge.

Fluorophores with high extinction coefficient and high quantum yield (at least in the high fluorescence state) will increase the photons available for detection. High intensity of exciting light can also increase sensitivity, but the usable intensity may depend on the photostability of assay components, particularly the fluorophore, together with the time that the irradiation must be maintained. In single-molecule experiments where laser excitation is used, many fluorophores are photobleached after a few seconds. On the other hand, when exciting by, for example, mercury lamps used in some fluorimeters and stopped-flow apparatuses, the same fluorophore may be stable for many minutes: with lower power xenon lamps stability may be many hours.

Finally, the size of the fluorophore might be an important factor. Fluorescent dyes that are bright and absorb and emit at long wavelengths are generally larger than shorter wavelength variants and this might affect the recognition element negatively. For example, a large (and hydrophobic) fluorophore may limit the solubility of the biosensor.

2.4.2 Recognition Element

In the case of a protein framework, in particular, a good starting point is using a protein which has a known high-resolution structure to aid design and is robust to ensure stability. Important general properties must also be considered, including the concentration range that a biosensor can measure, the sensitivity of the detection and how fast it can respond to changes in concentration. A titration under the experimental conditions of the actual assay will give information about these factors, as shown as an example in Fig. 2.1c for the P_i biosensor.

Another point to consider is whether other components in the assay may produce a response by interacting with the recognition element. The ADP biosensor mentioned above is an example in which this problem has been overcome to allow measurement of ADP in the presence of structurally similar ATP [6].

Different types of assay require measurements over different concentration ranges: this might be "as low as possible" for many measurements in vitro, but much higher to reproduce cellular conditions. In the case of protein-based biosensors, in particular, there is a rational approach to vary the concentration range by varying the affinity of the protein for the target. This was demonstrated, for example, for a P_i biosensor [52] or Ca^{2+} biosensors [42, 44]. An issue then becomes reagent usage and cost, versus sensitivity. A tight-binding biosensor may bind the target stoichiometrically under assay conditions, and it needs to be in excess over the target analyte. By weakening the affinity, the biosensor can be used sub-stoichiometrically, as it now responds to changes in target concentration on either side of the dissociation constant [56]. Going sub-stoichiometric reduces sensitivity but also reduces reagent usage and cost.

2.4.3 Time Resolution

For time-resolved measurements, the speed of response is important. For simple reagentless biosensors, based on ligand binding, the speed depends largely on the kinetics of association and dissociation. Real-time measurements of motor proteins require biosensors that have a rapid response: many motor proteins, for example, hydrolyze ATP at $>100\ s^{-1}$ and move in >100 steps s^{-1}. Ideally a biosensor should have a response severalfold faster than the process being assayed to ensure the observed rate is purely that of the motor protein. If a high speed of response is required, then characterization may need to include the rate of binding: an example trace is shown in Fig. 2.1d for P_i binding to the P_i biosensor in tens of milliseconds.

2.4.4 *Biosensors for Cellular and Single Molecule Studies*

Apart from hybridization probes, which are applied in situ, and genetically encoded sensors, most biosensors are currently used mainly in vitro. This has several practical reasons, including difficulty of getting a functional biosensor into a cell. The choice of label is furthermore limited by the fact that many different, physical environments exist within a single cell. These could affect the behaviour of fluorophores leading to high background and rendering environmentally sensitive fluorophores inapplicable. In addition, discrimination may be needed against many molecules present within the cytoplasm, and there may be a wide range of concentrations of the target molecule in different subcellular compartments. It thus remains a major challenge to develop probes for measuring precise concentrations in live cells.

For transferring assays to single-molecule formats, such as total internal reflection fluorescence microscopy (TIRFM), immobilization of the fluorophore on the surface may be sufficient to obtain the readout, rather than actual fluorescence enhancement. Such a strategy was used to measure helicase unwinding (Sect. 2.5.6), whereby the environmentally sensitive coumarin fluorophore used for solution studies was replaced by a cyanine dye with relatively constant high quantum yield for TIRFM [63]. This also had the advantage of two other properties favourable for TIRFM, namely, a much more stable fluorophore and a higher wavelength.

2.5 Applications to Motor Proteins

2.5.1 *Regulation: Calcium*

A distinct application of biosensors and similar probes to motor proteins has been to monitor levels of molecules involved in regulation. Possibly the most-used class of small molecule indicators are those for calcium ions. A series of papers from Tsien described a range of such indicators with improved performances, such as the indo and fura series [64, 65]. These are based on an aromatic core containing multiple carboxyl groups, which form a complex with the calcium ion, inducing a fluorescent conformation in the indicator molecule. A key feature is their selectivity for calcium over magnesium, together with an affinity for calcium that allows the indicators to measure concentrations relevant for cell signalling, ranging from sub-micromolar to micromolar. These probes have been widely used, and early examples of their application in the muscle field were to obtain time-resolved information, especially in smooth muscle, in order to understand the control of contraction by calcium [66, 67]. An example is work that combined the use of the calcium indicator, fluo-3 [68], with caged compounds (Fig. 2.3a). The timed sequence of activation events for permeabilized smooth muscle was measured

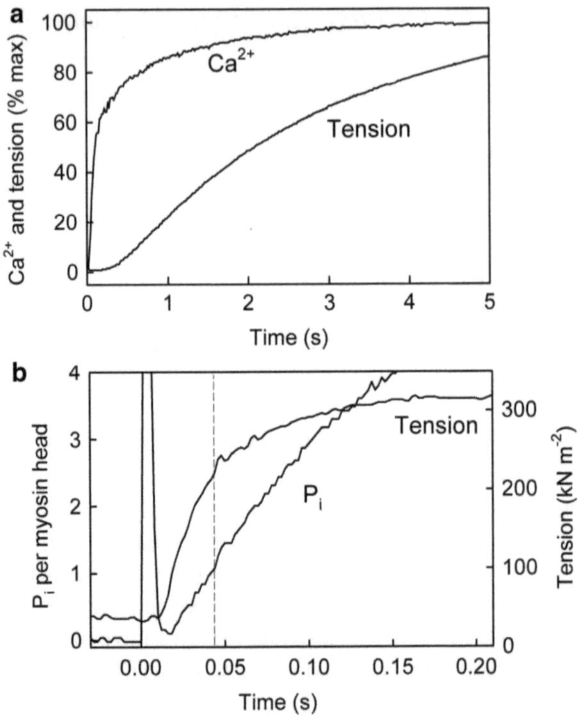

Fig. 2.3 Use of sensors to measure changes in target molecule concentration during activation of muscle contraction. (**a**) A Ca^{2+} sensor, fluo-3, was used to measure a Ca^{2+} transient leading to activation of smooth muscle force generation after release of inositol trisphosphate by photolysis of caged IP3 [69]. (**b**) A P_i biosensor, used to measure ATP usage during generation of force in striated muscle. Activation of the glycinerated muscle cell is by photolysis of caged ATP [70]. The spike at zero time is the flash photolysis artefact. The *vertical, dashed line* represents turnover of one ATP per myosin head (~150 μM) within the fibre, showing that most tension is generated concurrently with that first turnover

following rapid release of inositol trisphosphate (IP3) from its caged precursor [66, 69]. The IP3 concentration jump was followed by an increase in calcium, then in turn by muscle contraction.

2.5.2 *Mechanism*

Biosensors have also provided a way to get detailed, time-resolved information about motor protein mechanisms. Although some questions are likely to vary between different types of motors, there are some shared questions, particularly for linear motors, in which ATP hydrolysis drives movement along a track. A large number of such linear motors comprise those that move along DNA, such as helicases, and those that move along protein tracks, especially myosins on actin filaments, and kinesins and dyneins along microtubules. One common question for such linear motors is that of the processivity: how far the protein moves along the track before dissociation. A second question relates to the step size and how it is coupled to ATP hydrolysis. "Step size" can have various meanings, including what distance is moved for each discrete step: often the smallest discernible step comes out of single-molecule measurements. However, another type of step size describes how far the motor moves for each ATP used, and this can be designated as the "ATP-coupling step size". That leads to considerations of the efficiency of the motor and how movement is affected by load. An additional question, common to the different types of motor, linear or rotary, concerns mechanochemical coupling that understands how the ATPase cycle relates to movement, especially through protein conformation changes.

As described in Sect. 2.1, reagentless biosensors are particularly suited to measure reaction products, as they have only minimal impact on the system being studied. These types of measurements have played a significant part in understanding motor protein mechanism and defining their properties, particularly in respect to measuring the products of ATP hydrolysis. However, helicases unwind their nucleic acid track, making it possible to measure the extent of unwinding using a biosensor which binds the single-stranded DNA product of unwinding (see Sect. 2.5.6). In the case of protein tracks, such a measurement would be more difficult, because the motor proteins generally do not alter the track but also because it is more challenging to obtain protein tracks of uniform length in comparison with nucleic acids, where production of defined lengths is possible.

Because of its high speed, variants of the phosphate biosensor [1, 54] have been widely used to probe the ATP hydrolysis: under some conditions, its application gives the ATP usage relative to the extent of movement [5]. It also allows investigation of the P_i release step of the ATPase cycle of myosins [70, 4], kinesins [3] and helicases [71], which may be a key step in the transduction to produce movement and which otherwise is a difficult step to measure. Some specific examples of this detailed probing of ATP hydrolysis are discussed below.

2.5.3 Time-Resolved Measurements: Steady-State Kinetics

The simplest type of kinetic measurement is often to measure steady-state rates with the enzyme undergoing multiple turnovers. In the case of motor proteins, the concept is not quite straightforward as there is both movement and ATP hydrolysis that each can reach a steady state at different times. Although detailed interpretation of such measurements may be limited, steady-state measurements can provide a simple, reproducible measure of activity and a way to get basic kinetic parameters and information on the effectiveness of inhibitors. An example is given in Fig. 2.4a, in which the P_i biosensor is used to measure the ssDNA-activated ATPase activity of the helicase PcrA. The dependence on ATP concentration gives the basic

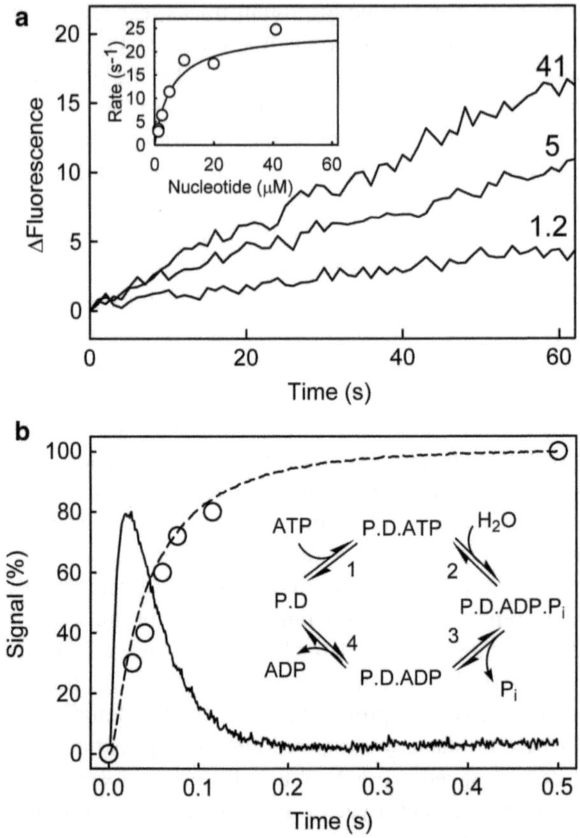

Fig. 2.4 P_i sensor used to measure PcrA helicase activity. (**a**) Steady-state ATPase assay with ssDNA, showing rate at micromolar ATP concentrations. The *inset* shows the concentration dependence of the ATPase rate, normalized to PcrA concentration (C.P. Toseland and M. R. Webb, unpublished). (**b**) Single-turnover measurement of mantATP binding (step 1 of inset with P. D being PcrA.DNA complex), cleavage (step 2), P_i release (step 3) and mantADP release (step 4) during dsDNA unwinding with RepD and PcrA. *Solid line*: mant fluorescence, showing mantATP binding (fluorescence increase) and subsequent mantADP release (decrease). *Circles*: quench-flow/HPLC measurement of mantATP cleavage. *Dashed line*: P_i release, measured using a P_i biosensor [71]

Michaelis–Menten parameters. The maximum rate gives a k_{cat} value of ~20 s^{-1}. However, this rate constant does not relate to ATPase activity during movement, but rather is limited by recycling the PcrA off the end of ssDNA and rebinding to another DNA strand. The ATP hydrolysis rate during translocation, obtained by a measurement analogous to that in Sect. 2.5.4, is severalfold faster [5].

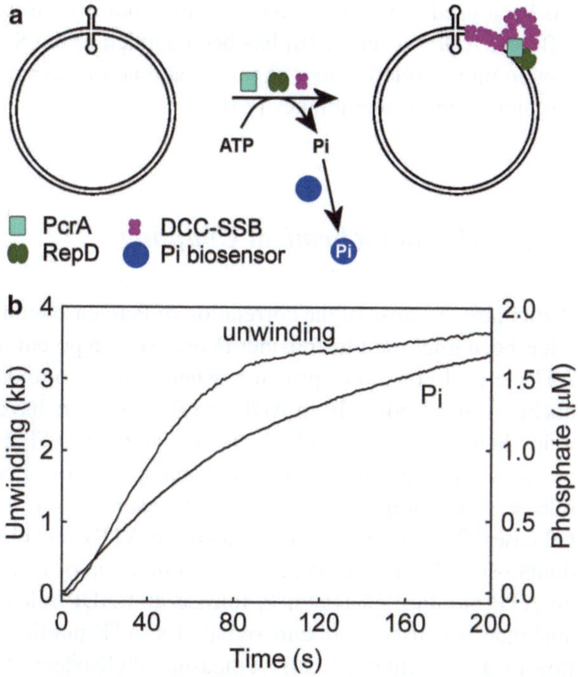

Fig. 2.5 P_i sensor used to measure PcrA unwinding plasmid. (**a**) *Cartoon* showing partial unwinding by PcrA and initiator protein, RepD in the presence of DCC-SSB and the P_i biosensor, MDCC-PBP [72]. (**b**) Single-turnover measurement of plasmid unwinding and P_i production

2.5.4 Single-Turnover Measurements

For reasons outlined above, well-defined tracks in terms of length can be produced for experiments involving nucleic acid motors. The phosphate biosensor [1] enabled the measurement of ATP usage in real time, while a DNA helicase, PcrA, translocated along a single-strand track of known length [5]. This method was then adapted to measure ATP usage during complete plasmid unwinding by this helicase (Fig. 2.5) [72]. The P_i produced can be correlated with the precise length of DNA: in this case the ATP-coupled step size is that one ATP is required for each base pair unwound.

A similar measurement cannot be done with a protein track, but nevertheless, measurement of ATP usage in a glycerinated muscle fibre was achieved [73]. In such preparations, the membrane of the muscle cell is disrupted, allowing introduction of the P_i biosensor and caged ATP. Both tension and P_i production could be measured in real time, following laser flash photolysis to release ATP (Fig. 2.3b). These types of measurement allowed the relationship between P_i release and force generation to be investigated: in skeletal muscle, there is a close correlation of these processes.

Although P_i measurements form the large majority of single-turnover measurements, ADP biosensors have also been used. Analogous to the use of the P_i biosensor, an ADP biosensor, or more correctly ADP/ATP ratio biosensor [35], has been used to measure ADP release from myosin within muscle fibres [74]. A different ADP biosensor [6] has been applied to the SecA kinetic mechanism: this protein moves relative to a peptide chain and is responsible for exporting proteins through bacterial membranes [75].

2.5.5 Mechanochemical Coupling

Particularly because of the correlation of P_i release with force generation in muscle, it has been important to consider P_i release as a potentially important process in the ATPase cycle of motor proteins, when aiming to understand the detailed mechanochemical coupling. Both ADP and P_i biosensors have been featured in studies of individual steps in the biochemical cycle of ATP hydrolysis. However, other methods such as fluorescent ADP analogues are available for measuring ADP release. An example of a P_i release measurement is shown in Fig. 2.4b for PcrA helicase. The experiment shows a single ATPase cycle of PcrA in the presence of double-stranded DNA, whereby a combination of measurements gave kinetic data on ATP binding, cleavage, P_i release and ADP release [71]. A fluorescent ATP analogue was used to obtain signals for ATP binding and ADP release. Quench-flow measurement gave a direct measure of cleavage. P_i release, in this case, had the same kinetics as cleavage, and ADP release was also similar, indicating that ATP cleavage is the rate-limiting step followed by rapid release of both products, ADP and P_i.

2.5.6 Measurement of Movement

As indicated above, biosensors are less often useful to measure motor protein movement. However, because helicases unwind double-stranded nucleic acids to give a single-stranded product, this change can be monitored. One way has been to use dsDNA-binding dyes to monitor this reaction [76], although there are potential problems as the substrate is altered, and the dye may affect the movement of the helicase. SSB binds highly selectively to ssDNA, a property which has been harnessed to probe DNA helicase activity. A fluorescent version, therefore, acts as an ssDNA biosensor, giving a fluorescence increase on binding to ssDNA [34]. This type of approach has been used to measure DNA unwinding in solution [72] (Fig. 2.5) and in a single-molecule format using TIRFM [63].

2.6 Summary and Future Directions

Fluorescent biosensors have facilitated new types of measurements for motor proteins, as well as for other systems, and so have greatly helped our understanding of such systems. However, often developments of biosensors have been driven by commercial applications, such as high-throughput drug screening, so the ease of preparation and the cost of production are likely to be important design considerations. Commercial applications, even more than in basic research, require simplicity and robustness, as well as affordability. Simplicity comes from both ease of application and large signal size. There are an ever increasing number of ways to design biosensors, some aimed at simplicity, but others, such as aptamers, having the advantage of being easily adapted to new targets. In most cases, there is unlikely to be, at least for a while, a universally applicable biosensor for any target, because different types of assays diverge greatly in what biosensor properties are most suitable.

Targets that are widely assayed are likely to continue to be addressed by new biosensors. Examples include important signalling molecules, such as glutamate, and important products of enzymes that are drug targets, such as ADP for kinases. Within the motility field, faster biosensors would aid real-time measurements. New biosensors for regulatory molecules other than calcium are desirable, especially if suitable for intracellular measurements. In the broader sense of motor proteins, assays for polymerases would benefit from a biosensor that is specific for pyrophosphate. While ingenuity can potentially lead to a wide array of biosensors with varying complexity, it is likely that the ones that can be applied simply and easily to assays for such molecules will also have widespread appeal.

Acknowledgements The authors thank the Medical Research Council (U117512742) and MRC Technology for financial support and the various co-workers who are authors on papers from this laboratory, cited herein.

References

1. Brune M, Hunter JL, Corrie JET, Webb MR (1994) Direct, real-time measurement of rapid inorganic phosphate release using a novel fluorescent probe and its application to actomyosin subfragment 1 ATPase. Biochemistry 33:8262–8271
2. Brune M, Hunter JL, Howell SA, Martin SR, Hazlett TL, Corrie JET, Webb MR (1998) Mechanism of inorganic phosphate interaction with phosphate binding protein from *Escherichia coli*. Biochemistry 37:10370–10380
3. Gilbert SP, Webb MR, Brune M, Johnson KA (1995) Pathway of processive ATP hydrolysis by kinesin. Nature 373:671–676
4. White HD, Belknap B, Webb MR (1997) Kinetics of nucleoside triphosphate cleavage and phosphate release steps by associated rabbit skeletal actomyosin, measured using a novel fluorescent probe for phosphate. Biochemistry 36:11828–11836

5. Dillingham MS, Wigley DB, Webb MR (2000) Demonstration of unidirectional single-stranded DNA translocation by PcrA helicase: measurement of step size and translocation speed. Biochemistry 39:205–212
6. Kunzelmann S, Webb MR (2009) A biosensor for fluorescent determination of ADP with high time resolution. J Biol Chem 284:33130–33138
7. Marvin JS, Hellinga HW (2001) Conversion of a maltose receptor into a zinc biosensor by computational design. Proc Natl Acad Sci USA 98(9):4955–4960
8. Looger LL, Dwyer MA, Smith JJ, Hellinga HW (2003) Computational design of receptor and sensor proteins with novel functions. Nature 423(6936):185–190
9. Frommer WB, Davidson MW, Campbell RE (2009) Genetically encoded biosensors based on engineered fluorescent proteins. Chem Soc Rev 38(10):2833–2841
10. VanEngelenburg SB, Palmer AE (2008) Fluorescent biosensors of protein function. Curr Opin Chem Biol 12(1):60–65
11. Keppler A, Gendreizig S, Gronemeyer T, Pick H, Vogel H, Johnsson K (2003) A general method for the covalent labeling of fusion proteins with small molecules in vivo. Nat Biotechnol 21(1):86–89
12. Gautier A, Juillerat A, Heinis C, Correa IR Jr, Kindermann M, Beaufils F, Johnsson K (2008) An engineered protein tag for multiprotein labeling in living cells. Chem Biol 15(2):128–136
13. Brun MA, Griss R, Reymond L, Tan KT, Piguet J, Peters RJ, Vogel H, Johnsson K (2011) Semisynthesis of fluorescent metabolite sensors on cell surfaces. J Am Chem Soc 133(40):16235–16242
14. Brun MA, Tan KT, Nakata E, Hinner MJ, Johnsson K (2009) Semisynthetic fluorescent sensor proteins based on self-labeling protein tags. J Am Chem Soc 131(16):5873–5884
15. Juskowiak B (2011) Nucleic acid-based fluorescent probes and their analytical potential. Anal Bioanal Chem 399(9):3157–3176
16. Morrison LE, Halder TC, Stols LM (1989) Solution-phase detection of polynucleotides using interacting fluorescent labels and competitive hybridization. Anal Biochem 183(2):231–244
17. Ellington AD, Szostak JW (1990) In vitro selection of RNA molecules that bind specific ligands. Nature 346(6287):818–822
18. Tuerk C, Gold L (1990) Systematic evolution of ligands by exponential enrichment: RNA ligands to bacteriophage T4 DNA polymerase. Science 249(4968):505–510
19. Fang X, Tan W (2010) Aptamers generated from cell-SELEX for molecular medicine: a chemical biology approach. Acc Chem Res 43(1):48–57
20. Keefe AD, Pai S, Ellington A (2010) Aptamers as therapeutics. Nat Rev Drug Discov 9(7):537–550
21. Iliuk AB, Hu L, Tao WA (2011) Aptamer in bioanalytical applications. Anal Chem 83(12):4440–4452
22. Liu Y, Matharu Z, Howland MC, Revzin A, Simonian AL (2012) Affinity and enzyme-based biosensors: recent advances and emerging applications in cell analysis and point-of-care testing. Anal Bioanal Chem 404(4):1181–1196
23. Yamana K, Ohtani Y, Nakano H, Saito I (2003) Bis-pyrene labeled DNA aptamer as an intelligent fluorescent biosensor. Bioorg Med Chem Lett 13(20):3429–3431
24. Yamana K, Iwai T, Ohtani Y, Sato S, Nakamura M, Nakano H (2002) Bis-pyrene-labeled oligonucleotides: sequence specificity of excimer and monomer fluorescence changes upon hybridization with DNA. Bioconjug Chem 13(6):1266–1273
25. Tyagi S, Bratu DP, Kramer FR (1998) Multicolor molecular beacons for allele discrimination. Nat Biotechnol 16(1):49–53
26. Wang RE, Zhang Y, Cai J, Cai W, Gao T (2011) Aptamer-based fluorescent biosensors. Curr Med Chem 18(27):4175–4184
27. Mascini M, Palchetti I, Tombelli S (2012) Nucleic acid and peptide aptamers: fundamentals and bioanalytical aspects. Angew Chem Int Ed 51(6):1316–1332
28. Song KM, Lee S, Ban C (2012) Aptamers and their biological applications. Sensors 12(1):612–631

29. Zeng X, Shen Z, Mernaugh R (2012) Recombinant antibodies and their use in biosensors. Anal Bioanal Chem 402(10):3027–3038
30. Terai T, Nagano T (2013) Small-molecule fluorophores and fluorescent probes for bioimaging. Pflugers Arch 465(3):347–359
31. Hirshberg M, Henrick K, Haire LL, Vasisht N, Brune M, Corrie JE, Webb MR (1998) Crystal structure of phosphate binding protein labeled with a coumarin fluorophore, a probe for inorganic phosphate. Biochemistry 37(29):10381–10385
32. De Lorimier RM, Smith JJ, Dwyer MA, Looger LL, Sali KM, Paavola CD, Rizk SS, Sadigov S, Conrad DW, Loew L, Hellinga HW (2002) Construction of a fluorescent biosensor family. Protein Sci 11(11):2655–2675
33. Dwyer MA, Hellinga HW (2004) Periplasmic binding proteins: a versatile superfamily for protein engineering. Curr Opin Struct Biol 14(4):495–504
34. Dillingham MS, Tibbles KL, Hunter JL, Bell JC, Kowalczykowski SC, Webb MR (2008) Fluorescent single-stranded DNA binding protein as a probe for sensitive, real time assays of helicase activity. Biophys J 95:3330–3339
35. Brune M, Corrie JET, Webb MR (2001) A fluorescent sensor of the phosphorylation state of nucleoside diphosphate kinase and its use to monitor nucleoside diphosphate concentrations in real time. Biochemistry 40:5087–5094
36. Berg J, Hung YP, Yellen G (2009) A genetically encoded fluorescent reporter of ATP:ADP ratio. Nat Methods 6(2):161–166
37. Nakai J, Ohkura M, Imoto K (2001) A high signal-to-noise Ca(2+) probe composed of a single green fluorescent protein. Nat Biotechnol 19(2):137–141
38. Baird GS, Zacharias DA, Tsien RY (1999) Circular permutation and receptor insertion within green fluorescent proteins. Proc Natl Acad Sci USA 96(20):11241–11246
39. Sakaguchi R, Endoh T, Yamamoto S, Tainaka K, Sugimoto K, Fujieda N, Kiyonaka S, Mori Y, Morii T (2009) A single circularly permuted GFP sensor for inositol-1,3,4,5-tetrakisphosphate based on a split PH domain. Bioorg Med Chem 17(20):7381–7386
40. Marvin JS, Borghuis BG, Tian L, Cichon J, Harnett MT, Akerboom J, Gordus A, Renninger SL, Chen TW, Bargmann CI, Orger MB, Schreiter ER, Demb JB, Gan WB, Hires SA, Looger LL (2013) An optimized fluorescent probe for visualizing glutamate neurotransmission. Nat Methods 10(2):162–170
41. Bizzarri R, Serresi M, Luin S, Beltram F (2009) Green fluorescent protein based pH indicators for in vivo use: a review. Anal Bioanal Chem 393(4):1107–1122
42. Miyawaki A, Llopis J, Heim R, McCaffery JM, Adams JA, Ikura M, Tsien RY (1997) Fluorescent indicators for Ca2+ based on green fluorescent proteins and calmodulin. Nature 388:882–887
43. Mank M, Reiff DF, Heim N, Friedrich MW, Borst A, Griesbeck O (2006) A FRET-based calcium biosensor with fast signal kinetics and high fluorescence change. Biophys J 90(5): 1790–1796
44. Heim N, Griesbeck O (2004) Genetically encoded indicators of cellular calcium dynamics based on troponin C and green fluorescent protein. J Biol Chem 279(14):14280–14286
45. Nikolaev VO, Bunemann M, Schmitteckert E, Lohse MJ, Engelhardt S (2006) Cyclic AMP imaging in adult cardiac myocytes reveals far-reaching beta1-adrenergic but locally confined beta2-adrenergic receptor-mediated signaling. Circ Res 99(10):1084–1091
46. DiPilato LM, Cheng X, Zhang J (2004) Fluorescent indicators of cAMP and Epac activation reveal differential dynamics of cAMP signaling within discrete subcellular compartments. Proc Natl Acad Sci USA 101(47):16513–16518
47. Honda A, Adams SR, Sawyer CL, Lev-Ram V, Tsien RY, Dostmann WR (2001) Spatiotemporal dynamics of guanosine 3',5'-cyclic monophosphate revealed by a genetically encoded, fluorescent indicator. Proc Natl Acad Sci USA 98:2437–2442
48. Nikolaev VO, Gambaryan S, Lohse MJ (2006) Fluorescent sensors for rapid monitoring of intracellular cGMP. Nat Methods 3(1):23–25

49. Fehr M, Frommer WB, Lalonde S (2002) Visualization of maltose uptake in living yeast cells by fluorescent nanosensors. Proc Natl Acad Sci USA 99(15):9846–9851
50. Fehr M, Lalonde S, Lager I, Wolff MW, Frommer WB (2003) In vivo imaging of the dynamics of glucose uptake in the cytosol of COS-7 cells by fluorescent nanosensors. J Biol Chem 278(21):19127–19133
51. Lager I, Fehr M, Frommer WB, Lalonde S (2003) Development of a fluorescent nanosensor for ribose. FEBS Lett 553(1–2):85–89
52. Gu H, Lalonde S, Okumoto S, Looger LL, Scharff-Poulsen AM, Grossman AR, Kossmann J, Jakobsen I, Frommer WB (2006) A novel analytical method for in vivo phosphate tracking. FEBS Lett 580(25):5885–5893
53. Imamura H, Nhat KP, Togawa H, Saito K, Iino R, Kato-Yamada Y, Nagai T, Noji H (2009) Visualization of ATP levels inside single living cells with fluorescence resonance energy transfer-based genetically encoded indicators. Proc Natl Acad Sci USA 106(37):15651–15656
54. Okoh MP, Hunter JL, Corrie JET, Webb MR (2006) A biosensor for inorganic phosphate using a rhodamine-labeled phosphate binding protein. Biochemistry 45:14764–14771
55. Kunzelmann S, Webb MR (2011) Fluorescence detection of GDP in real time with the reagentless biosensor rhodamine-ParM. Biochem J 440(1):43–49
56. Kunzelmann S, Webb MR (2010) A fluorescent, reagentless biosensor for ADP based on tetramethylrhodamine-labeled ParM. ACS Chem Biol 5(4):415–425
57. Miyawaki A, Griesbeck O, Heim R, Tsien RY (1999) Dynamic and quantitative Ca2+ measurements using improved cameleons. Proc Natl Acad Sci USA 96(5):2135–2140
58. Kerr R, Lev-Ram V, Baird G, Vincent P, Tsien RY, Schafer WR (2000) Optical imaging of calcium transients in neurons and pharyngeal muscle of C. elegans. Neuron 26(3):583–594
59. Palmer AE, Qin Y, Park JG, McCombs JE (2011) Design and application of genetically encoded biosensors. Trends Biotechnol 29(3):144–152
60. Mank M, Griesbeck O (2008) Genetically encoded calcium indicators. Chem Rev 108(5): 1550–1564
61. Brun MA, Tan KT, Griss R, Kielkowska A, Reymond L, Johnsson K (2012) A semisynthetic fluorescent sensor protein for glutamate. J Am Chem Soc 134(18):7676–7678
62. Masharina A, Reymond L, Maurel D, Umezawa K, Johnsson K (2012) A fluorescent sensor for GABA and synthetic GABA(B) receptor ligands. J Am Chem Soc 134(46):19026–19034
63. Fili N, Mashanov G, Toseland CP, Batters C, Wallace MI, Yeeles JTP, Dillingham MS, Webb MR, Molloy JE (2010) Visualizing DNA unwinding by helicases at the single molecule level. Nucleic Acids Res 38:4448–4457
64. Grynkiewicz G, Poenie M, Tsien R (1985) A new generation of Ca2+ indicators with greatly improved fluorescence properties. J Biol Chem 260(6):3440–3450
65. Tsien RY (1980) New calcium indicators and buffers with high selectivity against magnesium and protons: design, synthesis, and properties of prototype structures. Biochemistry 19(11): 2396–2404
66. Somlyo AP, Somlyo AV (1990) Flash photolysis studies of excitation-contraction coupling, regulation, and contraction in smooth muscle. Annu Rev Physiol 52:857–874
67. Himpens B, Somlyo AP (1988) Free-calcium and force transients during depolarization and pharmacomechanical coupling in guinea-pig smooth muscle. J Physiol 395(1):507–530
68. Minta A, Kao JP, Tsien RY (1989) Fluorescent indicators for cytosolic calcium based on rhodamine and fluorescein chromophores. J Biol Chem 264(14):8171–8178
69. Somlyo AV, Horiuti K, Trentham DR, Kitazawa T, Somlyo AP (1992) Kinetics of Ca2+ release and contraction induced by photolysis of caged D-myo-inositol 1,4,5-trisphosphate in smooth muscle. The effects of heparin, procaine, and adenine nucleotides. J Biol Chem 267 (31):22316–22322
70. He Z-H, Chillingworth RK, Brune M, Corrie JET, Trentham DR, Webb MR, Ferenczi MA (1997) ATPase kinetics on activation of rabbit and frog permeabilized isometric muscle fibres: a real time phosphate assay. J Physiol 501:125–148

71. Toseland CP, Martinez-Senac MM, Slatter AF, Webb MR (2009) The ATPase cycle of PcrA helicase and its coupling to translocation on DNA. J Mol Biol 392:1020–1032
72. Slatter AF, Thomas CD, Webb MR (2009) PcrA helicase tightly couples ATP hydrolysis to unwinding double-stranded DNA, modulated by the replication initiator protein, RepD. Biochemistry 48:6326–6334
73. He Z-H, Chillingworth RK, Brune M, Corrie JET, Webb MR, Ferenczi MA (1999) The efficiency of contraction in rabbit skeletal muscle fibres, determined from the rate of release of inorganic phosphate. J Physiol 517:839–854
74. West TG, Hild G, Siththanandan VB, Webb MR, Corrie JET, Ferenczi MA (2009) Time course and strain dependence of ADP release during contraction of permeabilized skeletal muscle fibers. Biophys J 96(8):3281–3294
75. Das S, Grady LM, Michtavy J, Zhou Y, Cohan FM, Hingorani MM, Oliver DB (2012) The variable subdomain of *Escherichia coli* SecA functions to regulate SecA ATPase activity and ADP release. J Bacteriol 194(9):2205–2213
76. Eggleston AK, Rahim NA, Kowalczykowski SC (1996) A helicase assay based on the displacement of fluorescent, nucleic acid-binding ligands. Nucleic Acids Res 24(7):1179–1186
77. Ledvina PS, Yao N, Choudhary A, Quiocho FA (1996) Negative electrostatic surface potential of protein sites specific for anionic ligands. Proc Natl Acad Sci USA 93:6786–6791

71. [illegible] and C.P. [illegible] (2005) The ATPase cycle of [illegible]
[illegible]

Chapter 3
Rapid Reaction Kinetic Techniques

Christopher P. Toseland and Michael A. Geeves

Abstract Most biochemical processes occur on sub-second time scales. Relaxation and rapid mixing methods allow reactions from microsecond time scales onwards to be monitored in real time. This chapter describes the instrumentation for these techniques and it discusses general topics of sample excitation and signal detection.

Keywords Kinetics • Stopped-flow • Fluorescence • Motor protein • Relaxation kinetics

Abbreviations

CCD	Charge-coupled device
Deac	Diethylaminocoumarin
DWR	Direct wave recording
FRET	Förster (or fluorescence) resonance energy transfer
k_{cat}	Catalytic turnover number
K_d	Equilibrium dissociation constant
K_m	Michaelis constant
k_{obs}	Observed rate constant
LED	Light-emitting diode
Mant	2′(3′)-*O*-(*N*-Methylanthraniloyl)-
NATA	*N*-Acetyl tryptophanamide

C.P. Toseland (✉)
Chromosome Organisation and Dynamics, Max-Planck Institute of Biochemistry, Martinsried 82152, Germany
e-mail: toseland@biochem.mpg.de

M.A. Geeves
School of Biosciences, University of Kent, Canterbury CT2 7NJ, UK
e-mail: m.a.geeves@kent.ac.uk

C.P. Toseland and N. Fili (eds.), *Fluorescent Methods for Molecular Motors*, Experientia Supplementum 105, DOI 10.1007/978-3-0348-0856-9_3,

P_i	Inorganic phosphate
PMT	Photomultiplier tube
λ_{ex}	Excitation wavelength

3.1 Introduction

Many biological processes occur on sub-second time scales. These can be specific processes such as DNA replication by DNA polymerase which can replicate hundreds of bases every second. Or they can be multiple processes which enable a bird to flap its wings 50 times a second. In order to characterise these processes, it is important to be able to determine the dynamic individual steps of these reactions. Steady-state measurements are useful for determination of global kinetic parameters such as the dissociation constant, K_d, the Michaelis constant, K_m, and the rate of catalytic turnover, k_{cat}, but these parameters are made up of many individual molecular events which are only revealed by transient kinetics. These events include ligand binding (bimolecular reactions) and dissociation, protein isomerisations (conformational changes) and catalytic events (bond breaking and bond making).

Most biological reactions are reversible and therefore reach an equilibrium, or steady state, under experimental conditions. This equilibrium can be perturbed through addition or removal of reactants or through changes in external parameters such as pressure and temperature [1]. Both the forward and reverse rate constants contribute to the rate at which equilibrium is achieved. Therefore, a more dynamic system, with faster forward and reverse rate constants, will reach equilibrium more quickly. It is this approach to equilibrium which is monitored using the methods to rapidly mix reactants or perturb the system. It is then possible to measure the forward and reverse rate constants of individual process to determine how rapidly the system can adapt.

Fluorescence is ideal to study these reactions because fluorescence itself is a rapid process, pico- to nanosecond for the absorption and re-emission of the photon, compared to the micro- to second time scale of protein-based biochemical reactions [2]. But, standard spectrophotometers are not sufficient to measure these processes, in which manual mixing and initiating the measurement takes several seconds. The stopped-flow technique allows measurements of reactions typically from <1 to 3 ms onwards. This apparatus rapidly mixes two solutions and triggers data acquisition automatically. Continuous flow can initiate the reaction in as little as 50 μs but with a much greater use of reactants [3]. Relaxation techniques, such as pressure jump and temperature jump, do not mechanically mix solutions, but they perturb the equilibrium through environmental changes [4, 5]. Therefore, these methods are able to measure reactions on fast time scales of 3–50 μs [1].

This chapter will highlight instrumentation available for fluorescent measurements using rapid mixing stopped-flow and relaxation methods, pressure and temperature jump. Firstly, common methods of sample illumination and detection will be discussed before reviewing the individual techniques.

3.2 Sample Illumination and Detection

3.2.1 Light Sources and Sample Excitation

Illumination is typically from arc lamps (xenon, mercury, xenon/mercury) to provide high-intensity sources. A typical light path for a stopped-flow apparatus is shown in Fig. 3.1a. Xenon arcs provide a smooth excitation from 200 to 800 nm, while mercury or xenon/mercury lamps have defined intensity spikes. These peaks are compatible for excitation of many common fluorophores such as mant, coumarin and Cy3. The peak illumination can be advantageous because the concentration of reactants can be reduced and/or the demands on the detectors can be reduced to increase signal-to-noise ratio. More recently, fast switchable LEDs are becoming bright enough to compete with arc lamps for many uses. Their stability, versatility and economy mean that they may replace arc lamps in all but the most exacting systems in the near future. Currently the wavelengths available are limited in the near UV but the range is constantly being expanded.

The excitation wavelength is typically selected using a monochromator but some systems use glass filters. Filters allow the most light to enter the system while monochromators typically cause a 30 % intensity decrease. However, filters can excite the sample over a relatively broad range, whereas monochromators allow a specific narrow wavelength to be selected. Therefore, a decision has to be made over illumination intensity and spectral purity. Monochromator systems have slits to control the amount of light entering the sample and adjust spectral purity. If the fluorophore has a large Stokes shift, then the slit widths can be large to increase the intensity and sensitivity. Small Stokes shift fluorophores will need narrow slits to reduce the direct excitation light reaching the detector. Reducing the slit widths, to reduce total light intensity, can also help to reduce photobleaching which is rarely a problem with rapid reaction measurements that are complete in less than a second but may occur over long reaction times. Alternatively, shutter systems can be configured with the excitation system. In this set-up the sample will only be illuminated at set intervals. Therefore, while time scales are not limited, classical spectrophotometers may be better for reactions on the order of minutes.

A fibre optic cable is then used to couple the light into the cuvette. This reduces noise from the monochromator mechanics upon the cuvette. A lens focuses the excitation light upon the cuvette to ensure light is efficiently acting upon the sample. For fluorescence observation cells are typically 1–2 mm square cross section and 5–10 mm long. Excitation along the long path length using a narrowly focused beam can be the best approach for very low concentrations of fluorophore, but at high concentrations of fluorophore, inner filter effects (in which the intensity

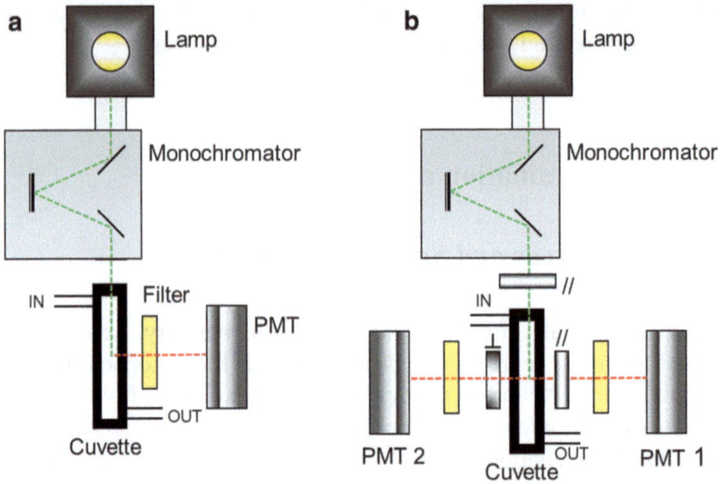

Fig. 3.1 Stopped-flow optical set-up. (**a**) Fluorescence intensity set-up. White light from a lamp passes through a monochromator, passing on to the solution at a chosen wavelength. Emission is measured at 90° through an emission cut-off filter, and the intensity measured using a photomultiplier tube (PMT). (**b**) Fluorescence anisotropy set-up. White light from a lamp passes through a monochromator, and a vertical polariser (||), passing on to the solution with vertically polarised light at a chosen wavelength. Emission is measured using two channels simultaneously, both at 90° to the excitation plane. Each channel has either a parallel (||) or perpendicular polariser (⊥), which polarises the emitted light before going through an emission cut-off filter, at the appropriate wavelength. The intensity is measured using a photomultiplier tube (PMT) for each channel

of the exciting light diminishes along the length of the cell) can reduce the signal strength. For high concentrations of fluorophore, excitation along the short path length may be preferred.

3.2.1.1 Fluorophore Selection

To have a strong signal, the fluorophores also have to be considered. Switching fluorophores is the easiest method to improve measurements compared to hardware alterations of the system. Proteins contain fluorescent amino acids, which include tryptophan, and these can be used to report upon biological processes. However, extrinsic fluorophores provide a greater variability for selecting the probes, but these modifications can perturb the activity of the protein. A greater discussion of these issues is presented in Chap. 1, but a few key properties are discussed below. There are three important factors to consider, the brightness, excitation and emission wavelengths, and stability.

The ease of observing a fluorophore is based upon its brightness. This is defined by two parameters, its extinction coefficient and quantum yield [2, 6]. The extinction coefficient is a measure of the ease in which a fluorophore absorbs light and the quantum yield is related to the probability of a photon being emitted through fluorescence. Therefore, a molecule with a high quantum yield and extinction coefficient is the brightest. Bright fluorophores allow lower concentrations of reactants to be used which can provide an advantage for concentration-dependent reactions and it can improve the signal-to-noise ratio. It also has to be noted that molecules with a high extinction coefficient will absorb more light and can increase inner filter effects at lower concentrations. The extinction coefficient and quantum yield are frequently affected by the environmental factors such as pH, solvents and viscosity. These properties can be harnessed to produce large fluorescent changes to measure binding events or conformation changes within the experimental reaction. To achieve this fluorescence change not only requires selection of a suitable fluorophore but also needs correct positioning of the fluorophore, as described in more detail in Chap. 1 and here [7].

The difference between the excitation and emission wavelength is the Stokes shift, and the degree of shift reflects the loss of energy due to non-radiative decay before the release of a photon [2, 6]. In general, larger Stokes shifts are useful because the emission peak is further from the excitation source. The slit widths of the excitation light source are important for fluorophores with a small Stokes shift, in order to prevent direct illumination of the detector. The excitation and emission wavelengths are more important for Förster (or fluorescence) resonance energy transfer (FRET) applications. FRET requires spectral overlap between the emission of one fluorophore (the donor) and the excitation of another (the acceptor) in order to generate a signal when the two fluorophores are spaced within 10 nm of each other [6].

Fluorophores will go through many cycles of excitation and emission but all will eventually be irreversibly deactivated or photobleached. During these cycles, the fluorophore can transfer from the singlet state to the highly reactive triplet state. It is in this state that the fluorophore can become permanently deactivated through reactions with oxygen and free radicals [8]. This mostly occurs with high-intensity light sources such as lasers for microscopy. Photobleaching is rarely a problem in reactions of less than 1 s. However, stability can be affected by reaction conditions, such as temperature and pH.

3.2.2 Detection and Data Recording

In a typical system, fluorescence detection is performed at 90° to the excitation. Photomultipliers (PMT) are used to detect light (approximately 160–800 nm) (Fig. 3.1a). Depending on the applications, it may be useful to use PMTs dedicated to UV, visible or near-red wavelengths to ensure the best sensitivity of detection.

Optical filters are positioned in front of the PMTs. The filters function to reduce excitation light and scattering from three sources: (1) Rayleigh at the excitation source wavelength, (2) the 1st harmonic of excitation source ($2 \times \lambda_{ex}$) and (3) Ramen from water ($1/\lambda_{ex} - 3.4 \times 10^{-4}$). Filters have to be carefully chosen to maximise the fluorescence intensity. Cut-off filters block all light below the selected wavelength but allow everything above to pass through to the detector. This is advantageous for detecting all emitted light but scattering can be an issue. Band-pass filters allow a selected range of wavelengths to pass through to the detectors. While scattering is reduced with these filters, they may cut the emission and thus decrease the fluorescence signal. Signal detection can be enhanced by selecting a fluorophore which will yield a large fluorescence change during the measurement. Furthermore, using fluorescent probes in the visible spectra will push the first harmonic scattering into the red spectra whereby photomultipliers are less responsive.

3.2.2.1 Improving Signal-to-Noise Ratio

Several modifications can be made to the detection methods to improve the signal-to-noise ratio. The signal can be affected by electronic noise, which is limited by the response time of the electronics (time constant). The time constant can be increased to remove this noise from the signal. It is important to have a time constant long enough to remove noise, but it must not be too long whereby it determines the observed signal. If the reaction observed is a single exponential process, then a simple rule of thumb of the time constant being <1 % of the reaction time required to capture the data is sufficient. If the reaction is more complex, then more care is required and the time constant must be <10 % of the relaxation time (1/observed rate constant) of the fastest event to be measured. Note that in some commercial systems, the time constant is selected automatically based on the collection time used. This may not always be appropriate for some types of reaction. If collecting data in a linear time scale, it is important to perform mixes over long and short time scales for accurate fitting of the data. However, for sample economy, collecting data on logarithmic time scales allows the initial data to be collected at short time intervals, and time intervals increase as the reaction progresses. This allows a sufficient number of data points to be collected for both fast and slow processes.

Data oversampling can also be used to improve the signal-to-noise ratio. Oversampling allows data to be acquired as fast as possible, but instead of displaying every data point, the data is averaged and presented with the selected number of data points. For example, a 2 s trace could be sampled every 1 ms but displayed with 512 data points. This means every data point is the average of 4 measurements. It is important to remember that the signal-to-noise ratio increases with the square root of the number of samples averaged.

Noise can be further reduced by using two photomultipliers in T format, one detector each side of the sample chamber, and averaging the signal from both detectors. Most system will already have two PMTs for performing fluorescence polarisation/anisotropy measurements.

3.3 Rapid Mixing

3.3.1 Stopped-Flow Apparatus

In a basic system, two reactants are driven from separate syringes into a mixing chamber before delivering the mixed sample to an observation chamber (Fig. 3.2a). As solution enters the observation chamber, the old solution is ejected into a stop syringe. Once filled to a set volume, the flow stops and triggers data acquisition. The reaction is then observed in real time as the transient approaches equilibrium. In this chapter we will only focus upon fluorescent acquisition, but this technique can be coupled to essentially any optical systems, e.g. UV/visible absorbance, CD, IR or any other measuring system with sufficient sensitivity and time resolution (NMR, ESR, calorimetry, conductivity).

3.3.1.1 Mixing

The two solutions are driven into the sample chamber by pneumatic drives or motors to rapidly mix the solutions. Typically, the two reactants are mixed with an equal volume, but this can be varied by using different size syringes or two syringes with independently driven motors. This can be advantageous for performing rapid buffer changes or dilutions which is often employed during protein folding experiments. The reproducibility of the drive mechanism is important to ensure the kinetic measurements are reliable. Stepper motors can deliver sub-microlitre precision, and now servo motors are being introduced to provide higher reproducibility and reliability. In general the motor-based systems provide a higher reliability over air-driven systems because the flow is not as affected by temperature and sample viscosity. However, motors will wear over time more readily than air-driven systems. Driving the solutions together causes a pressure increase within the flow circuits which has to be released during the experiment. The timing of this pressure release is important in relation to data acquisition in order to avoid data artefacts. This can be adjusted from a few milliseconds to a second after flow stops to ensure any artefact does not interfere with the reaction being observed.

Importantly, the flow speed must be limited to prevent cavitation whereby low pressure creates air bubbles within the flow circuits. Furthermore, high flow rates may generate shear effects upon long polymers, such as actin filaments, microtubules, DNA or lipid vesicles. Orientation of long molecules in the flow which then relax to a disordered state when flow stops can also lead to optical artefacts.

Efficient mixing is vital for the technique to work accurately. The efficiency of mixing itself is dependent upon flow speed, temperature and viscosity. Many commercial systems use a T-mixer, but specific systems use novel complex mixers to provide a greater mixing efficiency over a range of temperatures and viscosities. It is important to measure the mixing efficiency of the apparatus, especially as the

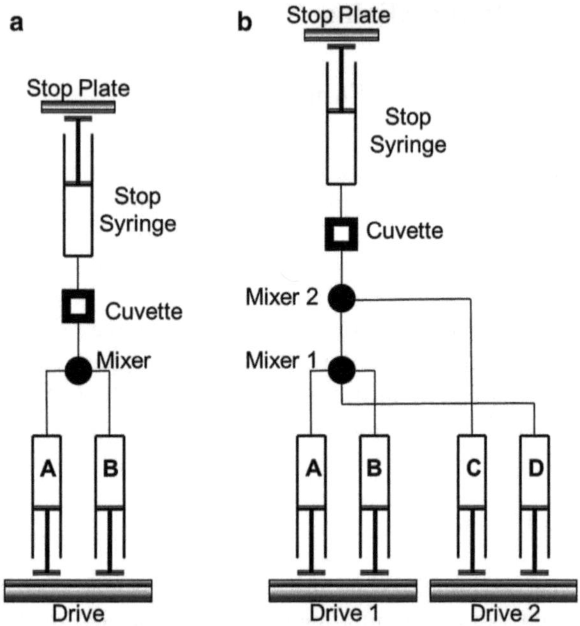

Fig. 3.2 Rapid mixing stopped-flow systems. (**a**) Single-mix system. *Reactants A and B* are rapidly mixed and the flow is stopped using the stopper plate. The fluorescence of the mixed solution is measured in the cuvette. (**b**) Double-mix system. *Reactants A and B* are rapidly mixed (Mixer 1). After a chosen time period, a second mix occurs (Mixer 2). The fluorescence following this mix is observed in the cuvette

machine ages. In order to perform this measurement, it is important to use a reaction which occurs essentially instantaneously (within the mixing time of the apparatus). The principle being, if mixing is efficient then a reaction transient is not observed, but if mixing is inefficient then a transient would be observed as the reaction is slowed due to poor mixing. Example test systems include pH indicators because proton transfer reactions are “instantaneous” compared to the mixing times.

3.3.1.2 Dead-Time

The mixing point and observation point are spatially distinct; therefore the initial reaction can be missed until the solution flows to the observation point. This time is referred to as the dead-time. For most lab systems, this time is typically 1–2 ms but new machines are capable of sub-millisecond dead-times. Reactions within or faster than this time frame (1,000 s^{-1}) cannot be accurately studied so relaxation methods maybe more applicable for these biological systems. For single exponential reactions, the observable signal amplitude can be calculated by

$$A/A_0 = e^{-kd} \tag{3.1}$$

where A_0 is the complete amplitude, k is the rate constant and d is the dead-time [9]. Therefore, for a machine with a dead-time of 1.5 ms, 98 % of the reaction will be visible for a rate constant of 10 s^{-1}. But, this decreases to 22 % for a rate constant of 1,000 s^{-1}. Figure 3.3 illustrates this effect for a dead-time of 3 ms. However, it

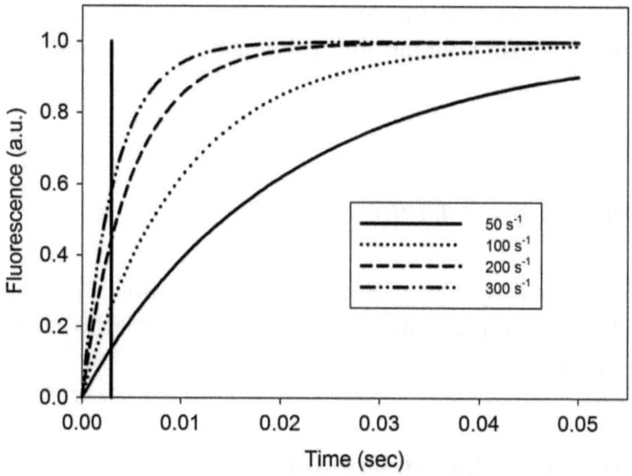

Fig. 3.3 Effect of dead-time on reaction amplitude. Simulated curves representing protein–ligand reactions at 50–300 s^{-1} are plotted. The solid line at 3 ms represents the dead-time. Note that the time scale represents the true start of the reaction and not that observed after the flow has stopped. Only the transients after the 3 ms cut-off would be observed in this theoretical stopped-flow. At 300 s^{-1}, the reaction loses 60 % of its total amplitude

has to be stated that reactions which begin, but are not over, within the dead-time can be studied. These processes will show decreased reaction amplitude due to the fact that a proportion of the reaction has occurred before data acquisition. But, as long as the process follows an exponential behaviour, it will be possible to fit the data and determine the reaction half-life to extract the rate constant [1].

There are several modifications that can be made to the stopped-flow system to decrease the dead-time. The cuvette volume can be reduced, typically from 20 to 5 μL, which reduces the mixing time because less solution has to be flowed through the system. However, due to the smaller reaction volume, there is a decrease in the fluorescence signal. Additionally, the flow speed can be increased but increasing this rate may cause cavitation or shearing of samples. However, an increased dead-time can reflect a problem with the drive mechanism, whereby a slowing in the flow speed can occur with more viscous samples or with blocked flow circuits.

The instrument dead-times, along with the mixing efficiency, are important parameters to measure in order to determine how well the system is working. This is especially important when modifications are made to the system. Dead-time measurements can be performed with any well-characterised second-order reaction. The reaction of *N*-acetyl tryptophanamide (NATA) with *N*-bromosuccinimide in 0.1 M sodium phosphate (pH 7.5) is typically used [1]. This reaction leads to a fluorescence decrease of NATA. 10 μM NATA is initially mixed against buffer to provide the initial starting fluorescence. The mixing of NATA against 100–600 μM *N*-bromosuccinimide under pseudo-first-order conditions reveals the dead-time. At higher a concentration of *N*-bromosuccinimide, the initial starting fluorescence is reduced—a decrease in the overall reaction amplitude. This occurs because the reaction started in the dead-time of the instrument. The precise dead-time is measured by extrapolating the reaction back to the starting fluorescence value.

3.3.2 *Stopped-Flow Variations*

The basic stopped-flow system can be modified to allow different fluorescence techniques and more complex mixing experiments in order to explore more complex biological processes.

3.3.2.1 Fluorescence Polarisation/Anisotropy

Anisotropy measurements are based upon the rotation (rotation correlation time) of a fluorescent species within its fluorescence lifetime, described in detail [10]. Two parameters are crucial for these measurements: the fluorescence lifetime and the size of the label. If the lifetime is too short, the population will appear highly anisotropic, whereas if it is too long, the species will have low anisotropy. Anisotropy measurements are particularly suited when one species is significantly smaller than the other, such as protein and ATP [11, 12]. When binding to the larger protein, the anisotropy of the smaller unit increases because the larger complex has a slower rotation correlation time. This provides a sensitive measurement of complex formation.

To perform the measurements, the stopped-flow is set up in T format (two detectors at 90° to the excitation source); polarisers are attached to excitation path (this must be able to rotate between vertically and horizontally polarised light) and both emission channels (Fig. 3.1b). The sample is excited by vertically polarised light; the intensity in the parallel and perpendicular light (in plane of excitation) is monitored. The two PMTs will respond differently to the intensity of the parallel and perpendicular light; therefore, the detectors have to be normalised [1]. To perform the normalisation, the fluorophore is excited with horizontally polarised light. In this format the amount of light depolarised on the parallel and perpendicular detectors is equal. Therefore, the voltage of the PMTs can be adjusted until the readout is equal.

Fluorescence anisotropy is then represented as a weighted average of "bound" versus "free" species:

$$r = I_{||} - I_{\perp}/I_{||} - 2I_{\perp} \qquad (3.2)$$

where $I_{||}$ represents emitted light in the parallel plane and $I\perp$ represents light emitted in the perpendicular plane. In practical terms, if all of the emitted light is in the parallel plane, then the molecules are fully polarised and $r = 1$. If the molecules are fully depolarised, then light is emitted equally in both planes and

$r = 0$. Measuring polarisation is performed in the same manner, but there is only a single contribution from the perpendicular intensity in the denominator.

Using polarisers causes a large loss of light; therefore, it is important to ensure that all optical components are correctly aligned. Furthermore, excitation slits may also have to be increased to enhance the intensity. The reduced sensitivity will also limit the minimum concentrations that can be used in the assays. Additional problems can result from scattering in the solution because all scattered light is polarised; therefore, this will compromise the signal. Degassing buffers will reduce air bubbles, which are a major source of scattered light.

3.3.2.2 Multi Mix: Double and Triple

The basic stopped-flow principle works by mixing two solutions together, but it is possible to perform multiple mix experiments whereby one or two mixes are made, prior to a final mix which is then observed in the cuvette (Fig. 3.2b). Importantly, the first mixes cannot be observed and this approach uses more reactants due to an increase in the number of mixes and flow path. This technique can be vitally important for understanding how steps relate to each other in complex reactions. For example, upon mixing a helicase RecD2, pre-bound to DNA, with ATP, a biphasic increase in fluorescence is observed [13] (Fig. 3.4a). The phases are related to binding followed by a hydrolysis-related conformation change. To test if DNA triggers this conformation change, a double-mix experiment was performed premixing RecD2 with mantATP before mixing with DNA. The second mix is delayed long enough for all of the mantATP to be bound before mixing with DNA. The fluorescent changes are identical to the second phase from the previous experiment, thereby showing the DNA induces this change (Fig. 3.4b).

3.3.2.3 Low Volume Cuvette/Low Dead Volume

While a stopped-flow reaction may require 50 μL of each reactant, typically five reactions would be collected for data averaging. This increases the total volume needed to 250 μL. However, additional sample is required to prime the system prior to observing the first reaction. This dead volume can increase the sample required to ~500 μL of each reactant. When using high concentrations this can be limited by a protein produced in low yield. The different commercial systems vary in the amount of priming solution required. This is essentially being reduced with newer systems. In other cases, adaptors can be built into the systems to preload samples in front of the cuvette. The system then pushes the sample with buffer into the observation chamber. This reduces the sample volume to 120 μL for 5 20 μL shots through a 5 μL observation cell.

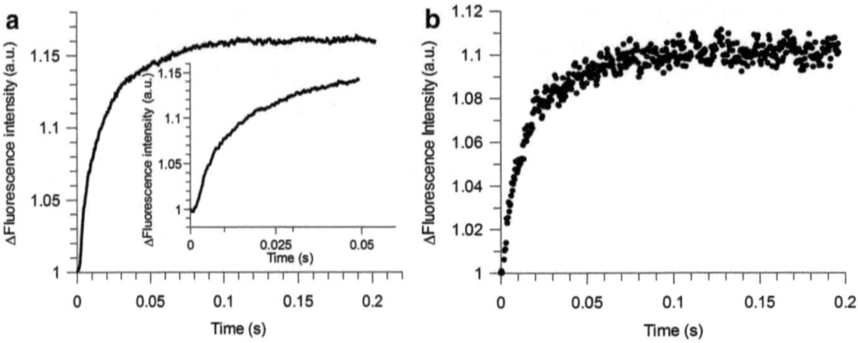

Fig. 3.4 Double-mix stopped-flow experiment. (**a**) Time course following mixing 12.5 μM RecD2 and 15 μM dT_{20} with 2.5 μM mantATP. Inset shows the initial increase in fluorescence. The trace was fitted by a double exponential. The first phase represented two-thirds of the amplitude had an observed rate constant of 250 s^{-1} and the second phase was 39 s^{-1}. (**b**) The time course of mant fluorescence upon mixing 2.5 μM mantATP and 12.5 μM RecD2 with 15 μM dT_{20} after a first mixing of 25 μM RecD2 and 5 μM mantATP and ageing for 0.1 s. The trace was fitted by a single exponential giving a rate constant of 42 s^{-1}

3.3.2.4 Dual Fluorescence and Fluorescence Emission Scanning

The use of two detectors can allow for measurements at two different fluorescence wavelengths. Filters will have to be correctly chosen to isolate each wavelength such as using a combination of band-pass and cut-off filters. This application can be advantageous for measuring the fluorescence from donor and acceptor fluorophores during a FRET measurement. However, this application would more greatly benefit from scanning the entire emission spectrum to directly monitor both donor and acceptor fluorophores. For this the detection system has to couple in a rapid scanning monochromator so that the spectra can be recorded in a time-dependent manner. There are two disadvantages to this approach, the first being that more light is removed from the system with a second monochromator. Secondly, the acquisition speed is limited by the speed of the detection system; therefore, this approach is not applicable to all reaction systems. Future application of diode arrays and CCD cameras will provide a greater sensitivity and acquisition speeds for these measurements.

3.3.2.5 Diode Array Detection

While this is strictly an absorbance-based measurement, it can be useful to consider that fluorophores can display environmental sensitivity resulting in changes in extinction coefficients, excitation/absorbance spectra. Therefore, it can be useful to monitor absorbance spectra during a reaction time course. A diode array detector can be situated in place of the PMT to provide rapid recording of absorbance over approximately 280–700 nm.

3.3.2.6 High-Throughput Screening

High-throughput screening has been applied to fluorescent systems, mostly through the application of plate readers. The AutoSF-120 from Kintek now allows for automated screening with stopped-flow measurements. These advances and developments will expand the use of transient kinetics for the pharmaceutical industry.

3.3.2.7 Stand-alone Rapid Mixing Devices

Simple rapid mixing devices can be purchased for coupling to standard fluorimeters for excitation/emission detection. The reactants are loaded into syringes and driven through an umbilical to a cuvette. Mixing occurs just before the cuvette in a similar manner to the stopped-flow apparatus. These devices have large volume requirements, mostly due to the umbilical, longer dead-times due to limited flow rates, while the detection speed is limited by the fluorimeters' capabilities. Therefore, reactions are limited to those with rate constant below 10–100 s^{-1}. These accessories provide a low cost option for performing stopped-flow measurements.

Such systems also allow for stopped-flow measurements to be performed with other cuvette-based optical systems. Such systems include circular dichroism and fluorescent lifetime measurements. The fluorescent lifetime of a fluorophore changes based upon the environment; therefore, upon a fluorescent ligand binding to a protein, an increase in lifetime can be observed. Additionally, if performing FRET measurements, the donor lifetime will decrease when in proximity to acceptor fluorophores. FRET lifetime measurements have the ability to detect the change in population of the lifetimes and therefore FRET distances not just the average FRET distance of an ensemble.

In addition, time-resolved fluorescence through direct waveform recording (DWR) provides sub-nanosecond temporal resolution. This is an advantageous approach over time-correlated single-photon counting methods with an average measurement time of 10 s. The DWR method can be used for different fluorescent applications including anisotropy and lifetime measurements [14, 15].

3.4 Relaxation Methods

The stopped-flow method perturbs the equilibrium by rapidly mixing solutions, but the equilibrium can also be perturbed by external factors such as temperature and pressure [16, 17]. A given reaction will result in thermodynamic changes in enthalpy (ΔH°) and molar volume (ΔV°), which are in turn related to changes in temperature and pressure. For temperature this is described by van't Hoff equation:

$$\Delta \ln K/\Delta T = \Delta H/\mathrm{RT}^2 \tag{3.3}$$

and for pressure:

$$\Delta K/K = \Delta V^\circ \cdot \Delta P/\mathrm{RT} \tag{3.4}$$

The relationship between perturbation and the observed change in reactant concentration depends upon K, along with the applied perturbation (ΔP) and thermodynamic parameter (ΔV°). For the reaction scheme below:

The amplitude of the observed reaction as a function of [L] directly reports on the equilibrium constant with the largest amplitude generated when the equilibrium constant is close to 1 [17]. For Scheme 3.1, when the change in [E] is plotted again K/[L], where $K =$ [E][L]/[EL], the change is largest when [L] $= K$; therefore [E]/[EL] $= 1$.

For Scheme 3.2:

Scheme 3.1 Reaction of Enzyme (E) and Ligand (L) forming the E.L complex.

$$\mathrm{E + L} \rightleftharpoons \mathrm{E{\cdot}L}$$

Scheme 3.2 Example reversible reaction of A to product B

$$\mathrm{A} \rightleftharpoons \mathrm{B}$$

If the concentration of A at time t is $[\mathrm{A}]_t$, then Δ[A] with time is

$$-\mathrm{d[A]}/\mathrm{d}t = k_{+1}[\mathrm{A}]_t - k_{-1}[\mathrm{B}]_t \tag{3.5}$$

Following integration:

$$\Delta[\mathrm{A}]_t = \Delta[\mathrm{A}]_0 \mathrm{e}^{-(k_{+1}+k_{-1})t} \tag{3.6}$$

The relaxation time (τ) is given by

$$1/\tau = k_{+1} + k_{-1} \tag{3.7}$$

where $1/\tau$ is identical to observed rate constant (k_{obs}) in rapid mixing experiments. For Scheme (3.1), the relaxation time would be

$$1/\tau = k_{+1}\left([\mathrm{A}]_{\mathrm{eq}} + [\mathrm{B}]_{\mathrm{eq}}\right) + k_{-1} \tag{3.8}$$

Relaxation methods work through the rapid application of temperature or pressure that change the equilibrium position [17]. The relaxation of the system to the new equilibrium can be followed using fluorescence or other spectral techniques. These changes are significantly faster than mixing reactions; therefore, measurements can be observed on significantly faster time scales. The equipment for both temperature and pressure jump essentially replaces the stopped-flow drive mechanism, while the same excitation and emission systems can be used. Therefore, all of the considerations outlined for the stopped-flow also apply here. However, as the equilibrium perturbations are small, then so is the signal change. Therefore, there is greater noise within these signals compared to rapid mixing reactions which puts greater importance upon the detection system. Fortunately, relaxation measurements do not consume reactants during the experiment; therefore, as long as reactants are stable, numerous repeat measurements can be performed upon a single sample. This leads to a large reduction in the random noise, thus improving the sensitivity. Furthermore, the stopped-flow variations such as fluorescence polarisation/anisotropy and the dual fluorescence detection are also viable options for these techniques.

3.4.1 Temperature Jump

As the name suggests, this method works through a rapid increase in temperature. A common set-up uses Joule heating. The temperature increases in a conducting solution due to the resistance to the electrical current flow through the solution. The capacitor discharge can heat a solution by 5 °C in 1–2 μs [1]. Alternatively, lasers can be used to induce heating of small volumes in time scales down to 10 ns [5]. Only rapid reactions can be observed with this technique because the solution will immediately (>0.1 s) begin to cool to ambient temperature, although methods for stabilising the new temperature are possible. Commercial equipment is available making this an easier method to attempt, but there are several disadvantages of this method over pressure jump [1]. (1) Measurements have to be performed at high ionic strength for the conductivity required in Joule heating. (2) It is hard to maintain temperature after the jump. (3) Jumps are unidirectional; samples cannot be rapidly cooled. (4) Fluorescence can be perturbed by these temperature changes; therefore the signal will artificially report upon processes.

3.4.2 Pressure Jump

Pressure-jump experiments require that the reaction occurs with a significant volume change, such as that seen in water packing around buried or exposed hydrophobic or charged groups [4]. These changes are common for most biological systems (ligand binding, protein conformational changes, dimerisations, assembly of polymers or protein folding reactions); therefore, these measurements can be attempted without significant prerequisites.

Devices using piezoelectric crystals can generate repeated, large pressure changes (500 bar), in small volumes (50 μl) and in 50 μs. The pressure can be rapidly increased and decreased in order to observe transients in both directions;

furthermore, the experiment can be repeated immediately. Importantly, the pressure can be maintained for long periods; therefore, slow transients can be observed as easily as fast transients.

3.5 Conclusions

The combination of relaxation and rapid mixing techniques allows biochemical pathways to be explored from microsecond time scales onwards. This has helped to reveal the important intermediates within biochemical processes. The application of fluorescence to these techniques has created a diverse range of assays with a high sensitivity. New technologies with faster acquisition times and greater sample efficiency will continue to drive the development of these machines.

References

1. Eccleston JF, Martin SR, Schilstra MJ (2008) Rapid kinetic techniques. Methods Cell Biol 84:445–477. doi:10.1016/S0091-679X(07)84015-5, S0091-679X(07)84015-5 [pii]
2. Lakowicz JR (2006) Principles of fluorescence spectroscopy, 3rd edn. Springer, Berlin
3. Shastry MC, Luck SD, Roder H (1998) A continuous-flow capillary mixing method to monitor reactions on the microsecond time scale. Biophys J 74(5):2714–2721. doi:10.1016/S0006-3495(98)77977-9
4. Pearson DS, Holtermann G, Ellison P, Cremo C, Geeves MA (2002) A novel pressure-jump apparatus for the microvolume analysis of protein-ligand and protein-protein interactions: its application to nucleotide binding to skeletal-muscle and smooth-muscle myosin subfragment-1. Biochem J 366(Pt 2):643–651. doi:10.1042/BJ20020462
5. Williams S, Causgrove TP, Gilmanshin R, Fang KS, Callender RH, Woodruff WH, Dyer RB (1996) Fast events in protein folding: helix melting and formation in a small peptide. Biochemistry 35(3):691–697. doi:10.1021/bi952217p
6. Shanker N, Bane SL (2008) Basic aspects of absorption and fluorescence spectroscopy and resonance energy transfer methods. Methods Cell Biol 84:213–242. doi:10.1016/S0091-679X(07)84008-8
7. Toseland CP (2013) Fluorescent labeling and modification of proteins. J Chem Biol 6(3):85–95. doi:10.1007/s12154-013-0094-5
8. Ha T, Tinnefeld P (2012) Photophysics of fluorescent probes for single-molecule biophysics and super-resolution imaging. Annu Rev Phys Chem 63:595–617. doi:10.1146/annurev-physchem-032210-103340
9. Johnson KA (1986) Rapid kinetic analysis of mechanochemical adenosine triphosphatases. Methods Enzymol 134:677–705
10. LiCata VJ, Wowor AJ (2008) Applications of fluorescence anisotropy to the study of protein-DNA interactions. Methods Cell Biol 84:243–262. doi:10.1016/S0091-679X(07)84009-X
11. Toseland CP, Martinez-Senac MM, Slatter AF, Webb MR (2009) The ATPase cycle of PcrA helicase and its coupling to translocation on DNA. J Mol Biol 392(4):1020–1032. doi:10.1016/j.jmb.2009.07.071
12. Toseland CP, Webb MR (2010) Fluorescence tools to measure helicase activity in real time. Methods 51(3):259–268. doi:10.1016/j.ymeth.2010.02.012

13. Toseland CP, Webb MR (2013) ATPase mechanism of the 5′-3′ DNA helicase, RecD2: evidence for a pre-hydrolysis conformation change. J Biol Chem 288(35):25183–25193. doi:10.1074/jbc.M113.484667
14. Muretta JM, Kyrychenko A, Ladokhin AS, Kast DJ, Gillispie GD, Thomas DD (2010) High-performance time-resolved fluorescence by direct waveform recording. Rev Sci Instrum 81(10):103101. doi:10.1063/1.3480647
15. Muretta JM, Behnke-Parks WM, Major J, Petersen KJ, Goulet A, Moores CA, Thomas DD, Rosenfeld SS (2013) Loop L5 assumes three distinct orientations during the ATPase cycle of the mitotic kinesin Eg5: a transient and time-resolved fluorescence study. J Biol Chem 288(48):34839–34849. doi:10.1074/jbc.M113.518845
16. Bernasconi CF (1976) Relaxation kinetics. Academic, New York
17. Geeves MA, Pearson DS (2013) Relaxation methods. In: Roberts GCK (ed) Encyclopedia of biophysics. Springer, Heidelberg

Chapter 4
Fluorescence to Study the ATPase Mechanism of Motor Proteins

Christopher P. Toseland

Abstract This chapter provides an overview of different methodologies to dissect the ATPase mechanism of motor proteins. The use of ATP is fundamental to how these molecular engines work and how they can use the energy to perform various cellular roles. Rapid reaction and single-molecule techniques will be discussed to monitor reactions in real time through the application of fluorescence intensity, anisotropy and FRET. These approaches utilise fluorescent nucleotides and biosensors. While not every technique may be suitable for your motor protein, the different ways to determine the ATPase mechanism should allow a good evaluation of the kinetic parameters.

Keywords Helicase • Myosin • Kinesin • Fluorescence • ATPase • Stopped-flow

Abbreviations

ATPγS	Adenosine 5′-*O*-(3-thio)triphosphate
Deac	Diethylaminocoumarin
dsDNA	Double-stranded deoxyribonucleic acid
FRET	Förster (or fluorescence) resonance energy transfer
k_{cat}	Catalytic turnover number
K_d	Equilibrium dissociation constant
K_i	Inhibition constant
K_m	Michaelis constant
Mant	2′(3′)-*O*-(*N*-methylanthraniloyl)-
MDCC	7-diethylamino-3-((((2-maleimidyl)ethyl)amino)carbonyl) coumarin

C.P. Toseland (✉)
Chromosome Organisation and Dynamics, Max-Planck Institute of Biochemistry, Martinsried 82152, Germany
e-mail: toseland@biochem.mpg.de

C.P. Toseland and N. Fili (eds.), *Fluorescent Methods for Molecular Motors*, Experientia Supplementum 105, DOI 10.1007/978-3-0348-0856-9_4,

MDCC-PBP	(A197C) Phosphate-binding protein adduct with MDCC
NADH	Nicotinamide adenine dinucleotide
P_i	Inorganic phosphate
ssDNA	Single-stranded deoxyribonucleic acid

4.1 Introduction

The conversion of ATP to ADP is a fundamental process for molecular motors. The ATP catalysis to ADP proceeds via cleavage of the β-γ phosphate bond to release inorganic phosphate (P_i). Hydrolysis of this bond results in -30.5 kJ mol^{-1} energy release but depending on the physiological conditions, this could be -57 kJ mol^{-1} [1]. The energy of ATP hydrolysis is coupled to conformation changes which modulate protein–track interactions and enable force production, both being the defining features of motor proteins.

Fluorescent nucleotides have been widely used to investigate the function of many molecular motors including myosins [2], kinesins [3] and helicases [4–6]. These nucleotides are employed to investigate the individual steps within the mechanistic cycle of these motors and to understand their relation to protein function. A minimal ATPase cycle mechanism will consist of binding, hydrolysis and product release steps (Scheme 4.1).

This is the "assumed" scheme when performing initial Michaelis–Menten steady-state measurements. These measurements provide important parameters such as the Michaelis constant (K_m) for ATP or the "track" and the inhibition constant (K_i) of ADP or non-hydrolysable analogues. This scheme becomes more elaborate with conformation changes associated with these core steps. The lifetime of the different biochemical intermediates can reveal how the protein functions and its regulation. This is particularly relevant to the myosin cytoskeletal motors, whereby differences in these lifetimes determine how the proteins function, for example, as a cellular transporter or a load-dependent sensor [7].

4.1.1 Fluorescent ATP

Tryptophan fluorescence changes upon nucleotide binding have proven to be a useful tool for investigating the ATPase kinetics of myosins [8]. However, these changes are not always present. Furthermore, it may be better to measure the ATP molecule and not the proteins response to the binding. Therefore, a fluorescent ATP is required.

ATPase rate constant varies widely from <1 s^{-1} to $>1{,}000$ s^{-1}. Therefore, it is important to have sensitive and rapid responses, which both are ensured by fluorescence methods. Furthermore, this is a safer alternative to radioactivity assays.

$$\text{Protein} + \text{ATP} \underset{k_{-1}}{\overset{k_{+1}}{\rightleftharpoons}} \text{P·ATP} \underset{k_{-2}}{\overset{k_{+2}}{\rightleftharpoons}} \text{P·ADP·P}_i \underset{k_{-3}}{\overset{k_{+3}}{\rightleftharpoons}} \text{P·ADP} \underset{k_{-4}}{\overset{k_{+4}}{\rightleftharpoons}} \text{P} + \text{ADP}$$

Scheme 4.1 A minimal ATPase cycle mechanism will consist of binding, hydrolysis and product release steps with associated forward and reverse rate constants

The sensitivity of fluorescence allows binding, product release and conformation changes to be reported in real time. The sensitivity is dependent upon the position of the fluorescent species, and therefore, this choice is important in order to avoid perturbation of the enzymatic activity by the fluorescent analogue.

The ATP structure can be modified at several locations as discussed in Chap. 1. Ribose modifications, such as mantATP [9], are frequently targeted. However, given that the modification is the two hydroxyl groups (2′ or 3′), ribose modifications result in the generation of two isomeric forms of the nucleotide, one of which may not be catalysed. Deoxy-ATP circumvents this problem by providing a single attachment site. It is important to note, increases in affinity have been reported, which can impede the overall catalytic cycle. For example, a slower dissociation of fluorescent-ADP delays the speed at which the next ATP can bind, which leads to an overall decrease in the catalytic rate constant [4, 5].

Overall, mantATP [9] and deac-aminoATP [10, 11] analogues (Chap. 1) are very useful for bulk solution methods, and examples with both nucleotides will be given in this chapter. Data will also be presented with Cy3-eda-ATP [12] which is suitable for single-molecule measurements.

It is vitally important to establish that the fluorescent ATP is a suitable analogue. Therefore, the Michaelis–Menten parameters should be similar to ATP (Sect. 4.2.1). Furthermore, it is important to perform a functional assay with the fluorescent nucleotide. For example, in the case of a helicase, DNA unwinding assays/translocation assays should be performed and compared to ATP [13].

Additionally, non-fluorescent modifications have also been successfully made, as in the case of ATPγS. This non-, or slowly, hydrolysable ATP analogue is particularly useful for separating out the kinetics steps of binding from hydrolysis. This is because any observed rate constants from ATP would consist of nucleotide binding and release (k_{+1} and k_{-1}), plus the forward rate constant (k_{+2}) of hydrolysis.

4.1.2 Types of Measurement

The fluorescent properties of the nucleotide are harnessed to measure intensity, anisotropy/polarisation and fluorescence/Förster resonance energy transfer (FRET). Intensity changes result from a change in the fluorophore's environment. This leads to changes in the quantum yield, and/or extinction coefficients can increase leading to large enhancements in the fluorescent signal, as discussed in Chap. 1. FRET is a mechanism of energy transfer between donor and acceptor fluorophores

[14]. Energy transfer occurs when two fluorophores with overlapping donor emission and acceptor excitation spectra are in close proximity (<10 nm). Anisotropy measurements are based upon the rotation (rotation correlation time) of a fluorescent species within its fluorescence lifetime, described in detail in Chap. 1 and [15]. Anisotropy measurements are particularly suited for nucleotide binding. When nucleotide binds to the larger protein, the anisotropy increases because the large complex has a slower rotation correlation time. This provides a sensitive measurement of complex formation.

This chapter will describe the application of fluorescent nucleotides in combination with stopped-flow (Chap. 3), quenched-flow and single-molecule methods (Chap. 7) to investigate the ATPase mechanism of motor proteins. Examples described here are focused upon the bacterial DNA helicases PcrA, RecG and RecD2. These enzymes interact with single-stranded DNA (ssDNA) and duplex DNA (dsDNA) to catalyse DNA unwinding in an ATP-dependent manner [4–6].

4.1.3 Pseudo-First-Order Kinetics

In nearly all experimental cases, either protein (P) or nucleotide (L) is in at least 10× molar excess, so that the concentration of one species can be considered constant during the reaction. This is called pseudo-first-order kinetics. These are important conditions for measuring bimolecular interactions because they allow each step of the mechanism to be described by a series of exponential functions. The basis of these conditions is defined below using adaptations from [16].

Firstly, let's consider a simple reversible first-order reaction (Scheme 4.2):

Scheme 4.2 A reversible first order reaction of P to P*

$$P \underset{k_{-1}}{\overset{k_{+1}}{\rightleftharpoons}} P^*$$

The reaction consists of unimolecular processes with first-order rate constants k_{+1} and k_{-1}. The rate $d[\mathrm{P}^*]/dt$ is shown below:

$$\frac{d[\mathrm{P}^*]}{dt} = k_{+1}[\mathrm{P}] - k_{-1}[\mathrm{P}] = k_{+1}[\mathrm{P_{TOTAL}}] - (k_{+1} + k_{-1})[\mathrm{P}^*] \quad (4.1)$$

The concentrations of P* can be expressed as a function of time:

$$[\mathrm{P}^*](t) = \mathrm{P}^*_{\mathrm{eq}} + \left(\mathrm{P}^*_0 - \mathrm{P}^*_{\mathrm{eq}}\right)\mathrm{e}^{k_{\mathrm{OBS}}t} \quad (4.2)$$

where P^*_0 is the initial concentration of P*, $\mathrm{P}^*_{\mathrm{eq}}$ is the equilibrium (or final) concentration of P and $k_{\mathrm{obs}} = k_{+1} + k_{-1}$. The larger the k_{obs}, the more rapid the system will equilibrate, and this is the process that is measured. It is also important to note that a reaction with the same k_{+1} but with different k_{-1} will equilibrate at different rates, with the reaction containing the largest k_{-1} equilibrating the fastest.

Scheme 4.3 A reversible second order reaction between Protein (P) and Ligand (L)

$$P + L \underset{k_{-1}}{\overset{k_{+1}}{\rightleftharpoons}} PL$$

Obviously, the reactions here focus upon protein binding to a ligand which leads to the formation of a protein–ligand (PL) complex. These reactions are termed second order, as shown in (Scheme 4.3).

The reaction consists of a bimolecular binding process k_{+1} (M^{-1} s^{-1}) and a unimolecular dissociation process k_{-1} (s^{-1}). The equilibrium dissociation constant (K_d) of this reaction is defined as k_{-1}/k_{+1} (M). The rate of $d[PL]/dt$ is shown below:

$$\begin{aligned}\frac{d[\mathrm{PL}]}{dt} &= k_{+1}[\mathrm{P}][\mathrm{L}] - k_{-1}[\mathrm{PL}] \\ &= k_{+1}([\mathrm{P_{TOTAL}}] - [\mathrm{PL}])([\mathrm{L_{TOTAL}}] - [\mathrm{PL}]) - k_{-1}[\mathrm{PL}] \end{aligned} \tag{4.3}$$

There is no simple analytical solution to this equation, as done for the first-order reactions. This is because the concentrations of both P and L vary in the reaction. In these situations kinetic simulation can be used to fit the data and extract the rate constants. However, if P or L is in a far larger excess, then, when the PL complex forms, the concentration of the excess component is essentially unchanged. So, in the case of excess L, $L_{TOTAL} - [PL] \approx L_{TOTAL}$. This is termed pseudo-first-order conditions. As a result, the formation of PL is in the same form as the first-order reaction, with $k_{obs} = k_{+1}[L_{TOTAL}] + k_{-1}$, for the example of excess ligand. This provides a simple approach to measure interactions without kinetic simulations.

4.2 Measurement of an ATPase Cycle

4.2.1 Michaelis–Menten Steady-State Kinetics

The catalytic turnover rate constant (k_{cat}) and the ATP K_m should be determined prior to performing transient kinetics to provide important parameters for subsequent measurements. The k_{cat} will provide an idea of the order of rate constants expected to be measured. The slowest individual step(s) determined from the transient kinetic measurements should relate to this rate constant. The ATP K_m provides an idea of the ATP affinity which is important to determine before performing binding measurements that have to be made above the equilibrium dissociation constant. It is also possible to perform competition assays to determine the inhibition constant (K_i), providing a measurement of the affinity for ADP and non-hydrolysable analogues. In addition, this will reveal if non-hydrolysable analogues are truly not a catalytic substrate and that they can efficiently compete with ATP for binding to the motor protein. The K_m for interactions with the track, for example, DNA [4] or filamentous actin [17], can also be determined with these methods. This is important to understand the affinity between protein and track because you may wish to preform this complex for ATPase measurements.

There are several methods available for measuring steady-state kinetics. The two most frequently employed are the enzyme-coupled assays monitoring the reduction of nicotinamide adenine dinucleotide (NADH) [18] and the fluorescent phosphate-binding protein (MDCC-PBP) [19]. MDCC-PBP will be discussed later and in Chap. 2. The MDCC-PBP provides the simplest method to directly measure P_i release during steady state. However, the NADH assay has the benefit of recycling ADP to ATP, and it does not sequester P_i; therefore, the concentrations do not change during the reaction. These measurements are performed in standard spectrometers or plate readers. NADH has an additional benefit by being an absorbance and fluorescent assay, thus allowing measurements to be performed with equipment based on either technique.

4.2.2 Nucleotide Binding and Release

These measurements are applicable to both ATP and ADP analogues. A diverse range of approaches can be taken here, such as intensity, FRET and fluorescence anisotropy/polarisation with either protein or nucleotide in excess under pseudo-first-order conditions. Each approach has its advantages for revealing the details of the ATPase mechanism, as highlighted below. A minimal mechanism is shown in Scheme 4.1 with four steps with the ATP and ADP association steps determined by the equilibrium of bound and unbound states (K_1 and K_4), respectively. The same mechanism is described under pseudo-first-order kinetics in Scheme 4.4 whereby each step is described by a series of exponentials.

4.2.2.1 Measurements with an Excess of Nucleotide

With an excess of nucleotide, direct excitation of the fluorophore results in a large background fluorescence. Considering pseudo-first-order kinetics requires a 10× excess of fluorescent nucleotide over protein, <10 % of the total nucleotide will form the protein-bound complex. This problem can be aided if the nucleotide yields a large signal change, as is often the case with deac-aminoATP, but sensitivity is always compromised. Additionally, photo-physical factors, such as the inner filter effect [20] (Chap. 3), begin to affect the signal at higher concentrations of the fluorescent species.

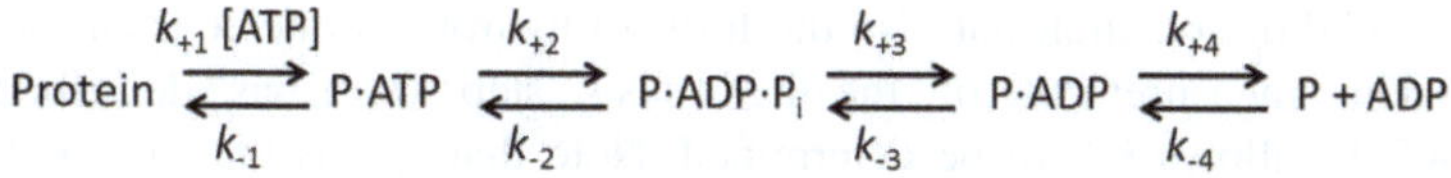

Scheme 4.4 A minimal ATPase cycle mechanism under pseudo-first order conditions

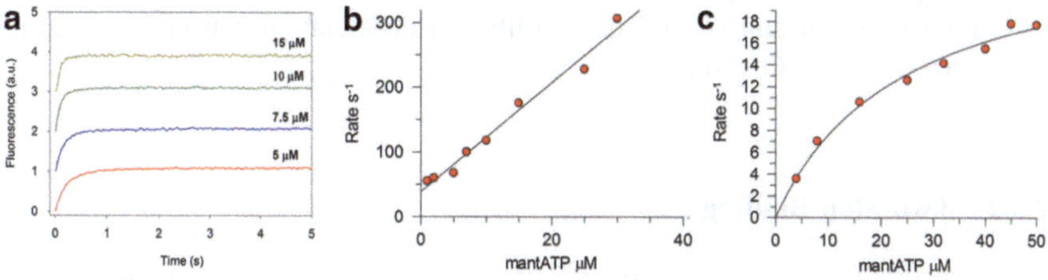

Fig. 4.1 ATP binding kinetics. (**a**) Example fluorescent transients from mixing mantATP at the micromolar concentrations shown with 0.5 μM RecG and 2.5 μM DNA. The fluorescence signal (arbitrary scale) was measured by exciting mant at 368 nm and measuring the emission through a 400 nm cut-off filter. Traces are offset for clarity. (**b**) A linear dependence between k_{obs} and [mantATP] for PcrA helicase. Individual traces were fit to single exponentials, and the dependence of the rate constants on concentration was then fitted to a *straight line*. The fit for mantATP gives a slope of 8.1 $\mu M^{-1}\ s^{-1}$ and an intercept of 43 s^{-1}. (**c**) The *traces* in (**a**) were fit to single exponentials and the dependence of the rate constants on concentration was then fitted to a hyperbola. The fit gives $1/K_1 = 24$ μM, $k_{+1b} + k_{-1b} = 26\ s^{-1}$ and k_{-1b} as $<1\ s^{-1}$

The high background fluorescence can be reduced by using FRET from tryptophan to mant nucleotides. This is dependent upon having a tryptophan close to the nucleotide binding site, and it limits the selection of fluorescent label on the nucleotide.

Typically experimental conditions are 0.5 μM protein and 5–50 μM fluorescent-ATP. The two reactants are rapidly mixed in the stopped-flow apparatus to follow the reaction in real time. The time course is described by an exponential function (Eq. 4.4) (Fig. 4.1a), but the time course may deviate from this, as discussed in Sect. 4.2.2.3:

$$y = Ae^{(-kt)} + \text{offset} \tag{4.4}$$

where A is the reaction amplitude and k is the observed rate constant (k_{obs}). In the case of ATP, the formation of P·ATP is described by

$$k_{OBS} = k_{+1}[\text{ATP}] + k_{off} \tag{4.5}$$

Performing the measurements at several concentrations of ATP allows the dependence to be plotted. A linear dependence between k_{obs} and ATP concentration ([ATP]) represents a simple bimolecular interaction (Fig. 4.1b). k_{+1} is determined from the slope and represents the second-order association rate constant and k_{off} from the intercept, giving the breakdown of P·ATP. Importantly, this includes not only the ATP dissociation but also the forward hydrolysis rate constant (k_{+2}). An independent measurement of the hydrolysis step (k_{+2}), as described later (Sect. 4.2.3), allows k_{-1} to be determined. Note that k_{obs} is the sum of k_{+1} and k_{off}. This is because the experiment describes the progress to equilibrium which can be reached with a large k_{off} value.

If the experiment was performed with ADP, then the intercept would give the dissociation rate constant (k_{+4}). The equilibrium dissociation constant, K_d, could then be extracted from the data k_{-4}/k_{+4}.

4.2.2.2 Two-Step Binding

The plot of k_{obs} against concentration may show a more complex nature by varying hyperbolically (Fig. 4.1c). Under these circumstances, it is assumed that a first-order isomerisation step occurs after a second-order association event (Scheme 4.5), or vice versa (Scheme 4.6).

Scheme 4.5 A first-order isomerisation step (P.ATP to P.ATP*) after a second-order association

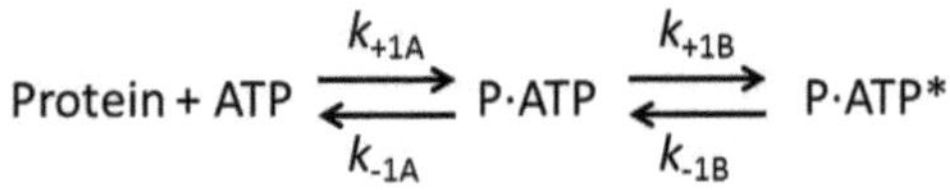

Scheme 4.6 A first-order isomerisation step (Protein to Protein*) prior to a second-order association event

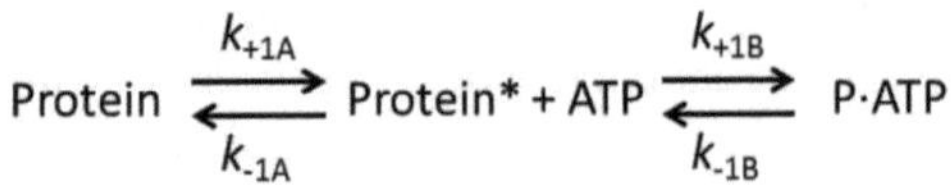

This is considered to represent ATP binding followed by a rapid isomerisation. Assuming the second step yields the fluorescence change, the data is fit to the following equation:

$$k_{OBS} = \frac{k_{+1B}\,[\mathrm{ATP}]}{[\mathrm{ATP}] + K_{d1A}} + k_{-1B} \tag{4.6}$$

which gives the affinity for the binding step ($1/K_{1A}$) and the maximum rate of the isomerisation ($k_{+1B} + k_{-1B} + k_{+2}$), whereas the intercept gives k_{off}, which is again dependent upon $k_{-1B} + k_{+2}$. The hydrolysis rate constant can be independently measured to calculate k_{-1B} (Sect. 4.2.3) which in turn allows k_{+1B} to be determined. This fitting assumes that the second step generates the fluorescence signal, and this assumption is only valid if fluorescence intensity is measured. If measurements are made using FRET or fluorescence anisotropy, then the signal is generated on the initial interaction. Therefore, it could be expected that a linear dependence would be observed while performing these measurements.

4.2.2.3 Measurements with an Excess of Protein: "The Single Turnover"

The high background fluorescence caused by an excess of nucleotide can also be reduced by switching to single-turnover conditions using an excess of protein. With this approach, all of the nucleotide is bound by the protein and then hydrolysed to yield the products. However, no further catalysis can occur. A large excess of protein, at least twice the K_d, is required to saturate the nucleotide in a single round of catalysis. Theoretically, this results in the maximum fluorescence signal change because 100 % of the nucleotide is bound, vastly improving the signal to noise. This approach is costly in terms of the amount of protein required. Furthermore, in combination with a weak ATP affinity (high K_m), quantitative complex formation will not be possible.

The single-turnover conditions are also beneficial because conformation changes and product dissociation can be revealed in the same trace following binding. Under these conditions, the traces will deviate from a single exponential to have a biphasic (or multi-phasic) behaviour. By performing the measurements at several protein concentrations, it is possible to establish the association rate constant (k_{+1}) from that of the subsequent steps. In the example below, only the first phase shows a concentration dependence. The second-order association rate constant is extracted from this process, as established with an excess of nucleotide. Importantly, this rate constant should be the same as that measured with an excess of nucleotide.

If the observed rate constant of the second exponential is independent of the protein concentration, then this can be described as a first-order process representing subsequent steps in the cycle. Where A_a is the reaction amplitude, k_a is the k_{obs} for the first phase, A_b and k_b are the reaction amplitude and k_{obs} for the second phase.

$$y = A_a e^{(-k_a t)} + A_b e^{(-k_b t)} + \text{offset} \tag{4.7}$$

Examples can be seen with RecG and RecD2 [5, 6]. The fluorescent traces are characterised by a biphasic fluorescence change (Fig. 4.2a). When fitted by a double exponential (Eq. 4.7), the observed rate constant can be determined. After performing the measurement at several concentrations, the first phase varies with protein concentration and the second phase is invariant (Fig. 4.2b).

4.2.2.4 Fluorescence Anisotropy/Polarisation

The very same conditions can be used to measure fluorescence polarisation/anisotropy [16]. Typically, only single-turnover measurements are performed. With an excess of nucleotide, there is likely to be problems with sensitivity due to the large unbound fraction. These measurements are particularly useful to aid in understanding of the fluorescent traces, because this approach directly measures complex formation. Therefore, both binding and dissociation measurements can be recorded. Fluorescence anisotropy will be referred to here given the simple mathematical

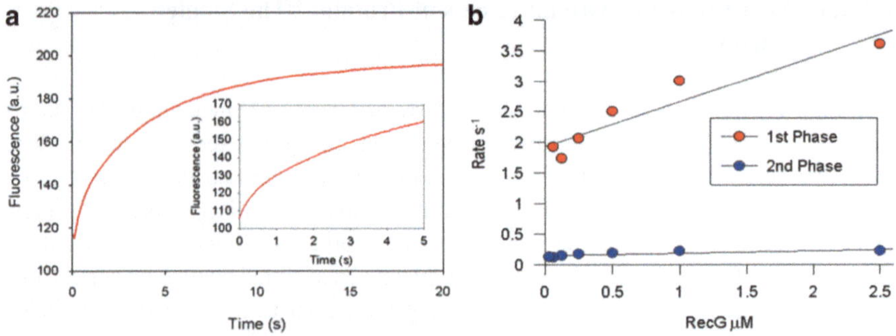

Fig. 4.2 Association kinetics with an excess of protein. (**a**) The change in mant fluorescence when 2.5 μM RecG·DNA junction was mixed with 0.5 μM mantATP (*inset* shows the short timescale). (**b**) Individual traces were fit to double exponentials, and the dependence of the rate constants on concentration was then fitted to a *straight line*. The fit for the first phase (*red circles*) gives a slope of 0.7 $\mu M^{-1}\,s^{-1}$ and an intercept of 1.9 s^{-1}. The second phase (*blue circles*) is independent to [RecG]

manipulations that can be performed due to the additive nature of anisotropy values, whereby the observed anisotropy represents free species plus the bound species (Chap. 1).

Emitted light is detected through polarisers, measuring the relative intensity in the parallel and perpendicular planes to that of the excited light. The ratio between the two intensities describes the rate of tumbling. Faster tumbling fluorophores will 'depolarise', increasing the relative intensity in the perpendicular plane, while, with slower tumbling fluorophores, the greater intensity will be emitted in the parallel plane and effectively remain 'polarised'. Fluorescence anisotropy (r) is expressed using Eq. (4.8):

$$r = I_{||} - I_{\perp}/I_{||} - 2I_{\perp} \tag{4.8}$$

where $I_{||}$ represents emitted light in the parallel plane and $I\perp$ represents light emitted in the perpendicular plane. In practical terms, if all of the emitted light is in the parallel plane, then the molecules are fully polarised and $r = 1$. If the molecules are fully depolarised, then light is emitted equally in both planes and $r = 0$ [21].

The raw fluorescence data is converted to anisotropy using Eq. (4.8). This is then fitted to a single exponential (Eq. 4.9). This expression takes into account a fluorescence change between protein bound and free states. If there is not a fluorescence change, then a normal single exponential function can be used to fit the data (Eq. 4.4).

$$r = \frac{(A_1 - A_2)}{(e^{(-kt)}(1 - D)) + D} \tag{4.9}$$

where k represents the observed rate constant, t represents the time and A_1 and A_2 represent the anisotropy at the start and end of the exponential, respectively. D is a correction factor representing the fluorescence intensity change between A_1 and A_2 and is calculated using the equation below:

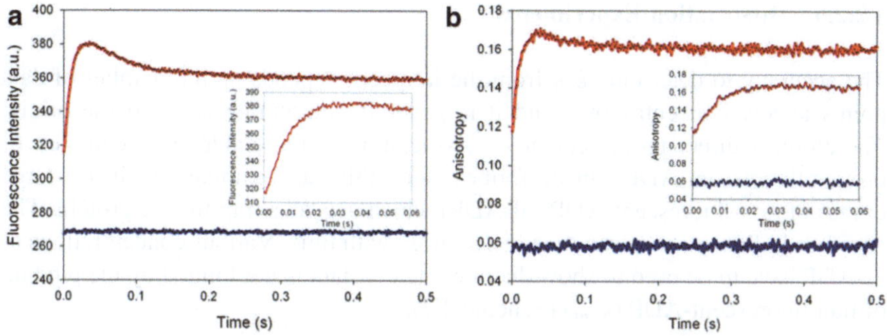

Fig. 4.3 Fluorescence anisotropy. Time course of mant fluorescence intensity (**a**) and fluorescence anisotropy (**b**) on mixing 2 μM mantATP with 8 μM PcrA·dT_{20}. The *inset* shows the first phase of the trace on a short timescale. *Blue traces* represent mixing mantATP against buffer

$$D = \frac{I_2}{I_1} \tag{4.10}$$

where I_1 and I_2 represent the fluorescence intensity at the start and end of the exponential, respectively.

In the case of PcrA [4], the intensity trace for the single-turnover measurement displayed two phases, an exponential increase followed by a partial decrease (Fig. 4.3a). Fluorescence anisotropy was used to confirm the binding results and importantly to confirm that the fluorescence decrease is related to a partial diphosphate dissociation. This is shown because the intensity and anisotropy traces are essentially the same (Fig. 4.3b).

4.2.2.5 Non-hydrolysable Analogues

Non-hydrolysable ATP analogues can be employed in the very same assays. These analogues help to relate the fluorescent traces to different steps in the cycle, apart from binding. A common nucleotide is ATPγS. In the example described below, an excess of RecD2·DNA was mixed with mantATP [6]. The trace shows a biphasic behaviour which can be fitted by a double exponential. The plots of k_{obs} versus [mantATP] showed that the first phase was a bimolecular binding event, while the second step was concentration independent showing a first-order process, as mentioned above. This first-order process could relate to a conformation change or steps following binding, such as hydrolysis. By using mantATPγS under the very same conditions, it was possible to determine that the first-order process must be a conformation change. This was established because mantATPγS had exactly the same behaviour as mantATP, while we were able to confirm by HPLC analysis that the nucleotide is not hydrolysed over the reaction time course.

4.2.2.6 Dissociation Experiments

The accuracy to determine k_{off} from the intercept depends on the number of data points at low concentrations, and it is generally inaccurate close to the origin. Therefore, a direct measurement of dissociation is preferable. If a complex of protein·fluorescent-ADP can be formed, then this can be mixed against a large excess of non-fluorescent-ADP (nf-ADP) which acts as a trap for the protein. The decrease in fluorescence can then be followed with time. Various concentrations of nf-ADP have to be used to show that the rate constant is not limited by the binding of non-fluorescent-ADP ($k_{+\text{nf}}$) (Scheme 4.7).

Scheme 4.7 Dissociation of fluorescent ADP from Protein and competing binding reaction on non-fluorescent ADP to the protein

$$\text{Protein·f-ADP} \underset{k_{-4}}{\overset{k_{+4}}{\rightleftharpoons}} \text{P + f-ADP} \qquad \text{P + ADP} \underset{k_{-\text{nf}}}{\overset{k_{+\text{nf}}}{\rightleftharpoons}} \text{P·ADP}$$

Scheme 4.8 Reaction consisting of an ADP-bound isomerisation step prior to ADP dissociation

$$\text{P·ADP·P}_i \underset{k_{-3}}{\overset{k_{+3}}{\rightleftharpoons}} \text{P·ADP} \underset{k_{-4a}}{\overset{k_{+4a}}{\rightleftharpoons}} \text{P·ADP*} \underset{k_{-4b}}{\overset{k_{+4b}}{\rightleftharpoons}} \text{P + ADP}$$

The above example works on the assumption that preforming a protein–ADP complex represents exactly the same species following hydrolysis. ADP release could occur in two steps, whereby the species following hydrolysis cannot be formed by mixing with ADP in an association experiment (Scheme 4.8). Therefore, this state can only be observed by measuring ADP release from hydrolysed ATP. This can be achieved by performing a similar measurement to the above but in a double-mix stopped-flow experiment. Firstly, protein and fluorescent ATP are mixed together. The delay before the second mixing with an excess of non-fluorescent-ADP allows for hydrolysis (details to measure the hydrolysis time course are shown in the next Sect. 4.2.3) to occur. Therefore, this approach directly measures ADP release from the hydrolysed protein complex. The trace may be characterised by a double exponential decrease in fluorescence due to two ADP states. The rate constant from the second step should be similar to that obtained by pre-incubating the protein with ADP, as described at the start of this section. Furthermore, by increasing the delay time in the double-mix experiment, it would be possible to isolate the second step if the system has enough time to reach ADP state 1 (ADP*) before mixing with the excess of non-fluorescent-ADP.

4.2.3 Hydrolysis Step

The hydrolysis step can be determined using a combination of quenched-flow and HPLC. Quenched-flow is a similar technique to stopped-flow, whereby two

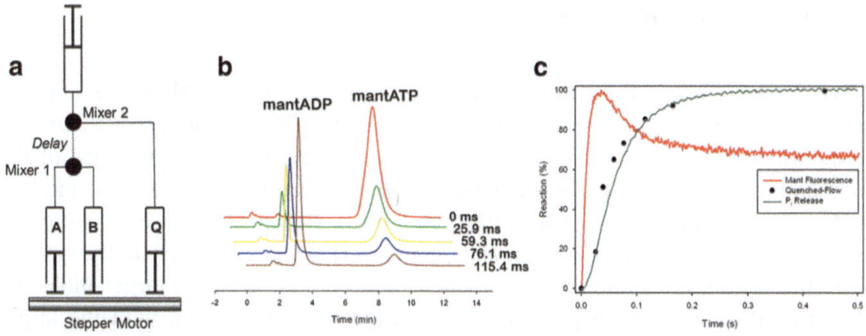

Fig. 4.4 Quenched-flow measurements of ATP hydrolysis. (**a**) Reagents A, B and quench (Q) are pushed simultaneously by the stepper motor. Reagents A and B mix and move into the delay loop. The delay loop of a defined length which governs the duration of the reaction before being mixed with quench. The quenched sample is then removed for analysis using the sample collector. (**b**) The HPLC analysis of quenched samples from mixing 2 μM mantATP, 8 μM PcrA and 16 μM dT_{20}. The reaction was allowed to proceed for the indicated amount of time before quenching with 10 % perchloric acid. (**c**) Time course of mantADP formation. The *circles* represent single time points for a quenched-flow measurement of mantADP formation. The mant fluorescence trace, performed under the same conditions, is added for comparison to binding kinetics; the trace was set to transition between 0 and 100 %. In addition, P_i release kinetics are also added so that each step of the cycle, binding, hydrolysis and P_i release can be compared within a single experiment. The measurements of Pi release were supplemented with 10 μM MDCC-PBP

solutions can be rapidly mixed with one another, followed by a second mixing with a quench agent (Fig. 4.4a) [22, 23]. This is typically an acid to denature the ATPase, which therefore stops the reaction at chosen time points [4]. Other quench agents include a large excess of EDTA to chelate Mg^{2+}, thereby reducing the hydrolysis. However, chelating kinetics has to be fast enough to halt the reaction. Furthermore, this would not stop the slow hydrolysis of $ATP{\cdot}Mg^{2+}$ already bound to the protein. The solution can then be removed and analysed by HPLC to give a ratio of ATP/ADP (Fig. 4.4b). Thereby, an ADP time course can then be determined. Fluorescent nucleotides provide a greater sensitivity to perform measurements at low concentrations of nucleotide, such as single-turnover conditions. These conditions allow hydrolysis to be directly related to mantATP binding under the same conditions (Fig. 4.4c). This not only gives the rate constant of the cleavage step, but it can also identify what processes fluorescent changes refer to.

The data can be analysed by fitting the trace to two exponentials (Eq. 4.7). The first exponential has no amplitude, and this represents the lag phase, corresponding to processes prior to hydrolysis, particularly binding in this case. The second exponential represents the rate of hydrolysis.

4.2.4 *Product Release*

Two products result from the ATPase reaction, ADP and P_i. There are coupled enzyme assays that measure ADP or P_i, for example [24, 25]. However, these may require addition of several components, increasing the possibility that one of the additions interferes with the biological system. Therefore, the best method to measure these products is reagentless biosensors (Chap. 2). These measure the formation of products in real time, without interfering significantly with the system. The biosensors are fluorescently labelled proteins that detect a target molecule (such as P_i). The protein provides the specificity by interacting with the target molecule, and the fluorophore gives a signal when that interaction occurs. For measurement of ADP and P_i, the biosensor protein only interacts with its target, once that product has dissociated from the motor protein.

The single-turnover kinetics employed during binding and hydrolysis can be used again here. Importantly, it is possible to relate the multiple fluorescent changes observed with a fluorescent ATP binding to steps within the ATP cycle (Fig. 4.4c). Typically, the single-turnover measurements show a trace with a small lag followed by an exponential increase in fluorescence. The data can be analysed by fitting the trace to two exponentials, as described for the hydrolysis measurement.

4.2.4.1 ADP Release

The ADP sensor is based upon *Escherichia coli* ParM [26]. The biosensor responds to ADP binding with a large fluorescence change by MDCC. The sensor has sub-micromolar ADP affinity while ATP affinity is 400× weaker. However, ATP concentration is limited to 200 μM because of binding to the sensor. Furthermore, it is important to ensure that the ATP substrate is a pure as possible, at least 99 %. The association rate constant is 0.65 $\mu M^{-1}\ s^{-1}$ which limits ADP binding to 13 s^{-1} under normal protein conditions. Therefore, this biosensor is not applicable to fast reactions. The fluorescent signal is calibrated through titrations of ADP and ATP. A rhodamine–ParM is also available for combination with coumarin-labelled nucleotides [27].

4.2.4.2 P_i Release

Two versions of the PBP have been used, one with a coumarin fluorophore (MDCC-PBP) and the other with two rhodamines (rhodamine-PBP) [19, 28]. Based upon the fluorescent nucleotide being used in the system, it is possible to select the relevant biosensor. With respect to MDCC-PBP, binding of phosphate causes a rapid conformational change responsible for the sixfold fluorescence increase. PBP is very sensitive to P_i; therefore. a "P_i Mop" has to be used to clean all equipment of contaminating P_i prior to use. This mop consists of 0.01 U ml^{-1} purine nucleoside phosphorylase and 200 μM 7-methylguanosine [19]. A calibration has to be performed to convert the fluorescent signal into μM P_i. The calibration is performed by measuring the fluorescent intensity of PBP in the present of several concentrations of P_i. A linear fit to the plot of fluorescence intensity versus P_i concentration gives the fluorescent change per μM P_i.

4.2.5 Kinetics of Non-fluorescent Nucleotides

These are important measurements given that there can be different affinities between labelled and unlabelled nucleotides. These differences would be first revealed in the steady-state analysis. Through these measurements, the mechanism described with a fluorescent ATP can be related to that with native ATP. It is important to remember that although the use of fluorescent nucleotides allows the mechanism to be revealed, it is possible that the absolute rate constants between nucleotide substrates do vary.

4.2.5.1 Competition Experiments

Fluorescent nucleotides can be used to measure association kinetics of native non-fluorescent nucleotides. In these experiments, protein is rapidly mixed with fluorescent-ATP (f-ATP) and non-fluorescent ATP (nf-ATP), which are both in a large excess over protein (Scheme 4.9). The experiments are performed at various concentrations of nf-ATP but fixed protein and f-ATP. The observed rate constant is

$$k_{\mathrm{OBS}} = k_{+1}[\mathrm{f}-\mathrm{ATP}] + k_{+2}[\mathrm{nf}-\mathrm{ATP}] \tag{4.11}$$

A plot of k_{OBS} versus nf-ATP gives a slope with k_{+2}, the ATP second-order association rate constant and an intercept of k_{+1}[f-ATP].

Scheme 4.9 Competitive binding between fluorescent and non-fluorescent ATP for protein

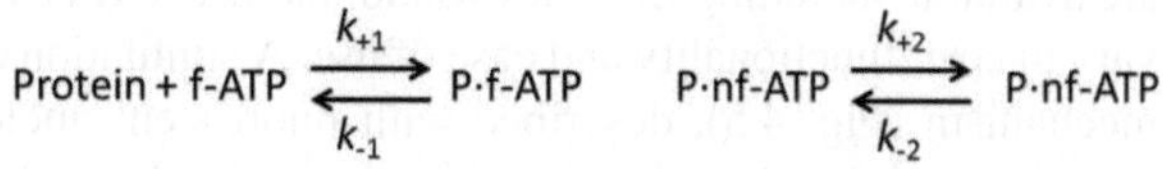

4.2.5.2 Multi-turnover P_i Release

The phosphate biosensor was described above in order to measure the P_i release kinetics from a single-turnover experiment. The biosensor can also be used in multiple turnover (MTO) experiments. The MTO approach is advantageous because

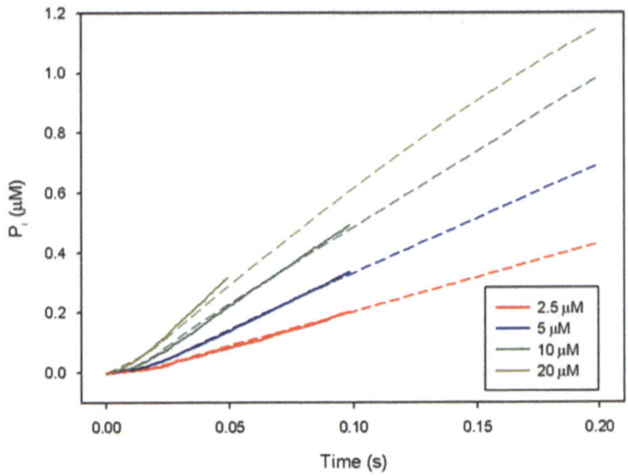

Fig. 4.5 Measurements with non-fluorescent nucleotides. The MDCC-PBP fluorescence traces (*solid lines*) for mixing ATP at the micromolar concentrations shown with 0.5 μM PcrA and 2.5 μM dT_{20} in the presence of 10 μM MDCC-PBP. Traces were calibrated to the amount of P_i produced. The model of the ATPase cycle was used fit the traces (*dashed lines*)

it does not require a large excess of the protein. The concept of this approach is to determine individual rate constants for the ATPase mechanism by monitoring the rate of P_i release following mixing with various concentrations of ATP [4, 6]. Pseudo-first-order conditions are once again employed with an excess of ATP. The traces are characterised by a slight lag, followed by a fairly linear rate of P_i release (Fig. 4.5). Both phases are dependent on ATP concentration. The lag duration decreases, while the rate of the linear phase increases with greater concentrations of ATP. The lag is due to the steps prior to P_i release, namely, ATP binding and cleavage during the first cycle of hydrolysis. Faster ATP binding at higher concentrations decreases the duration of the lag phase. The linear phase shows the steady-state turnover. The linear phase has a faster rate at higher ATP concentrations, due to the rapid rebinding of ATP. If the trace showed a burst phase before the linear phase, then this would suggest that ADP release is the rate-limiting step in the ATPase mechanism. Such a burst phase would result from the relatively fast P_i release of the initial cycle before the P_i release of subsequent cycles was limited by slow ADP release.

Simulations and modelling will not be discussed in detail here. Several software are available including Berkeley Madonna, DYNAFIT and KinTek Explorer. These vary in cost, functionality and ease of use. A simulation can be made based upon the mechanism (Fig. 4.5), described with fluorescent nucleotides. A global fit to the experimental data based upon a kinetic model derived from the fluorescent nucleotides can then be performed. Rate constants which fit to the entire dataset can then be extracted, assuming that the mechanism (i.e. the number of steps) is equivalent to the fluorescent ATP.

4.2.5.3 Diphosphate Dissociation Experiments

It may be possible to perform a direct measurement of ADP dissociation. This requires the tight binding of ADP which allows quantitative formation. This can be determined by steady-state measurements of K_i. In the example below, an excess of RecG·DNA was pre-bound to ADP before mixing with a large excess of mantADP in the stopped-flow apparatus [5]. This is the opposite approach to that employed in Sect. 4.2.2.6 with fluorescent dissociation. The very tight binding of mantADP to RecG made that a suitable trap for free RecG·DNA, after dissociation of ADP. The rate constant for dissociation was determined from a single exponential fit as $11.4\ s^{-1}$.

4.2.6 Single-Molecule PcrA ATPase

Measurements of ATPase activity are not restricted to bulk studies. Single-molecule measurements can also be performed to provide important information relating to the heterogeneity of the sample [29]. It is also possible to measure the ATP cycle directly coupled to movement of the motor protein [30]. This provides a direct visualisation of the chemo-mechanical coupling.

Either the motor protein or the protein's track (such as F-actin or DNA) has to be immobilised on an inert surface. In the case of helicases, the same conditions to study DNA unwinding can be used here [31–33]. If there is a significant fluorescent enhancement upon ATP binding, then it is easy to distinguish protein-bound ATP. Otherwise, background binding must be reduced as much as possible. Total internal reflection fluorescence (TIRF) microscopy is required to provide a high signal to noise which allows visualisation of individual fluorophores (Chap. 7). The ATP also has to be labelled with bright, stable fluorophores. Deac (coumarin based) and Cy3 have successfully been applied (Chap. 1). In the case of Cy3, a linker was used to space the large fluorophore away from the nucleotide (eda-ATP) [12].

Unless there is a significant change in fluorescence between ATP- and ADP-bound states, fluorescent spots will abruptly appear, remain stationary (if the protein is immobilised) and then disappear (Fig. 4.6a, b). The fluorescence trace will be characterised by a single increase in fluorescence with a plateau lasting for the duration of the ATP cycle (dwell time). The imaging conditions have to be optimised for single molecules where an image is acquired every 50 ms. This means that this assay is suited to slower ATPase motors with cycles lasting hundreds of milliseconds (or seconds). Therefore, each event would be observed over several frames which help to distinguish these events from background noise. It is important to confirm that the fluorescence decrease is due to ADP dissociation and does not result from photobleaching. This can be achieved by immobilising the fluorophore on the surface and measuring the dwell time. In addition, the measurement can be performed at different laser intensities, so that the dwell times should be invariant to this change.

The ATPase kinetics are determined from the distribution of dwell times (Fig. 4.6c). The distribution is fitted by a single exponential, whereby the reciprocal of the mean lifetime ($1/\tau$) gives the ATPase rate. Measurements with fluorescent-ADP allows the ADP dissociation constant to be measured. As with the above measurements, the dwell times are plotted before single exponential fitting to determine $1/\tau$.

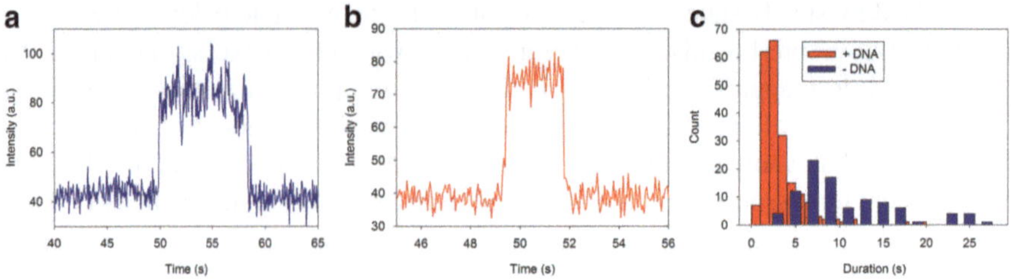

Fig. 4.6 Single-molecule ATPase of PcrA and DNA. (**a**) Example fluorescent time course of PcrA alone and with DNA (**b**). The lifetimes (dwell times) are plotted in (**c**). The mean lifetime + DNA was 3.1 s which gave an ATPase rate of 0.3 s^{-1}. With PcrA alone, the ATPase rate was 0.06 s^{-1}. Under the conditions of the experiment, the mean Cy3-eda-ATP photobleaching time was 24.3 s

4.3 Conclusions

Fluorescent nucleotides allow the molecular mechanism of motor proteins to be dissected through a variety of approaches. This flexibility should allow one to tailor a method to his or her specific needs. Further development of fluorescent nucleotides will aid this. An interesting development would be a specific fluorescence change between ATP and ADP bound states. It is most important to remember that modifications of any substrate can perturb its catalysis and that interpreting fluorescence signals in order to understand what gives rise to them can be challenging. Overall, there are now more tools available than ever before to decipher these complex biochemical mechanisms in more and more detail.

References

1. Berg JM, Tymoczko JL, Stryer L (2002) Biochemistry, 5th edn. W.H. Freeman, New York
2. De La Cruz EM, Wells AL, Rosenfeld SS, Ostap EM, Sweeney HL (1999) The kinetic mechanism of myosin V. Proc Natl Acad Sci USA 96(24):13726–13731
3. Friel CT, Howard J (2011) The kinesin-13 MCAK has an unconventional ATPase cycle adapted for microtubule depolymerization. EMBO J 30(19):3928–3939. doi:10.1038/emboj.2011.290
4. Toseland CP, Martinez-Senac MM, Slatter AF, Webb MR (2009) The ATPase cycle of PcrA helicase and its coupling to translocation on DNA. J Mol Biol 392(4):1020–1032. doi:10.1016/j.jmb.2009.07.071
5. Toseland CP, Powell B, Webb MR (2012) ATPase cycle and DNA unwinding kinetics of RecG helicase. PLoS One 7(6):e38270. doi:10.1371/journal.pone.0038270
6. Toseland CP, Webb MR (2013) ATPase mechanism of the 5'-3' DNA helicase, RecD2: evidence for a pre-hydrolysis conformation change. J Biol Chem 288(35):25183–25193. doi:10.1074/jbc.M113.484667
7. Bloemink MJ, Geeves MA (2011) Shaking the myosin family tree: biochemical kinetics defines four types of myosin motor. Semin Cell Dev Biol 22(9):961–967. doi:10.1016/j.semcdb.2011.09.015

8. Trentham DR, Eccleston JF, Bagshaw CR (1976) Kinetic analysis of ATPase mechanisms. Q Rev Biophys 9(2):217–281
9. Hiratsuka T (1983) New ribose-modified fluorescent analogs of adenine and guanine nucleotides available as substrates for various enzymes. Biochim Biophys Acta 742(3):496–508
10. Webb MR, Corrie JE (2001) Fluorescent coumarin-labeled nucleotides to measure ADP release from actomyosin. Biophys J 81(3):1562–1569. doi:10.1016/S0006-3495(01)75810-9
11. Webb MR, Reid GP, Munasinghe VR, Corrie JE (2004) A series of related nucleotide analogues that aids optimization of fluorescence signals in probing the mechanism of P-loop ATPases, such as actomyosin. Biochemistry 43(45):14463–14471. doi:10.1021/bi0486334
12. Toseland CP, Webb MR (2011) Fluorescent nucleoside triphosphates for single-molecule enzymology. Methods Mol Biol 778:161–174. doi:10.1007/978-1-61779-261-8_11
13. Toseland CP, Webb MR (2010) Fluorescence tools to measure helicase activity in real time. Methods 51(3):259–268. doi:10.1016/j.ymeth.2010.02.012
14. Ha T (2001) Single-molecule fluorescence resonance energy transfer. Methods 25(1):78–86. doi:10.1006/meth.2001.1217
15. James NG, Jameson DM (2014) Steady-state fluorescence polarization/anisotropy for the study of protein interactions. Methods Mol Biol 1076:29–42. doi:10.1007/978-1-62703-649-8_2
16. Eccleston JF, Martin SR, Schilstra MJ (2008) Rapid kinetic techniques. In: Correia JJ, Detrich III HW (eds) Methods in Cell biology, vol 84. Elsevier, Amsterdam
17. Batters C, Woodall KA, Toseland CP, Hundschell C, Veigel C (2012) Cloning, expression, and characterization of a novel molecular motor, Leishmania myosin-XXI. J Biol Chem 287(33):27556–27566. doi:10.1074/jbc.M112.381301
18. De La Cruz EM, Ostap EM (2009) Kinetic and equilibrium analysis of the myosin ATPase. Methods Enzymol 455:157–192. doi:10.1016/S0076-6879(08)04206-7
19. Brune M, Hunter JL, Howell SA, Martin SR, Hazlett TL, Corrie JE, Webb MR (1998) Mechanism of inorganic phosphate interaction with phosphate binding protein from Escherichia coli. Biochemistry 37(29):10370–10380. doi:10.1021/bi9804277
20. Shanker N, Bane SL (2008) Basic aspects of absorption and fluorescence spectroscopy and resonance energy transfer methods. Methods Cell Biol 84:213–242. doi:10.1016/S0091-679X(07)84008-8
21. Lakowicz JR (2006) Principles of fluorescence spectroscopy, 3rd edn. Springer, New York
22. Eccleston JF, Martin SR, Schilstra MJ (2008) Rapid kinetic techniques. Methods Cell Biol 84:445–477. doi:10.1016/S0091-679X(07)84015-5, S0091-679X(07)84015-5 [pii]
23. Barman TE, Bellamy SR, Gutfreund H, Halford SE, Lionne C (2006) The identification of chemical intermediates in enzyme catalysis by the rapid quench-flow technique. Cell Mol Life Sci 63(22):2571–2583. doi:10.1007/s00018-006-6243-z
24. Webb MR (1992) A continuous spectrophotometric assay for inorganic phosphate and for measuring phosphate release kinetics in biological systems. Proc Natl Acad Sci USA 89(11): 4884–4887
25. Charter NW, Kauffman L, Singh R, Eglen RM (2006) A generic, homogenous method for measuring kinase and inhibitor activity via adenosine 5'-diphosphate accumulation. J Biomol Screen 11(4):390–399. doi:10.1177/1087057106286829
26. Kunzelmann S, Webb MR (2009) A biosensor for fluorescent determination of ADP with high time resolution. J Biol Chem 284(48):33130–33138. doi:10.1074/jbc.M109.047118
27. Kunzelmann S, Webb MR (2010) A fluorescent, reagentless biosensor for ADP based on tetramethylrhodamine-labeled ParM. ACS Chem Biol 5(4):415–425. doi:10.1021/cb9003173
28. Okoh MP, Hunter JL, Corrie JE, Webb MR (2006) A biosensor for inorganic phosphate using a rhodamine-labeled phosphate binding protein. Biochemistry 45(49):14764–14771. doi:10.1021/bi060960j
29. Oiwa K, Eccleston JF, Anson M, Kikumoto M, Davis CT, Reid GP, Ferenczi MA, Corrie JE, Yamada A, Nakayama H, Trentham DR (2000) Comparative single-molecule and ensemble myosin enzymology: sulfoindocyanine ATP and ADP derivatives. Biophys J 78(6): 3048–3071. doi:10.1016/S0006-3495(00)76843-3

30. Sakamoto T, Webb MR, Forgacs E, White HD, Sellers JR (2008) Direct observation of the mechanochemical coupling in myosin Va during processive movement. Nature 455(7209): 128–132. doi:10.1038/nature07188
31. Chisty LT, Toseland CP, Fili N, Mashanov GI, Dillingham MS, Molloy JE, Webb MR (2013) Monomeric PcrA helicase processively unwinds plasmid lengths of DNA in the presence of the initiator protein RepD. Nucleic Acids Res 41(9):5010–5023. doi:10.1093/nar/gkt194
32. Fili N, Mashanov GI, Toseland CP, Batters C, Wallace MI, Yeeles JT, Dillingham MS, Webb MR, Molloy JE (2010) Visualizing helicases unwinding DNA at the single molecule level. Nucleic Acids Res 38(13):4448–4457. doi:10.1093/nar/gkq173
33. Fili N, Toseland CP, Dillingham MS, Webb MR, Molloy JE (2011) A single-molecule approach to visualize the unwinding activity of DNA helicases. Methods Mol Biol 778: 193–214. doi:10.1007/978-1-61779-261-8_13

Chapter 5
Use of Pyrene-Labelled Actin to Probe Actin–Myosin Interactions: Kinetic and Equilibrium Studies

Nancy Adamek and Michael A. Geeves

Abstract Studying the dynamics of the interaction between actin and myosin and how this is modulated by ATP and other nucleotides is fundamental to any understanding of myosin motor protein activity. The fluorescent label pyrene, covalently attached to actin (at Cys 374), has been one of the most useful optical probes to report myosin binding to actin. The unique spectral features of pyrene make it sensitive to changes in the microenvironment of the probe and allow to monitor processes such as conformational changes and protein–protein interactions. Here we describe how to make and use pyrene-labelled actin and describe a set of fluorescence stopped-flow measurements that allow the actin–myosin interaction to be explored at protein concentrations from μM to nM for many of the known myosin motors.

Keywords Pyrene • Actin • Kinetics

Abbreviations

A	Actin
Actin.myosin	Actin–myosin complex
D	ADP
F-actin	Actin in its filamentous form
G-actin	Actin in its globular form (depolymerised)
M	Myosin (or S1)
Pyr-actin	Pyrene-labelled actin
Pyr-actin.S1	Pyrene-labelled actin-S1 complex
S1	Motor domain fragment of myosin (subfragment 1)
T	ATP

N. Adamek • M.A. Geeves (✉)
School of Biosciences, University of Kent, Canterbury CT2 7NJ, UK
e-mail: m.a.geeves@kent.ac.uk

C.P. Toseland and N. Fili (eds.), *Fluorescent Methods for Molecular Motors*, Experientia Supplementum 105, DOI 10.1007/978-3-0348-0856-9_5,

5.1 Introduction

The study of myosin, like most long-term studies, has progressed in fits and starts over its long history, since muscle myosin was first separated from actin in the 1940s, through to the first crystal structure of a myosin motor domain in 1993 and now onto the huge family of myosin motors that we are still exploring. Since those early days, it has been understood that the binding of myosin to actin was central to its biological role. Finding ways to study this interaction between actin and myosin has been a long slow process. We still have no atomic resolution structure of an actin–myosin complex. Studies of the dynamics of the interaction, the rate of formation and the breakdown of the complex in response to ATP and other nucleotides have progressed. Early studies in solution used light scattering to follow the formation of an actin–myosin complex or its dissolution by ATP. But the light scattering signal is not very sensitive for concentrations of less than 1 μM and it is subject to artefacts from other scattering particles in solution. Researchers therefore sought ways to label either actin or myosin with a reporter group—normally a fluorescence tag—that could detect the presence of the binding partner. Actin in particular has a cysteine group on its surface (Cys374) that is accessible to modification. It is close to the myosin-binding site on actin but not directly involved in the actin–myosin contact. Cys374 can therefore be modified without altering the affinity of actin for myosin, while the environmental sensitivity of a probe can report the local perturbation caused by myosin binding.

Many fluorescence groups have been used that can report myosin binding, but the one that is now most commonly used is a pyrene that is reacted via an iodoacetamide group with Cys 374 (see Fig. 5.1a). In addition to reporting myosin binding, the group has also been widely used to report polymerisation of actin from the globular to filamentous form [1]. F-actin is the form that normally interacts with myosin. The advantages of the pyrene label over others that have been used include the following:

- It is relatively simple to label actin to >90 % on a single specific site.
- The labelled protein is stable for 2–3 weeks at 4 °C.
- The label does not appear to alter the behaviour of actin in either its polymerisation reaction or its interaction with myosin.
- The signal change on binding myosin is large, >50 % quench of intensity.
- The fluorescence intensity is high allowing studies of pyr-actin at nM concentrations.

There are two drawbacks to the pyrene fluorophore: the excitation wavelength is in the UV and photo bleaching can be a problem for long exposure times. The excitation and emission wavelengths are quite close together (see Fig. 5.1), and therefore, very sharp cut-off filters are required to cleanly separate the scattered excitation light from the emission. We normally use a Schott KV399 filter.

The combination of properties makes pyr-actin a very versatile tool for exploring actin–myosin interaction in solution. We will describe here the most common

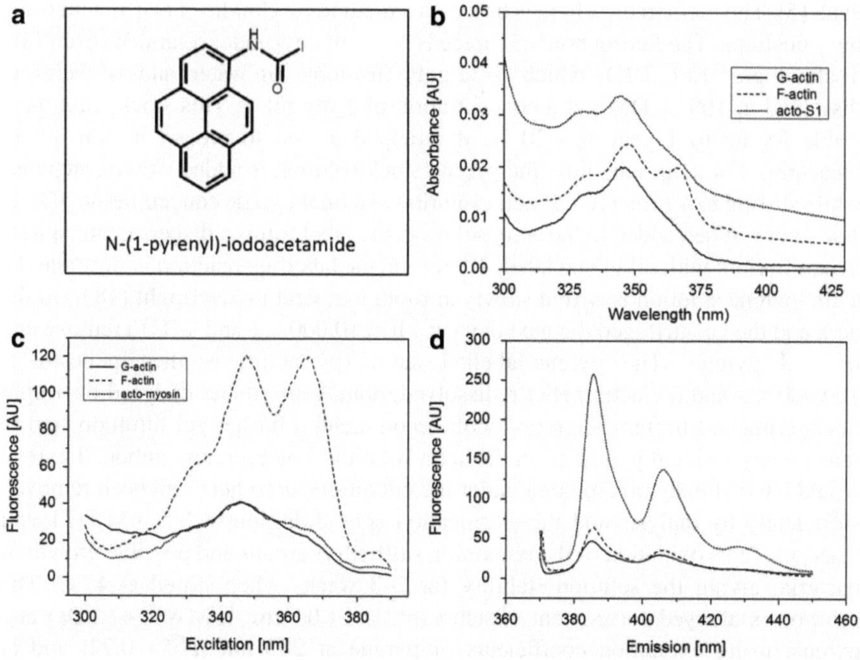

Fig. 5.1 Spectra of pyrene-labelled actin and acto-myosin complexes. (**a**) Pyrene iodoacetamide structure. (**b**) Absorbance spectra of pyrene-labelled G-actin, F-actin and actin.S1 at 1 μM concentration. (**c**) Fluorescence excitation spectra of pyrene-labelled G-actin, F-actin and actin. S1 at 1 μM concentration; emission wavelength: 407 nm. (**d**) Fluorescence emission spectra of pyrene-labelled G-actin, F-actin and actin.S1 at 1 μM concentration; excitation wavelength: 365 nm. (*Buffers*: F-actin and actin.S1 were measured in 20 mM MOPS pH 7.0, 100 mM KCl, 5 mM $MgCl_2$; G-actin was measured in depolymerising buffer containing 5 mM Tris pH 7.5, 0.2 mM $CaCl_2$, 1 mM NaN_3)

assays that can be used with any of the myosin family of proteins, and in combination they can reveal a great deal about how a myosin is tuned for its specific physiological role. These include assays of the rate and equilibrium constants for the formation of the actin–myosin complex and how this is altered by the presence of nucleotide on the myosin.

5.2 Preparation of Pyrene-Labelled Actin

The preparation of F-actin is based on a method by Lehrer and Kerwar [2] as modified by Spudich and Watt [3], where actin is extracted from a muscle acetone powder and purified through a series of polymerising and depolymerising steps. The labelling of actin at cysteine 374 with pyrene was first reported by Kouyama and Mihashi [4] and further developed by Cooper et al. [1] and Criddle et al. [5]. The method used here is based on a method by Criddle et al. [5] with some modifications. The fluorophore of choice is *N*-(1-pyrenyl)iodoacetamide (from Life Technologies Ltd, UK), which is largely insoluble in water and is therefore dissolved in 100 % DMF at a concentration of 5 mg ml^{-1}. This stock solution is stable for up to 1 year at −20 °C if wrapped in foil to protect it from photo bleaching. For actin labelling, the pyrene stock solution is added slowly, stepwise while stirring to a 1 mg ml^{-1} F-actin solution to a final pyrene concentration of 0.8–1 % (w/w). When added to the actin solution, the label forms a dispersed precipitate in equilibrium with dissolved label. Therefore, the labelling reaction is slow and the actin–pyrene solution is stirred slowly at room temperature overnight (18 h) in the dark and then centrifuged the next day for 1 h at 10,000 × *g* and 4 °C to remove any residual pyrene. The pyrene-labelled actin (pyr-actin) is then pelleted at 100,000 × *g* and the actin pellet redissolved, homogenised and dialysed overnight in experimental buffer. Some protocols recommend a further gel filtration step to remove any residual pyrene from the actin solution; however, any unbound pyrene would have already precipitated under the conditions used here and been removed sufficiently by dialysis and the clarification spin. Addition of 0.5 mM DTT and NaN_3 prevents oxidation of the remaining sulfhydryl groups and possible growth of bacteria, giving the solution stability for 2–3 weeks when stored at 4 °C. The solution is assayed for content of actin ($\varepsilon^{1\%} = 11.04$ cm^{-1}, MW = 42 kDa) and pyrene, using extinction coefficients of pyrene at 280 nm ($\varepsilon^{1\%} = 0.22$) and at 344 nm ($\varepsilon^{1\%} = 0.233$). The efficiency of labelling with pyrene is usually 90–100 %. If over- or under-labelling is observed, the (w/w) ratio of the pyrene stock solution used can be adjusted accordingly.

5.2.1 *Phalloidin-Stabilised Actin*

In solution, F-actin is in equilibrium with G-actin and the fraction of G-actin becomes significant at a total actin concentration below 1 μM. To maintain F-actin in experiments using lower protein concentrations, actin is stabilised in its filamentous form by the addition of phalloidin [6]. Phalloidin-stabilised actin is widely used in motility assays and in single molecule studies of actin and myosins. These and solution kinetic studies reported no significant effects of phalloidin on the actin–myosin interactions to date [7]. A 10 μM F-actin stock solution is prepared containing an equimolar amount of phalloidin (from *Amanita phalloides*, from Sigma Aldrich). The mix is incubated at 4 °C for several hours (ideally overnight) before use and is stable at 4 °C for about 1 week.

The absorbance of pyrene-labelled actin in the G- and F-forms is shown in Fig. 5.1b together with the excitation and emission spectra (Fig. 5.1c, d). The absorbance spectrum of G-actin shows a broad peak with a peak at 344 nm and a shoulder at 332 nm. On polymerisation, an additional peak appears at 365 nm with a general increase in scattering at all wavelengths. The addition of myosin S1 results in a further increase in light scattering and a loss of the 365 nm peak. Similar changes are observed in the excitation spectra (panel c), while the emission spectra (panel d) show a double peak at 385 and 405 nm. The fluorescence intensity is high for F-actin and the fluorescence is quenched by ~70 % in either G-actin or on binding of S1 to F-actin. This large signal change has proved very useful for monitoring actin polymerisation as well as the binding of myosin to F-actin.

5.3 ATP-Induced Dissociation of Actin.S1 by Stopped Flow Using Fast Skeletal Myosin S1

In stopped-flow experiments, the pyrene fluorescence is excited at 365 nm and the emission monitored using a 399 nm cut-off filter (KV399). The binding of ATP to the actin.S1 is measured by mixing the pyrene-labelled actin-S1 (A.M) complex with an excess concentration of Mg.ATP (T). Upon binding of ATP and subsequent dissociation of actin from myosin, the pyrene fluorescence is quenched by 70 %. The reaction can be described by the model of Millar and Geeves [8] shown in Scheme 5.1.

$$\mathrm{A{\cdot}M + T} \underset{}{\overset{K_1}{\rightleftharpoons}} \mathrm{A{\cdot}M{\cdot}T} \overset{k_{+2}}{\rightleftharpoons} \mathrm{A{\sim}M{\cdot}T} \xrightarrow{k_{dissn}} \mathrm{A + M{\cdot}T}$$

Scheme 5.1 ATP-induced dissociation of the actin-myosin complex

K_1 defines the equilibrium constant for the formation of the A·M·T collision complex, which is followed by isomerisation of the complex to the ternary complex A~M·T in which actin is more weakly bound and ATP more tightly bound with the rate constant of k_{+2}. This conformational change, step 2, is associated with the exponential change in pyrene fluorescence observed (as shown in the example transient in Fig. 5.2a). The following step, actin dissociation from myosin, can be followed by light scattering and shows no lag behind the fluorescence change. It is therefore assumed to be a very fast process following the conformational change. At low ATP concentrations ($K_1{\cdot}[\mathrm{ATP}] \ll 1$), the observed rate constant shows linear dependency on the ATP concentration as expected for a second-order reaction, and the slope of a plot of k_{obs} versus [ATP] is the apparent second-order rate constant $K_1{\cdot}k_{+2}$. At higher ATP concentration, k_{obs} shows hyperbolic dependence upon the ATP concentration (Eq. 5.1):

$$k_{obs} = [\mathrm{ATP}]k_{+2}/([\mathrm{ATP}] + K_{0.5}) \tag{5.1}$$

The dissociation reaction reaches a saturation plateau at a rate constant k_{max} which corresponds to k_{+2}. The point of half saturation defines the equilibrium constant of the dissociation from the complex $K_{0.5}$ (=1/K_1). This gives the affinity of the nucleotide in the collision complex (A.M-T).

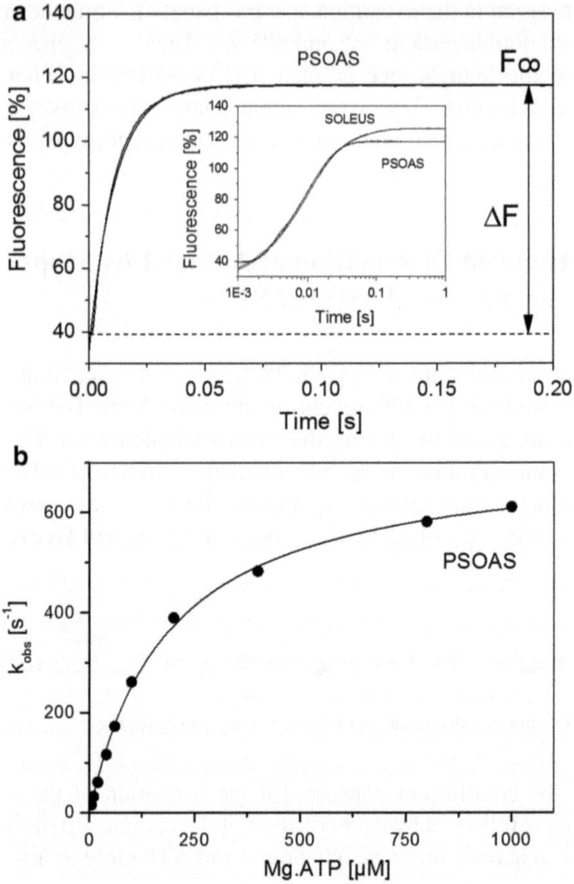

Fig. 5.2 ATP-induced dissociation of the acto-myosin complex. (**a**) Stopped-flow transient of 0.5 μM pyr-actin.*Psoas*S1 mixed with 25 μM ATP. The change in pyrene fluorescence is monitored over time and the best fit to a single exponential equation is superimposed $\Delta F_t = \Delta F_0 \exp(-k_{obs}t) + F_\infty$ where ΔF_0 is the amplitude of the fluorescence change. The observed rate constant (k_{obs}) = 90.4 s^{-1} and amplitude of 71.2 % for Psoas. The Soleus transient (Insert) is biphasic and the best fit to a double exponential equation ($\Delta F_t = \Delta F_1 \exp(-k_{obs1} \cdot t) + \Delta F_2 \exp(-k_{obs2} \cdot t) + F_\infty$) and gave $k_{obs1} = 87.1$ s^{-1} with an amplitude of 66.3 % and $k_{obs2} = 14.3$ s^{-1} with an amplitude of 8.0 %. (**b**) Plot of observed rate constant as a function of [ATP]. A hyperbolic fit to the data gives $1/K_1 = 185 \pm 9$ μM and $k_{+2} = 720 \pm 12$ s^{-1} (in 20 mM MOPS pH 7.0, 30 mM KCl, 5 mM $MgCl_2$). Data re-plotted from Iorga et al. [13]

This method of measuring the ATP-induced dissociation of actin.myosin has been applied to many different types of myosin, as most exhibit a large pyrene fluorescence change on binding and releasing myosin. There are a few examples where the pyr-actin signal change is very weak or non-existent: *Drosophila* indirect flight muscle myosin [9] and myosin IXb [10]. In such cases, light scattering can be used, but it requires higher protein concentrations and/or use of flash photolysis with caged ATP to generate high-quality data. Even where the pyrene signal is very weak for a particular myosin, it is possible to probe how it binds to pyr-actin by using a competition between a standard rabbit muscle myosin and the second myosin [11].

Some myosins show distinct behaviour compared to the fast muscle myosin shown in Fig. 5.2: Slow skeletal muscle myosin is characterised by a biphasic dissociation reaction, where an ATP-independent slow phase is observed in addition to the ATP-dependent fast phase as shown in Fig. 5.2 insert [12, 13]. Similar slow phases have also been observed for skeletal smooth muscle myosin II [14], mammalian non-muscle myosin II [15], Myosin 1b and 1c [16, 17], *Dictyostelium* Myosin 1e [18], as well as for myosin V and VI [19, 20].

Scheme 5.2 ATP-induced dissociation of actin-myosin with two forms of myosin

In the model proposed by Geeves et al. [16] shown in Scheme 5.2, actin.myosin (A.M) exists in two conformations: A.M (open) to which ATP (T) can bind, giving rise to the fast phase, and A.M' (closed), which has to isomerise to the open form before ATP can bind giving rise to the slow phase. The amplitude of each phase is proportional to the amount of each complex present, and therefore, the ratio of amplitudes is proportional to [A.M]/[A.M'] = $K_\alpha(=k_{+\alpha}/k_{-\alpha}$; see Scheme 5.2). The k_{obs} for the slow phase is defined by $k_{+\alpha}$, provided it is much slower than the following steps—a slow phase k_{obs} independent of ATP concentration is a diagnostic of this case. The one exception reported is for the mammalian myosin 1c in which both phases show ATP concentration dependence and unusually the slow phase has the largest amplitude [21]. The analysis is more complicated and the reader is referred to the original paper. Two similar complexes (A.M.D and A.M'. D) are also thought to exist in the presence of ADP (D), and in this case the isomerisation may be part of the strain-sensing mechanism (see Scheme 5.3) [22, 23].

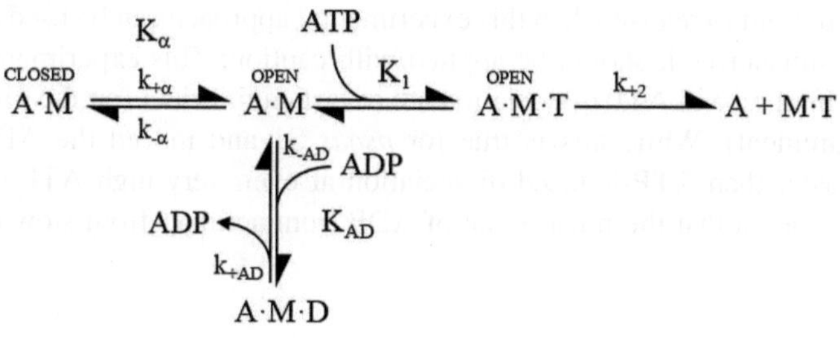

Scheme 5.3 Competition between ATP- and ADP-binding to actin-myosin

5.4 Measuring the Affinity of ADP for Actin.Myosin (K_{AD})

In the presence of ADP, both nucleotides compete for the binding to the actin–myosin complex, and since ADP binding does not normally dissociate the actin–myosin complex, the ATP-induced reaction is observed to apparently slow down in the presence of ADP. However, one of the characteristics of slow muscle myosin isoforms compared to fast isoforms is their much tighter binding of ADP to actin. myosin [13, 24, 25]. Assessing the ADP-binding properties therefore provides a good tool to define the nature of a novel myosin. There are several ways in which to assess the ADP-binding properties of a myosin isoform, but each approach is associated with its own problems regarding the slow isoforms, as will be shown.

5.4.1 Competitive Inhibition of the ATP-Induced Dissociation of Soleus Actin.S1 by ADP

If we assume that ADP binding is in rapid equilibrium with actin.S1, compared to the rate constant of ATP-induced dissociation of actin.S1, then ADP will act as a competitive inhibitor for the dissociation reaction, i.e. $k_{+AD}+[ADP]$ $k_{-AD} \gg K_1k_{+2}[ATP]$. Such an experimental set-up was first used by Siemankowski et al. [25] and is shown in Fig. 5.3 using myosin from rabbit *psoas* and *soleus* muscle [13]: pyr-actin.S1 from *psoas* and *soleus* muscle is mixed with 20 μM ATP, which was pre-incubated with increasing concentrations of ADP. Here the single exponential change in pyrene fluorescence was monitored (as shown in Fig. 5.3a) and the observed rate constants were normalised (to zero [ADP] value, $k_0 = K_1k_{+2}[ATP]$) and plotted as a function of ADP concentration (Fig. 5.3b).

As the ADP concentration is increased, the observed rate constant is slowed down and a fit of the data is described by a hyperbola (Eq. 5.2) giving an affinity of ADP for actin.S1 (K_{AD}) of 124 μM for *psoas*S1 and 8.5 μM for *soleus*S1, where k_0 is the k_{obs} in the absence of ADP.

$$k_{obs} = k_0/(1 + ([ADP]/K_{AD})) \tag{5.2}$$

It is important to realise when this experimental approach can be used and under which circumstances it should be applied with caution: This experiment assumes that ADP binding and ATP dissociation are at fast equilibrium (on the timescale of the measurement). While this is true for *psoas* S1 and indeed the ADP release remains faster than ATP-induced dissociation at even very high ATP concentrations, it is known that the release rate of ADP from actin.S1 from slow or cardiac

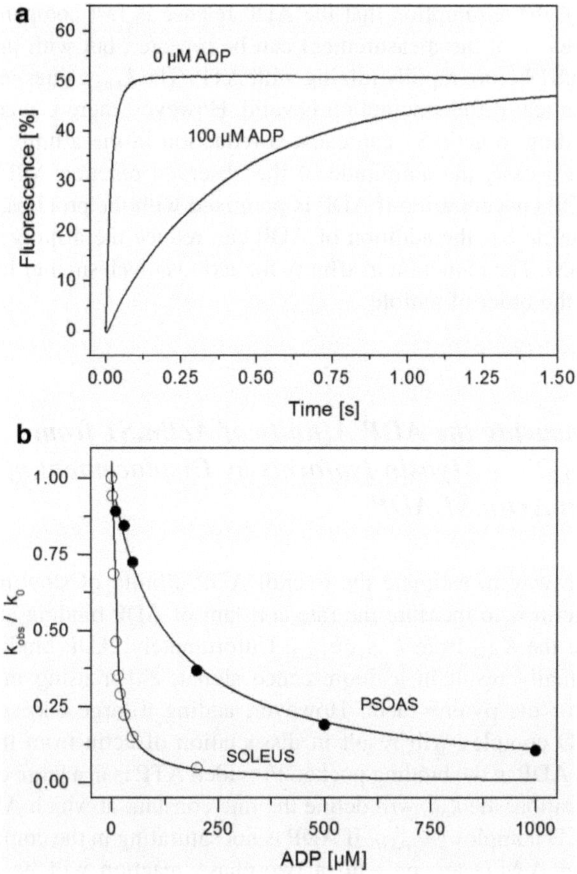

Fig. 5.3 Affinity of ADP for *soleus* and *psoas* actin-S1. (**a**) Stopped-flow transients of 0.5 µM pyr-actin.*soleus*S1 mixed with 20 µM Mg.ATP + 0 µM ADP or 100 µM ADP. The change in pyrene fluorescence is monitored over time and fitted to a single exponential equation (best fit superimposed). The k_{obs} was 45.0 s^{-1} without ADP and 3.1 s^{-1} in the presence of ADP, with an amplitude of 48.3 and 47.5 %, respectively. (**b**) ADP Affinity plot: 0.5 µM pyr-actin.S1 is mixed with 20 µM Mg.ATP pre-incubated with increasing concentrations of ADP and the single exponential change in pyrene fluorescence is monitored. The observed rate constants were normalised (to zero [ADP] value, k_0) and plotted as a function of [ADP] (*open circle* = *soleus*; *filled circle* = *psoas*). The data sets were fitted to a hyperbola ($k_{obs} = k_0/(1 + ([ADP]/K_{AD}))$) to obtain the K_{AD}: 124 ± 10 µM for *psoas* and 8.5 ± 0.9 µM for *soleus* with $k_0 = 65.8 \pm 1.2$ s^{-1} and 49.3 ± 1.8 s^{-1} respectively (in 20 mM MOPS pH 7.0, 100 mM KCl, 5 mM $MgCl_2$). Data re-plotted from Iorga et al. [13]

muscle myosin is relatively slow [12, 13, 24, 25]. Therefore, the above assumption may not be correct for slow myosins unless very low ATP concentrations are used. Ideally the ADP release rate constant from A.M.D needs to be determined, but this is not always possible.

As a test of the assumption that the ADP release is fast compared to the ATP dissociation reaction, the measurement can be repeated but with premixing ADP with the actin.S1 before rapidly mixing with ATP. The k_{obs} values in the two cases should be identical if the assumption is valid. However, there is a complication in that ADP binding to actin.S1 can cause a reduction in the affinity of myosin for actin. In such a case, the amplitude of the observed reaction will decrease with increasing ADP concentration if ADP is premixed with the proteins. In the case of rabbit fast muscle S1, the addition of ADP can reduce the affinity for actin from ~30 nM to 2 μM. The reduction in affinity for actin is much smaller for slow muscle myosins—of the order of tenfold.

5.4.2 *Measuring the ADP Affinity of Actin.S1 from Slow-Type Myosin Isoforms by Displacement of ADP from Actin.S1.ADP*

An alternative way to estimate the overall ADP affinity of slow myosins in the presence of actin is to measure the rate constant of ADP binding and release and then calculate the K_{AD} from k_{-AD}/k_{+AD}. Unfortunately, ADP binding to actin.S1 does not normally result in a fluorescence signal, either using intrinsic protein fluorescence or the pyrene label. However, adding a large excess of ATP to a ternary A.M.D complex will result in dissociation of actin from the complex as ATP replaces ADP in the binding pocket. Provided ATP is in a large excess over the ADP concentration, the k_{obs} will define the rate constant at which ADP is released from the A.M.D complex (k_{-AD}). If ADP is not saturating in the complex, such that both A.M and A.M.D are present, a two-phase reaction will be observed (see Fig. 5.4a): a fast phase as ATP binds to A.M and dissociates actin, as discussed previously, and a slow phase as ATP replaces ADP in the A.M.D complex (at a $k_{obs} = k_{-AD}$) and then dissociates actin (see Scheme 5.3). The fast and slow amplitudes of the two-phase reaction are proportional to A.M and A.M.D, respectively. While the two observed rate constants are relatively independent of ADP concentration present, the amplitudes show hyperbolic dependence on ADP concentration (Fig. 5.4b), and a fit of Eqs. (5.3) and (5.4) (for the fast and slow phase, respectively) to the data yields the affinity of ADP (K_{AD}) for the complex.

$$\text{fast phase}: y = (A_{max} * K_{AD})/(K_{AD} + x) + A_{min} \tag{5.3}$$

$$\text{slow phase}: y = (A_{max} * x)/(K_{AD} + x) + A_{min} \tag{5.4}$$

For some myosins, the analysis of the slow phases in this experimental set-up is complicated further by the presence of additional components in the transient at zero and high ADP concentrations. Careful analysis of each component and its dependence on ADP and ATP concentration will normally resolve the different component.

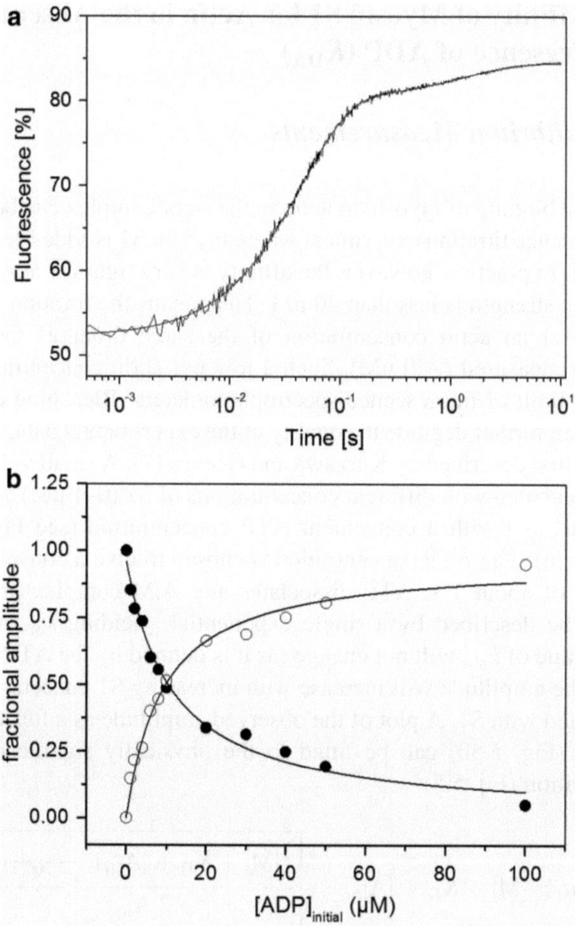

Fig. 5.4 ADP displacement from actin.myosin. (**a**) Stopped-flow transient: 0.1 µM pyr-actin.-β-myosinS1 pre-incubated with 100 µM ADP was mixed with 2 mM ATP and the change in fluorescence monitored over time (and fitted to a double exponential equation, best fit superimposed): $k_{obs}1 = 60.4\ s^{-1}$, $A1 = 38\ \%$ and $k_{obs}\ 2 = 0.78\ s^{-1}$ and $A2 = 4\ \%$. Data reproduced with permission from [37]. (**b**) ADP affinity K_{AD} of acto-*soleus* S1 0.2 µM pyr-actin.*soleus* S1, pre-incubated with ADP at different concentrations, was mixed with 250 µM Mg·ATP, and the change in pyrene fluorescence was monitored (ADP-displacement method). The biphasic transients were fitted to a double exponential equation and the amplitudes of the fast phase (*filled circle*) and slow phase (*open circle*) were plotted as a function of ADP concentration present. The data sets were fitted to a hyperbola and the obtained ADP affinity K_{ADP} of the fast and slow phase were 11.3 ± 1.0 µM and 9.2 ± 1.0 µM, respectively (in 20 mM MOPS pH 7.0, 100 mM KCl, 5 mM $MgCl_2$). Data re-plotted from Iorga et al. [13]

5.5 The Affinity of Myosin S1 for Actin in the Absence (K_A) and Presence of ADP (K_{DA})

5.5.1 Equilibrium Measurements

In principle the binding of myosin to actin in the rigor complex can be assessed in a simple fluorescence titration experiment where myosin S1 is added sequentially to a stirred cuvette. In practice, however, the affinity is very tight for most myosins and at normal ionic strength is less than 40 nM. This means the titration should ideally be performed at an actin concentration of the same order as the dissociation constant to be measured (~40 nM). Such a low pyr-actin concentration can be at the resolution limit of fluorescence spectrophotometers. Bleaching of the fluorescence signal can further degrade the quality of the experimental data. An alternative approach was first described by Kurzawa and Geeves [7]: A small volume of 30 nM pyr-actin is incubated with different concentrations of S1 (0–1 μM) and then mixed in the stopped flow with a convenient ATP concentration (see Fig. 5.5a, using human β-myosin). The ATP concentration is chosen to give a convenient monitoring timescale of about 1 s. ATP dissociates any A.M complex present and the transient can be described by a single exponential, yielding k_{obs} and amplitude values. The value of k_{obs} will not change (as it is defined by the ATP concentration used), while the amplitude will increase with increasing S1 concentration until the actin is saturated with S1. A plot of the observed amplitude as a function of free S1 concentration (Fig. 5.5b) can be fitted to the physically significant root of the quadratic equation (Eq. 5.5):

$$\alpha = [\mathrm{M}] + K_\mathrm{D} + [\mathrm{A}]_0 - \sqrt{\frac{\left([\mathrm{M}] + K_\mathrm{D} + [\mathrm{A}]_0\right)^2 - \frac{4}{[\mathrm{M}][\mathrm{A}]_0}}{2[\mathrm{A}]_0}} \qquad (5.5)$$

where α = fraction of actin with S1 bound ($\alpha = (F_0 - F)/(F_0 - F_\infty)$), [M] = [S1], $[\mathrm{A}]_0$ = total actin concentration and K_D is the dissociation constant of S1 for actin.

The more S1 is present, the larger the observed amplitude of the reaction will be; hence, the measurement depends critically on the concentration of active protein in the sample. Since it is well known that the K_A is very strongly dependent on ionic strength (tightening as ionic strength is decreased), careful consideration must also be given to the experimental buffer to be used [26–28].

The same method can be applied to determine the K_{DA}, the actin affinity for S1 in the presence of saturating concentration of ADP. The main difference in the experimental set-up is the use of a higher ATP concentration in order to compete with the ADP present. For example, 30 nM pyr-actin pre-incubated with varying amounts of S1 and 100 μM ADP is mixed with 500 μM ATP (Fig. 5.5b). The presence of ADP can significantly weaken the actin affinity for many myosins from 1- to 300-fold. This loss in affinity for actin (K_A) in the presence of ADP (K_{DA}) is

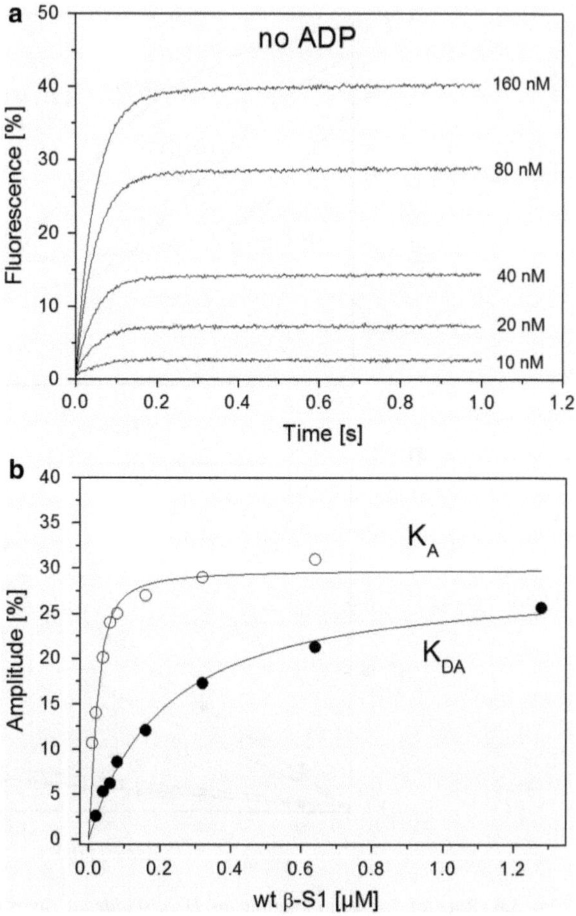

Fig. 5.5 Measurement of K_A and K_{DA}. (**a**) Example transients of 30 nM phalloidin-stabilised pyr-actin pre-incubated with increasing amounts of β-myosin and mixed with 20 μM ATP. (**b**) Plot of amplitudes as a function of S1 concentration for human β-myosinS1 in the absence and presence of ADP. The best fit was determined using the quadratic equation describing the binding isotherm (Eq. 5.5), resulting in $K_A = 8$ nM and $K_{DA} = 190$ nM (in 20 mM MOPS pH 7.0, 100 mM KCl, 5 mM $MgCl_2$). Data reproduced with permission from Bloemink et al. (2014)

reflected in the loss of affinity of ADP for S1 (K_D) compared to actin.S1 (K_{AD}). This is described as the thermodynamic coupling constant $K_A/K_{DA} = K_D/K_{AD}$ and is one of the key indicators, along with duty ratio, that show the type of mechanical activity a myosin is capable of [23].

5.6 Kinetics of Actin Binding to Myosin S1

Pyrene fluorescence can also be utilised to measure the rate at which actin binds to and dissociates from myosin. Figure 5.6 shows the transient observed for the binding of pyr-actin to different *Dictyostelium* myosin motor domain constructs (myosin residues 1–765; wild type and with an addition of 4 or 8 residues in loop 2, [29]). In this experiment, an excess of pyr-actin (1.5 μM) was mixed with

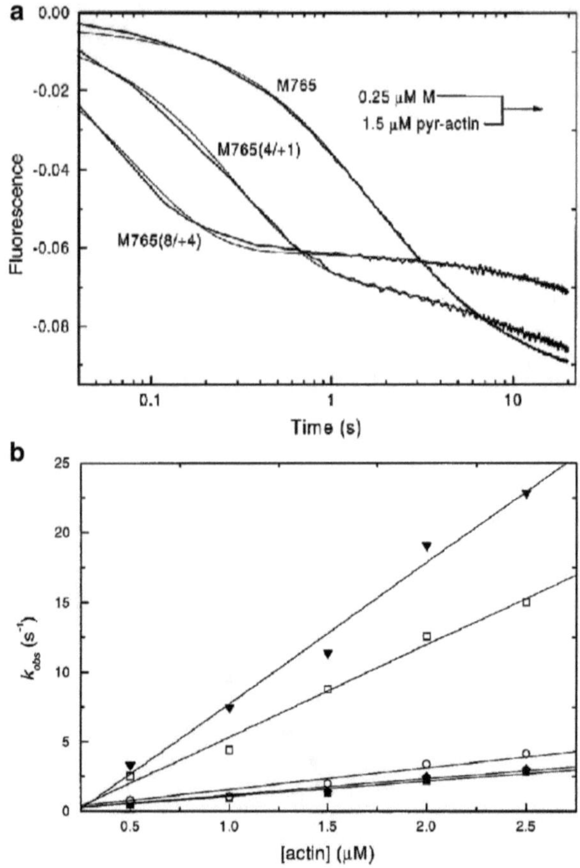

Fig. 5.6 Rate of pyr-actin binding to *D. discoideum* myosin motor domain constructs. (**a**) Stopped-flow transient of mixing 1.5 μM pyr-actin with 0.25 μM myosin motor domain construct. The time course of the fluorescence decreases after mixing myosin constructs and pyr-actin fits a double exponential. (**b**) Plot of the observed rate constants as a function pyr-actin concentration. The data were fitted to a straight line, the slope of which gives a second-order binding constant (k_{+A}). The intercepts are all near zero since the rate of dissociation is slower than the rate of association determined by these plots. Values of 1.34 M^{-1} s^{-1} [M765(wt)], 1.40 M^{-1} s^{-1} [M765 (11/0)], 2.16 M^{-1} s^{-1} [M765(4/+1)], 7.11 M^{-1} s^{-1} [M765(8/+4)] and 10.89 M^{-1} s^{-1} [M765 (20/+2)] were determined from the slopes of the linear fits. The symbols correspond to the following *D. discoideum* myosin constructs: M765(wt) (*filled square*), M765 (5/−1) (*filled diamond*), M765 (4/+1) (*open circle*), M765 (8/+4) (*open square*) and M765 (20/+12) (*filled inverted triangle*) (in 20 mM MOPS pH 7.0, 100 mM KCl, 5 mM $MgCl_2$). Data reproduced with permission from Furch et al. [29]

0.25 μM myosin and the reduction in fluorescence intensity monitored over 20 s on a log timescale. The fluorescence decreased in a single exponential with a slower bleaching of signal (Fig. 5.6a).

As expected the k_{obs} values were linearly dependent upon actin concentration (Fig. 5.6b: $k_{obs} = k_{+A}[pyrA] + k_{-A}$), and in terms of Scheme 5.4, the slope defines the second-order rate constant for binding (k_{+A}) and the intercept the dissociation rate constant (k_{-A}). As seen in Fig. 5.6b, the intercept is close to zero on the y-axis and so is poorly defined. The rate constant for actin dissociation (k_{-A}) can be measured by displacing pyr-actin bound to myosin by addition of a 20-fold excess of unlabelled actin (as shown in Fig. 5.7 using *D. discoideum* myosin motor domain mutants). The resulting pyrene fluorescence increase is single exponential, and provided the unlabelled actin is in a large excess, the observed rate constant corresponds directly to the actin release rate constant k_{-A}. The dissociation equilibrium constant K_A for actin binding can then be calculated from the ratio of the two rate constants as $K_A = k_{-A}/k_{+A}$. On some occasions, double exponential transients may be observed in the displacement experiments, which usually are caused by contamination of protein solutions with traces of nucleotide, giving rise to an additional fast phase. In those cases, pre-incubation of stock solutions with apyrase should alleviate the problem. Conversely, both measurements can be repeated in the presence of saturating ADP to define the rate constants of formation and breakdown of the A.M.D complex from actin and M.D.

$$\overset{\mathbf{K_A}}{\mathrm{A + M} \underset{k_{-A}}{\overset{k_{+A}}{\rightleftharpoons}} \mathrm{A{\cdot}M}}$$

Scheme 5.4 Actin-binding to myosin

The measurements of the association reaction described (excess pyr-actin mixed with myosin) can in principle be repeated, but mixing pyr-actin with an excess myosin instead, which, in theory, should give the same second-order rate constant when plotted against myosin concentration. In practice, however, the reaction becomes distinctly biphasic with an additional slow component. This is thought to be due to myosin binding to a free actin site between two sites already occupied by myosin being slowed by the steric constraints of the large myosin heads bound to the smaller actin monomer. Evidence for this comes from experiments pre-loading the actin filaments with myosin heads, as first described by Trybus and Taylor [30]. Pre-binding actin with up to 50 % saturation with myosin reduces the amplitude of the fast component with little effect on the slow component. Further increases in myosin binding reduce the amplitude of both phases. Measurement of the fast component corresponds closely to the phase seen with excess actin and gives a reliable estimate of the true rate of myosin binding. A complete description of the binding transient requires a more complex model in which the rate of binding alters as the degree of saturation increases. This can be done and the reader is referred to the relevant literature [31, 32].

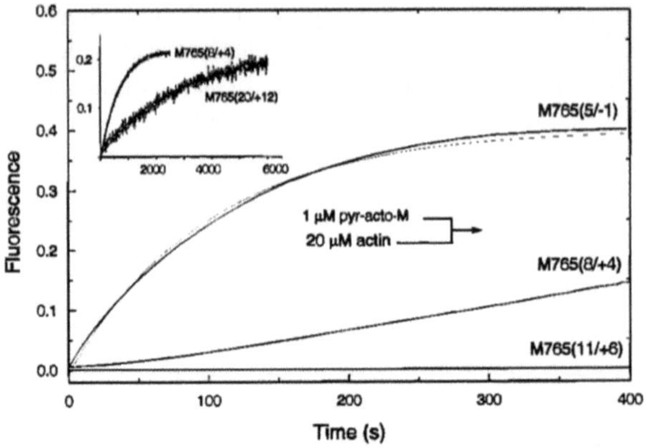

Fig. 5.7 Determination of the rate constant of actin displacement from actin–myosin using *D. discoideum* myosin motor domain constructs. The dissociation rate constants (k_{-A}) were determined from the rate of fluorescence enhancement when excess unlabelled actin was added to a sample of 0.5 μM pyr-actin equilibrated with an equimolar amount of mutant myosin construct. The stopped-flow and spectrofluorimeter transients of M765(5/−1), M765(8/+4) and M765(11/+6) are shown. The best fit to a single exponential is superimposed (in 20 mM MOPS pH 7.0, 100 mM KCl, 5 mM $MgCl_2$). Data reproduced with permission from Furch et al. [29]

5.7 Final Word

The above description outlines some of the simple elegant measurements that can be made using pyrene-labelled actin. The list is not exhaustive and many other measurements can be made. However, the above description defines the basic measurements that can be completed for almost any myosin. Some other experiments are more specialised, such as using actin with tropomyosin and troponin, in which the rate at which myosin binds to actin will depend upon the calcium concentration present in the assay [33, 34]. Similarly, some myosins have regulatory domains in the light chains (mammalian myosin 1c [21], smooth muscle [14], scallop muscle myosin [35]) which modulate how the actin-binding reaction takes place. For any novel myosin, the expectation is that the assays described here can be completed as described. As always the point of discovery begins when we find something unexpected. For a novel myosin, the binding to actin may not conform to all of the expectations set out here. Once the quality of the protein preparation has been confirmed and the equipment is working as expected, then trust the data and learn something new. One such recent example is the mouse myosin 18 which binds weakly to actin in an ATP-independent manner [36].

References

1. Cooper JA, Walker SB, Pollard TD (1983) Pyrene actin: documentation of the validity of a sensitive assay for actin polymerization. J Muscle Res Cell Motil 4(2):253–262
2. Lehrer SS, Kerwar G (1972) Intrinsic fluorescence of actin. Biochemistry 11(7):1211–1217
3. Spudich JA, Watt S (1971) The regulation of rabbit skeletal muscle contraction. I. Biochemical studies of the interaction of the tropomyosin-troponin complex with actin and the proteolytic fragments of myosin. J Biol Chem 246(15):4866–4871
4. Kouyama T, Mihashi T (1981) Fluorimetry Study of N-(1-Pyrenyl)iodoacetamide-Labelled F-Actin. Eur J Biochem 38:33–38
5. Criddle AH, Geeves MA, Jeffries T (1985) The use of actin labelled with N-(1-pyrenyl) iodoacetamide to study the interaction of actin with myosin subfragments and troponin/tropomyosin. Biochem J 232(2):343–349
6. Dancker P, Löw I, Hasselbach W, Wieland T (1975) Interaction of actin with phalloidin: polymerization and stabilization of F-actin. Biochim Biophys Acta 400(2):407–414
7. Kurzawa SE, Geeves MA (1996) A novel stopped-flow method for measuring the affinity of actin for myosin head fragments using microgram quantities of protein. J Muscle Res Cell Motil 17(6):669–676
8. Millar NC, Geeves MA (1983) The limiting rate of the ATP-mediated dissociation of actin from rabbit skeletal muscle myosin subfragment 1. FEBS Lett 160(1–2):141–148
9. Silva R, Sparrow JC, Geeves MA (2003) Isolation and kinetic characterisation of myosin and myosin S1 from the Drosophila indirect flight muscles. J Muscle Res Cell Motil 24(8):489–498
10. Nalavadi V, Nyitrai M, Bertolini C, Adamek N, Geeves MA, Bähler M (2005) Kinetic mechanism of myosin IXB and the contributions of two class IX-specific regions. J Biol Chem 280(47):38957–38968
11. Coluccio LM, Geeves MA (1999) Transient kinetic analysis of the 130-kDa myosin I (MYR-1 gene product) from rat liver. A myosin I designed for maintenance of tension? J Biol Chem 274 (31):21575–21580
12. Bloemink MJ, Adamek N, Reggiani C, Geeves MA (2007) Kinetic analysis of the slow skeletal myosin MHC-1 isoform from bovine masseter muscle. J Mol Biol 373(5):1184–1197
13. Iorga B, Adamek N, Geeves MA (2007) The slow skeletal muscle isoform of myosin shows kinetic features common to smooth and non-muscle myosins. J Biol Chem 282(6):3559–3570
14. Cremo CR, Geeves MA (1998) Interaction of actin and ADP with the head domain of smooth muscle myosin: implications for strain-dependent ADP release in smooth muscle. Biochemistry 37(7):1969–1978
15. Wang F, Kovacs M, Hu A, Limouze J, Harvey EV, Sellers JR (2003) Kinetic mechanism of non-muscle myosin IIB: functional adaptations for tension generation and maintenance. J Biol Chem 278(30):27439–27448
16. Geeves MA, Perreault-Micale C, Coluccio LM (2000) Kinetic analyses of a truncated mammalian myosin I suggest a novel isomerization event preceding nucleotide binding. J Biol Chem 275(28):21624–21630
17. Batters C, Arthur CP, Lin A, Porter J, Geeves MA, Milligan RA, Molloy JE, Coluccio LM (2004) Myo1c is designed for the adaptation response in the inner ear. EMBO J 23 (7):1433–1440
18. Tsiavaliaris G, Fujita-Becker S, Dürrwang U, Diensthuber RP, Geeves MA, Manstein DJ (2008) Mechanism, regulation, and functional properties of Dictyostelium myosin-1B. J Biol Chem 283(8):4520–4527
19. De La Cruz EM, Wells AL, Rosenfeld SS, Ostap EM, Sweeney HL (1999) The kinetic mechanism of myosin V. Proc Natl Acad Sci USA 96(24):13726–13731
20. Veigel C, Wang F, Bartoo ML, Sellers JR, Molloy JE (2002) The gated gait of the processive molecular motor, myosin V. Nat Cell Biol 4(1):59–65
21. Adamek N, Coluccio LM, Geeves MA (2008) Calcium sensitivity of the cross-bridge cycle of Myo1c, the adaptation motor in the inner ear. Proc Natl Acad Sci USA 105(15):5710–5715

22. Nyitrai M, Geeves MA (2004) Adenosine diphosphate and strain sensitivity in myosin motors. Philos Trans R Soc Lond B Biol Sci 359(1452):1867–1877
23. Bloemink MJ, Geeves MA (2011) Shaking the myosin family tree: biochemical kinetics defines four types of myosin motor. Semin Cell Dev Biol 22(9):961–967
24. Marston EW, Taylor SB (1980) Comparison of the myosin and actomyosin ATPase mechanisms of the four types of vertebrate muscles. J Mol Biol 139(4):573–600
25. Siemankowski RF, Wiseman MO, White HD (1985) ADP dissociation from actomyosin subfragment 1 is sufficiently slow to limit the unloaded shortening velocity in vertebrate muscle. Proc Natl Acad Sci USA 82(3):658–662
26. Konrad M, Goody RS (1982) Kinetic and thermodynamic properties of the ternary complex. Eur J Biochem 5(124):547–555
27. Marston SB, Weber A (1975) The dissociation constant of the actin-heavy meromyosin subfragment-1 complex. Biochemistry 14(17):3868–3873
28. Coates JH, Criddle AH, Geeves MA (1985) Pressure-relaxation studies of pyrene-labelled actin and myosin subfragment 1 from rabbit skeletal muscle. Biochem J 232:351–356
29. Furch M, Geeves MA, Manstein DJ (1998) Modulation of actin affinity and actomyosin adenosine triphosphatase by charge changes in the myosin motor domain. Biochemistry 37 (18):6317–6326
30. Trybus KM, Taylor EW (1982) Transient kinetics of adenosine 5'-diphosphate and adenosine 5'-(beta, gamma-imidotriphosphate) binding to subfragment 1 and actosubfragment 1. Biochemistry 21(6):1284–1294
31. Mijailovich SM, Kayser-Herold O, Li X, Griffiths H, Geeves MA (2012) Cooperative regulation of myosin-S1 binding to actin filaments by a continuous flexible Tm-Tn chain. Eur Biophys J 41(12):1015–1032
32. Geeves M, Griffiths H, Mijailovich S, Smith D (2011) Cooperative [Ca^2+]-dependent regulation of the rate of myosin binding to actin: solution data and the tropomyosin chain model. Biophys J 100(11):2679–2687
33. Mckillop DFA, Geeves MA (1993) Regulation of the interaction between actin and myosin subfragment 1: evidence for three states of the thin filament. Biophys J 65(2):693–701
34. Head JG, Ritchie MD, Geeves MA (1995) Characterization of the equilibrium between blocked and closed states of muscle thin filaments. Eur J Biochem 227(3):694–699
35. Nyitrai M, Stafford WF, Szent-Györgyi AG, Geeves MA (2003) Ionic interactions play a role in the regulatory mechanism of scallop heavy meromyosin. Biophys J 85(2):1053–1062
36. Guzik-Lendrum S, Heissler SM, Billington N, Takagi Y, Yang Y, Knight PJ, Homsher E, Sellers JR (2013) Mammalian myosin-18A, a highly divergent myosin. J Biol Chem 288 (13):9532–9548
37. Bloemink et al, 2014, JBC 289(8), pp 5158–67

Chapter 6
Fluorescent Methods to Study Transcription Initiation and Transition into Elongation

Aishwarya P. Deshpande*, Shemaila Sultana*, and Smita S. Patel

Abstract The DNA-dependent RNA polymerases induce specific conformational changes in the promoter DNA during transcription initiation. Fluorescence spectroscopy sensitively monitors these DNA conformational changes in real time and at equilibrium providing powerful ways to estimate interactions in transcriptional complexes and to assess how transcription is regulated by the promoter DNA sequence, transcription factors, and small ligands. Ensemble fluorescence methods described here probe the individual steps of promoter binding, bending, opening, and transition into the elongation using T7 phage and mitochondrial transcriptional systems as examples.

Keywords Fluorescence • Transcription initiation • RNA polymerase • Kinetics • 2-Aminopurine • Anisotropy • FRET

Abbreviations

2-AP	2-Aminopurine fluorescence
bp	Base pair
DNA	Deoxyribonucleic acid
ds	Double stranded
EC	Elongation complex
FRET	Förster resonance energy transfer
IC	Initiation complex
K_d	Equilibrium dissociation constant

*These authors made equal contribution to this work.

A.P. Deshpande • S. Sultana • S.S. Patel (✉)
Department of Biochemistry and Molecular Biology, Robert Wood Johnson Medical School, Rutgers, The State University of New Jersey, Piscataway, NJ 08854, USA
e-mail: patelss@rwjms.rutgers.edu

C.P. Toseland and N. Fili (eds.), *Fluorescent Methods for Molecular Motors*, Experientia Supplementum 105, DOI 10.1007/978-3-0348-0856-9_6,

mt	Mitochondrial
NTP	Nucleoside triphosphate
RNA	Ribonucleic acid
RNAP	RNA polymerase
ss	Single stranded
TAMRA	Tetramethylrhodamine

6.1 Introduction

The past few years have witnessed enormous growth in the understanding of the mechanism and regulation of DNA transcription through numerous structural snapshots of the RNA polymerase (RNAP) complexes from various organisms and through genome-wide gene expression studies and mutational studies [1–11]. Nevertheless, most of the quantitative research to elucidate the detailed mechanism of the transcriptional pathway has been conducted using kinetic approaches that monitor RNA synthesis by high-resolution radiometric gel assays [12] or intermediates by real-time fluorescent assays [13–15, 2, 16]. The focus of this chapter is to discuss how ensemble fluorescence measurements can be combined with transient-state kinetic methods to probe and quantify the various steps of transcription initiation and transition into elongation, the two stages of transcription that critically regulate the efficiency of RNA synthesis.

DNA-dependent RNA polymerases are motor enzymes that read the sequence of the DNA and copy the information into RNA molecules using ribonucleotide triphosphates (NTP) as substrates in a process called transcription. Transcription occurs in distinct stages including initiation, elongation, and termination. During initiation, the RNAP binds to a specific promoter DNA to form a closed and open complex in which a short stretch of DNA near the +1 transcription start site is melted to expose the template for initiation of RNA synthesis (Fig. 6.1). The initiating NTPs bind in the active site of the RNAP and base pair with the melted +1 and +2 template bases to form the initially transcribing complex. Phosphodiester bond formation between the two NTPs results in the synthesis of the first 2-mer RNA product, which is then extended to longer RNAs by sequential addition of NTPs coupled to DNA unwinding. During this phase, short RNAs from 2 to 12 mer dissociate into solution as abortive products. After 8–12 mer RNA synthesis, the RNAP is able to clear the promoter and make the transition into the elongation phase, where full-length RNA is made by processive translocation, DNA unwinding, and RNA synthesis.

This chapter is designed to follow the individual steps of transcription initiation (Fig. 6.1) and to empower researchers entering the field of transcription to ask targeted questions about particular steps and employ appropriate fluorescence techniques to address them. We also describe how these methods can be used to study the conformational changes in DNA and to determine the exact stage when

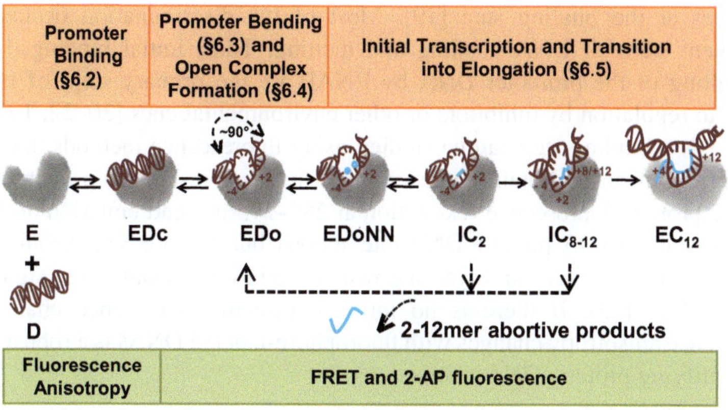

Fig. 6.1 Fluorescence methods probe the specific steps of transcription initiation. RNA polymerase (E) binds to the promoter DNA (D) to form a closed complex (EDc), which is converted to an open complex (EDo), where the DNA is sharply bent and locally melted. The initiating NTPs bind at the active site of the RNAP and base pair with +1 and +2 template bases to form the initially transcribing complex (EDoNN) that undergoes reaction to form the 2-mer RNA bound to the initiation complex (IC2). The 2-mer RNA is elongated to 8-mer and 12-mer in the initiation phase (IC8–12), and then the RNAP makes the transition into the elongation complex (EC12). Short 2–12 mer RNAs readily dissociate in solution to accumulate as abortive products

the RNAP transitions into elongation. We use studies of the well-characterized single-subunit T7 phage RNAP and the homologous yeast mitochondrial (mt) RNAP to elaborate on how fluorescence methods can be combined with transient-state kinetics to elucidate the mechanism of transcription initiation. We give a general overview of the theory behind each fluorescence technique, explain the rationale in designing the experiment, and expound on how to quantitatively analyze the data. The methods described here are general and applicable to studies of other RNA polymerases.

6.2 Promoter Binding

Transcription begins with RNAP binding to the promoter DNA. In T7 phage, the single-subunit T7 RNAP recognizes the consensus −17 to +6 promoter sequence without requiring any transcription factors. In yeast (*Saccharomyces cerevisiae*) mitochondria, transcription is carried out by a single-subunit class of enzymes related to T7 RNAP, but the core Rpo41 polymerase requires the transcription factor Mtf1 for specific transcription [17]. T7 RNAP discriminates against the non-promoter sequences at the initial binding step, showing ~10^5-fold tighter affinity for promoter relative to non-promoter sequences [18]. In contrast, Rpo41–Mtf1 shows little discrimination between promoter and non-promoter sequences at the binding step [19]. Most of the discrimination occurs at the subsequent steps of DNA bending and melting. Thus, initial binding, bending, and melting of the promoter DNA by RNAP are the primary steps of initiation subject to regulation by inhibitors or other environmental cues [20–22, 19].

The DNA binding step can be studied using fluorescence methods that rely on signals from intrinsic protein fluorescence or fluorophore-labeled protein or DNA. Intrinsic protein fluorescence (excitation at 280–295 nm and emission at 340 nm) can be used to quantify protein–DNA interactions, but it depends on the number and placement of tryptophan and tyrosine residues and on a quantifiable signal upon DNA binding [23]. If there is no intrinsic protein fluorescence change, then fluorescence anisotropy changes with fluorophore-labeled DNAs is a robust method for quantifying protein–DNA interactions.

6.2.1 Fluorescence Anisotropy

Fluorescence anisotropy-based titrations allow one to monitor protein–DNA interactions with a high precision, in real-time, and under true-equilibrium conditions and are a preferred technique over the traditional filter-binding or electrophoretic mobility shift assays [24]. The basic principle of fluorescence anisotropy is that when a group of fluorescently labeled molecules are excited with polarized light, the molecules with their excitation dipoles oriented along the incident light will get excited and further emit polarized light depending on the location, size, and movement of the fluorophore [25]. A fluorescently labeled DNA will freely tumble in solution and depolarize the emitted light. However, when the DNA is bound by the large RNAP, tumbling will be reduced and emission will retain more polarization, resulting in an increase in anisotropy (Fig. 6.2a). Fluorescence anisotropy change serves as a signal to carry out both equilibrium and kinetic studies of RNAP–DNA complexes.

6.2.2 Fluorophore-Labeled DNA Substrates

Fluorophore-labeled oligodeoxynucleotides can be purchased with a wide selection of fluorophores which absorb and fluoresce above the absorbance maxima of the DNA and protein (>400 nm), thus minimizing background fluorescence and inner-filter effects. The parameters that need to be considered for designing the DNA substrates are as follows: (1) the preferred length of the DNA is the minimal promoter length that binds RNAP with 1:1 stoichiometry, (2) the sequence at the DNA ends such as GC bp that quench the fluorescence of the fluorophore or stack the fluorophore [26], (3) the extinction coefficient and quantum yield of the fluorophore, (4) the length of the linker between the fluorophore and the DNA, and (5) the distance between the fluorophore and the RNAP binding site. The

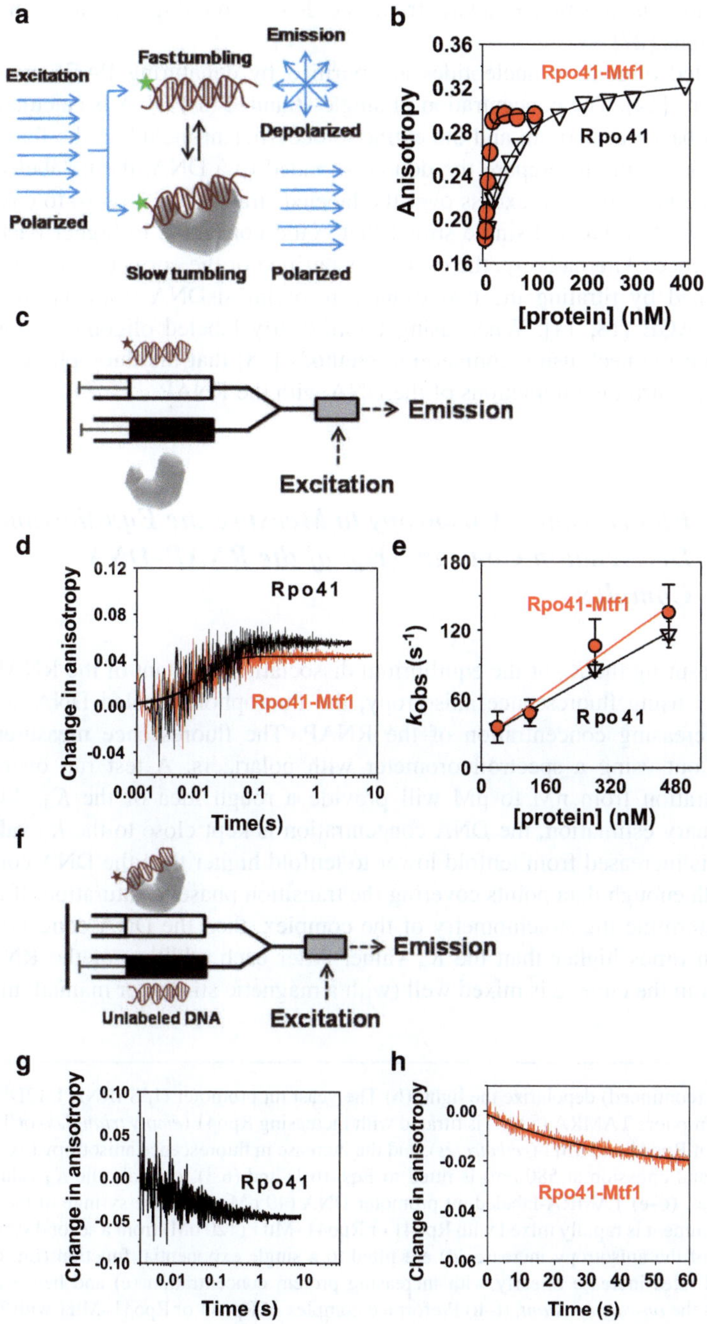

Fig. 6.2 Fluorescence anisotropy methods measure the K_d and kinetic *on-* and *off-rate constants.* **(a)** Fluorescence anisotropy is lower when the fluorophore-labeled DNA is not bound to RNAP and increases when the DNA binds to RNAP, because the larger complex does not readily

fluorophore is positioned away from the RNAP binding site to avoid direct interactions [27].

Labeled oligodeoxynucleotides are purified by denaturing PAGE under dark conditions [14]. The concentration of single-stranded (ss) DNA is calculated from its absorbance at 260 nm and the extinction coefficient including the fluorophore. During annealing to prepare the double-stranded (ds) DNA, the unlabeled strand should be kept in slight excess over the labeled strand (1.1: 1 ratio) to ensure that there is no free labeled single strand that could contribute to higher background fluorescence. Alternatively, the dsDNA is purified or the correct annealing ratio is determined by titrating the two strands, resolving dsDNA from the ssDNA by native PAGE [18, 14]. When using fluorescently labeled oligonucleotides, it is important to check using competition methods [28] that the fluorophore does not greatly perturb the interactions of the DNA with the RNAP.

6.2.3 *Fluorescence Anisotropy to Measure the Equilibrium Dissociation Constant* (K_d) *of the RNAP–DNA Complex*

For measuring the K_d or the equilibrium dissociation constant of the RNAP–DNA complex using fluorescence anisotropy, the fluorophore-labeled DNA is titrated with increasing concentration of the RNAP. The fluorescence measurement is carried out using a spectrofluorometer with polarizers. A test run over RNAP concentration from nM to μM will provide a rough idea of the K_d. After this preliminary estimation, the DNA concentration is kept close to the K_d value, and RNAP is increased from tenfold lower to tenfold higher than the DNA concentration with enough data points covering the transition phase to saturation. If the goal is to determine the stoichiometry of the complex, then the DNA concentration is kept ten times higher than the K_d value. After each addition of the RNAP, the solution in the cuvette is mixed well (with a magnetic stir bar or manual mixing by

---

Fig. 6.2 (continued) depolarize the light. (**b**) The yeast mt promoter (15S rRNA U12D8) labeled with fluorophore TAMRA (5 nM) is titrated with increasing Rpo41 (*empty triangles*) or 1:1.2 ratio mixture of Rpo41 and Mtf1 (*red circles*), and the increase in fluorescence anisotropy (excitation at 555 nm and emission at 580 nm) is fitted to Eqs. (6.1) and (6.3) to obtain the K_d values of the complexes. (**c–e**) TAMRA-labeled mt promoter DNA (40 nM) from one syringe of the stopped-flow instrument is rapidly mixed with Rpo41 or Rpo41–Mtf1 (120 nM) from a second syringe. The kinetics of the anisotropy increase (**d**) are fitted to a single exponential function (Eq. 6.4). The observed rates increase linearly with increasing protein concentration (**e**) and hence the slope estimates the *on-rate constant*. (**f–h**) Preformed complex of Rpo41 or Rpo41–Mtf1 with TAMRA-labeled mt promoter (40 nM) from one syringe of the stopped-flow instrument is mixed with unlabeled mt promoter (1 μM) from the other syringe. The decrease in fluorescence anisotropy with time at 100 nM Rpo41 (**g**) and 100 nM Rpo41–Mtf1 (**h**) is fit to a single exponential function (Eq. 6.4) to obtain the *off-rate* [19] pipette) and incubated at a constant temperature for a predetermined time interval to reach equilibrium before exciting with sample with the excitation wavelength of the fluorophore. For all fluorescence measurements, we recommend using the highest grade spectroscopic reagents. Additionally, distilled water stored in plastic containers should be avoided for making buffers, as the plastic material tends to leach fluorescent species over time.

The titration of tetramethylrhodamine (TAMRA)-labeled promoter DNA with increasing concentration of Rpo41 and Rpo41–Mtf1 is shown (Fig. 6.2b) [19]. The sample was excited at 555 nm, and fluorescence intensities ($I_{\text{parallel}(\parallel)}$ and $I_{\text{perpendicular}(\perp)}$) were measured at 580 nm. The observed anisotropy, $r_{\text{obs}} = (I_{\parallel} - I_{\perp})/(I_{\parallel} + 2I_{\perp})$, is a function of anisotropies of both free DNA (r_{f}) and the RNAP-bound DNA (r_{b}) (Eq. 6.1):

$$r_{\text{obs}} = r_{\text{b}} f_{\text{b}} + r_{\text{f}}(1 - f_{\text{b}}) \tag{6.1}$$

where f_{b} is the fraction of the total DNA ([D_t]) in the RNAP–DNA complex. In case the fluorescence intensity of the DNA-labeled dye is quenched by a factor Q, the anisotropy is corrected using Eq. (6.2):

$$r_{\text{obs}} = \frac{Q r_{\text{b}} f_{\text{b}} + r_{\text{f}}(1 - f_{\text{b}})}{1 - f_{\text{b}}(1 - Q)} \tag{6.2}$$

Formation of the RNAP–DNA complex will increase the anisotropy of labeled DNA without significantly changing the total fluorescence intensity ($I_{\text{total}} = I_{\parallel} + 2I_{\perp}$). However, if fluorescence intensity changes are greater than ~10 %, then the anisotropy is corrected (Eq. 6.2). Alternatively, both anisotropy and emission intensity data are simultaneously fit using SPECTRABIND [29]. If the intensity change is significant, then fluorescence intensity itself can be used to determine the K_{d} value.

The corrected anisotropy is plotted as a function of total RNAP concentration ([E_t]) to obtain the K_{d} values. The data are fit to the quadratic equation (Eq. 6.3) using the least squares method with softwares such as SigmaPlot (Systat software, Inc. San Jose, CA) [22]:

$$f_{\text{b}} = \frac{K_{\text{d}} + [\text{E}_{\text{t}}] + [\text{D}_{\text{t}}] - \sqrt{(K_{\text{d}} + [\text{E}_{\text{t}}] + [\text{D}_{\text{t}}])^2 - 4[\text{E}_{\text{t}}][\text{D}_{\text{t}}]}}{2[\text{D}_{\text{t}}]} \tag{6.3}$$

Anisotropy-based titrations showed that Rpo41-promoter DNA complex has a K_{d} of 58 nM, which indicated that Rpo41 itself binds to the promoter DNA with a high affinity. However, in the presence of its transcription factor Mtf1, promoter binding became ~300-fold tighter (K_{d} of 0.2 nM) [19]. The two K_{d} values were measured using 2–8 nM 5′-end TAMRA-labeled promoter DNA titrated with increasing protein concentration. Similar titrations determined that Rpo41–Mtf1 binds non-promoter DNA with only a fivefold weaker affinity (K_{d} ~ 1 nM) than the promoter DNA. This is in contrast to T7 RNAP, where competition assays showed that a non-promoter DNA binds ~10^5-fold more weakly than the promoter DNA [18]. Thus, the two RNAPs use different mechanisms for choosing promoter over non-promoter DNAs during transcription initiation.

6.2.4 Kinetic Studies of RNAP–DNA Complex Formation

Fluorescence anisotropy changes can be monitored in real time to determine the dynamics of the RNAP–DNA complex. The *on- and off*-rates show how fast RNAP binds and falls off the DNA, and studies with transcription factors determine how the stability of the complexes is modulated by other factors. Such measurements demonstrated that the high affinity of Rpo41–Mtf1 for the promoter DNA is because Mtf1 decreases the *off-rate* and not due to an increase in *on-rate*. The *on-rate* and *off-rate* can also be used to determine K_d values (*off-rate*/*on-rate*) of extremely tight RNAP–DNA complexes such as T7 RNAP with pre-melted promoter (K_d of 5 pM).

6.2.4.1 Measurement of the On-Rate of RNAP–DNA Complex

To measure the *on-rate* of RNAP–DNA complex, fluorescence anisotropy is recorded after mixing fluorescently labeled promoter DNA with RNAP in a stopped-flow instrument (Fig. 6.2c). Here, automated motor-driven syringes help rapid mixing of RNAP with DNA at a constant temperature with continuous fluorescence emission measurement at a particular wavelength. The anisotropy changes are measured as a function of time after mixing at constant DNA and various RNAP concentrations under pseudo-first-order conditions (where the RNAP concentration is always in tenfold excess over the labeled DNA). Multiple time traces (at least 7–8 shots) are averaged for each RNAP concentration and the averaged observed anisotropy (r_{obs}) versus time is fit to the exponential equation (Eq. 6.4) to obtain the observed rate constant (k_{obs}):

$$r_{obs} = r_f + Ae^{-k_{obs}t} \tag{6.4}$$

where r_f is the anisotropy value at the start, A is the change in anisotropy, and t is time. Representative time traces of increase in anisotropy of TAMRA-labeled promoter DNA upon addition of Rpo41 and Rpo41–Mtf1 are shown (Fig. 6.2d) [19]. The observed rates increase linearly with increase in Rpo41 and Rpo41–Mtf1 concentrations (Fig. 6.2e), indicating that binding of RNAP to promoter DNA is a single-step process (Scheme 6.1). Therefore, the dependency can be fit to a linear equation (Eq. 6.5) to obtain the *on-rate* (k_1) and *off-rate* (k_{-1}):

$$\mathrm{E} + \mathrm{D} \underset{k_{-1}}{\overset{k_1}{\rightleftarrows}} \mathrm{ED_c}$$

Scheme 6.1 One-step binding of RNAP to DNA

$$k_{\mathrm{obs}} = k_{-1} + k_1[\mathrm{E}] \tag{6.5}$$

where the slope provides the *on-rate* and the *y*-intercept is the *off-rate*. If the *off-rate* is slow, then it is best determined more directly from chase experiments (Sect. 6.4.2.2). Such measurements with Rpo41 and Rpo41–Mtf1 showed that each binds to the promoter DNA with similar *on-rates* (2–2.5 × 10^8 M^{-1} s^{-1}), which indicates that the transcription factor Mtf1 does not affect the kinetics of complex formation [19]. Thus, the lower K_d of Rpo41–Mtf1–DNA complex must be due to slower *off-rate*.

Similar anisotropy binding experiments showed that T7 RNAP forms a closed complex with its promoter DNA with an *on-rate* of 1.9 × 10^8 M^{-1} s^{-1} and an open complex with a pre-melted promoter with similar *on-rate* of 3 × 10^8 M^{-1} s^{-1} [30]. Thus, the lower K_d of T7 RNAP-pre-melted promoter complex is due to its slower off-rate. Once both *on* and *off-rates* are obtained, then the ratio provides an independent measure of the equilibrium dissociation constant (K_d) for the RNAP–DNA complex, which should match the K_d from the equilibrium titrations (Sect. 6.2.3).

6.2.4.2 Measurement of the Off-Rate of RNAP–DNA Complex

The *off-rate* of the RNAP–DNA complex is directly measured using a chase experiment, where a preformed complex of RNAP and fluorescently labeled DNA is mixed with a large molar excess (10- to 20-fold) of the same unlabeled DNA (Fig. 6.2f). When the labeled DNA dissociates from the RNAP, it is diluted in the pool of excess of unlabeled DNA. The fluorescence anisotropy of the free labeled DNA is lower, and therefore a decrease in anisotropy is observed over time with a rate equal to the *off-rate*. A TAMRA-labeled promoter DNA complexed with Rpo41 (Fig. 6.2g) or Rpo41–Mtf1 (Fig. 6.2h) was chased with excess of unlabeled promoter DNA [19]. The exponential decrease in anisotropy with time is fit to Eq. (6.4) to obtain the off-rate from the k_{obs}.

The *off-rate* of the promoter DNA from Rpo41 is 16 s^{-1} and from Rpo41–Mtf1 is 0.02 s^{-1}. Hence, Mtf1 stabilizes the ternary complex by decreasing the *off-rate* by 800-fold without significantly affecting the *on-rate*, as shown above. Similar methods were used to measure the *off-rates* of T7 RNAP complexed with duplexed or pre-melted promoter. T7 RNAP has a very high affinity for the pre-melted promoter such that the K_d could not be measured by equilibrium binding assays.

Therefore, the K_d of the pre-melted promoter complex was determined from the ratio of the *off-rate*/*on-rate* [18]. The *off-rate* of the duplex promoter from T7 RNAP is 0.8 s^{-1} and of the pre-melted promoter is 0.0015 s^{-1} [30]. Thus, the ratio of the *off-rate*/*on-rate* provided a K_d of 5 pM for the pre-melted promoter-T7 RNAP complex and 4 nM for the duplex promoter complex. The kinetic experiments show that the tighter binding of the pre-melted promoter is not due to its faster *on-rate* on T7 RNAP (1.6-fold faster than duplexed promoter), but due to its slower *off-rate* (530-fold slower).

6.3 Promoter Bending

RNA polymerases bend the promoter DNA during open complex formation, and the bending angle changes as the RNA gets elongated during the initiation stage and when the RNAP makes the transition into elongation. DNA conformational changes like bending can be detected using structural methods such as X-ray crystallography [31–33], nuclear magnetic resonance [34], electron microscopy [35], atomic force microscopy [36, 37], electrophoretic mobility assays like circular permutation assay [38–40] and helical phasing assay [39, 41, 40], ring closure methods [39, 42, 40], and fluorescence methods including Förster resonance energy transfer (FRET). FRET methods are increasingly used in transcription for monitoring DNA bending changes and for monitoring movements of the RNAP relative to the DNA [19, 43, 30, 22, 44]. With the recent advances in single-molecule FRET [45–48], these events can be monitored in real time on single molecules. Both ensemble and single-molecule FRET have been successfully used to study bending and unbending dynamics of T7 RNAP and yeast mt RNAP promoter complexes [22, 19, 49, 30].

6.3.1 Förster Resonance Energy Transfer

FRET relies on two-fluorophore system, donor, and acceptor, chosen carefully such that there is spectral overlap between the emission spectrum of the donor and the excitation spectrum of the acceptor (Fig. 6.3a). On exciting the donor, energy is transferred through dipole–dipole coupling to the acceptor that emits at a higher emission wavelength. The FRET efficiency (E_{FRET}) depends on the distance between the fluorophores (Eq. 6.9) and therefore serves as a molecular ruler with a range of ~10–100 Å, ideal for monitoring protein–DNA conformational changes and DNA bending angles in complexes. The E_{FRET} also serves as a signal to measure the K_d of RNAP–DNA complexes and the kinetics of DNA bending and unbending.

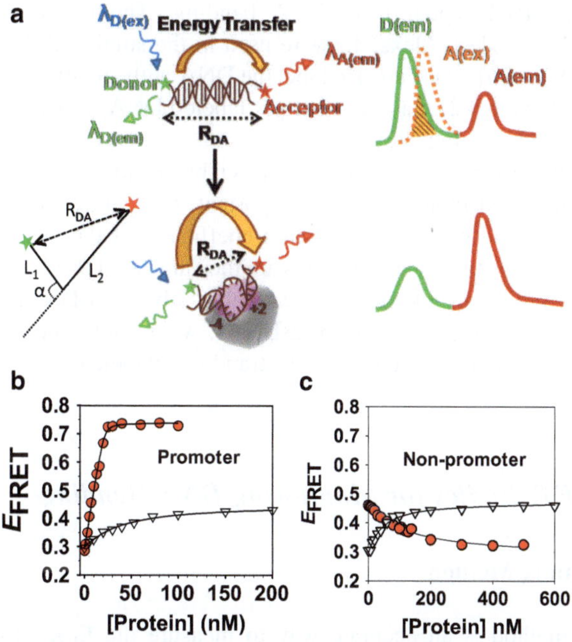

Fig. 6.3 FRET methods to study promoter bending and unbending. (**a**) The spectral overlap (*yellow-striped region*) between the donor (*green*) and acceptor (*red*) fluorophores attached to the DNA enable FRET measurements. When the RNAP binds to the promoter DNA and induces a bend, the distance (R_{DA}) between the two fluorophores decreases, leading to an increased E_{FRET} monitored as a decrease in donor's emission or increase in acceptor's emission intensity. The bending angle α is calculated from the R_{DA} value and the assumed upstream L_1 and downstream L_2 DNA lengths based on a single-kink model of DNA. Yeast mt promoter (**b**) and non-promoter DNAs (**c**), doubly labeled with TAMRA and Alexa Fluor 647, are titrated with increasing concentration of Rpo41 (*empty triangles*) or Rpo41–Mtf1 (*red circles*). The E_{FRET} provides the DNA bending angles and through titrations the K_d values of complexes [19]

6.3.2 DNA Substrates for FRET Studies

FRET measurements require DNAs that are doubly labeled with donor and acceptor fluorophores. Commercial DNA synthesis facilities covalently link the fluorophore through a linker to the 5′-end or the 3′-end of the ssDNA or to the dT base within DNA. A donor and acceptor pair must be chosen, and several are available with Förster distance (R_0) ranging from 50 to 85 Å. The R_0 is the distance (r) between the fluorophores when E_{FRET} is 50 %. For example, the R_0 of Fluorescein–TAMRA or Cy3–Cy5 is ~55 Å, TAMRA–Alexa647 or Alexa488–Alexa546 is ~65 Å, and Alexa594–Alexa647 is 85 Å. Thus, depending on the size of the consensus promoter, appropriate pairs and attachment sites on the duplex DNA are chosen to maximize the FRET signal from DNA bending. The FRET pairs TAMRA–Alexa647 and Alexa488–Alexa546 were used in the studies of T7 and mt RNAP [22, 30, 47, 50, 19, 26]. One can also label the DNA with an intrinsic 2-aminopurine probe (2-AP) (Sect. 6.4.2) making it a triply labeled DNA, which can monitor both DNA bending and opening [22].

The fluorophore-labeled ssDNAs are purified by denaturing PAGE that separates labeled and unlabeled molecules. It is important to accurately determine concentration of each strand based on extinction coefficient of the DNA as well as the fluorophore [51]. The two labeled strands are then annealed in a 1:1 ratio and PAGE purified (Sect. 6.2.2). In addition to the doubly labeled dsDNA, one also needs donor-only and acceptor-only control dsDNAs, where only one strand is fluorescently labeled, while the complementary strand is unlabeled.

6.3.3 FRET Studies for Monitoring DNA Bending

6.3.3.1 (Ratio)$_A$ Method

The (ratio)$_A$ method is an accurate way to measure the E_{FRET} [52, 22]. In this method, the acceptor emission after donor excitation is divided by the acceptor emission when it is directly excited, and the resulting ratio is used to determine the E_{FRET}. Thus, the experiment includes measuring three different steady-state fluorescence spectra: (a) acceptor emission in doubly labeled DNA after donor excitation $F_{DA}(ex_D, em_A)$, (b) acceptor emission in doubly labeled DNA when the acceptor is directly excited $F_{DA}(ex_A, em_A)$, and (c) donor emission in donor-only DNA excited at its own excitation wavelength $F_D(ex_D, em_D)$. The value of (ratio)$_A$ is calculated from Eq. (6.6):

$$(\text{ratio})_A = \frac{F_{DA}(ex_D, em_A) - \alpha \cdot F_D(ex_D, em_A)}{F_{DA}(ex_A, em_A)} \tag{6.6}$$

where α is a normalization factor of donor fluorescence in the absence and presence of FRET, i.e., $F_{DA}(ex_D, em_D)/F_D(ex_D, em_D)$. Depending on the FRET fluorophore pairs used and their spectral overlap, the donor excitation wavelength slit is between 1 and 2 nm to minimize direct excitation of the acceptor. E_{FRET} is linearly dependent on the measured (ratio)$_A$ by Eq. (6.7):

$$(\text{ratio})_A = E_{FRET} d^{+} \frac{\epsilon^{D}_{ex\ D}}{\epsilon^{A}_{ex\ A}} + \frac{\epsilon^{A}_{ex\ D}}{\epsilon^{A}_{ex\ A}} \tag{6.7}$$

where d^{+} is the labeling ratio with the donor dye, $\varepsilon_D(ex_D)$ and $\varepsilon_A(ex_A)$ are the extinction coefficients of the donor and acceptor at their respective excitation wavelengths, and $\varepsilon_A(\text{ex}_D)$ is the extinction coefficient of the acceptor at the donor's excitation wavelength. Ideally, labeled DNA should be PAGE purified from the unlabeled DNA, resulting in d^+ as 1.

Since the observed E_{FRET} depends on the concentration of the complex, the doubly labeled DNA is titrated with increasing amounts of RNAP to determine the maximal E_{FRET}. The binding fraction of DNA is related to the measured E_{FRET} by Eq. (6.8):

$$f_b = \frac{E - E_0}{E_b - E_0} \tag{6.8}$$

where E_0 and E_b are starting and ending E_{FRET} values, respectively, in the titration of DNA with increasing RNAP. One can also obtain the RNAP promoter K_d by plotting the observed E_{FRET} as a function of RNAP concentration and fitting it to a quadratic equation (Eq. 6.3).

In studies of mt RNAP, the promoter DNAs were 20 bp in length (−12 to +8) with TAMRA and Alexa647 attached to the two 5′ ends. Representative traces of change in E_{FRET} of promoter (Fig. 6.3b) and non-promoter DNA (Fig. 6.3c) with increasing concentrations of Rpo41 and Rpo41–Mtf1 are shown [19]. The E_{FRET} of free DNAs is ~0.3 which increased to ~0.45 upon addition of Rpo41 (Fig. 6.3b, c, empty triangles). This indicates that Rpo41 induces a slight bend in both the promoter and non-promoter DNAs. In contrast, similar experiments with Rpo41–Mtf1 resulted in a higher E_{FRET} of ~0.74 with the promoter DNA, but not with the non-promoter DNA (Fig. 6.3b, c respectively, circles). This indicates that Rpo41–Mtf1 recognizes the promoter sequence and induces a selective severe bending in the promoter DNA. The non-promoter sequence actually showed a final decrease in E_{FRET}, which indicates DNA unbending. The titrations also provide K_d values, which are consistent with the K_d values obtained from the fluorescence anisotropy-based titrations (Sect. 6.2.3) [19].

FRET studies of mt RNAP promoter under single-molecule conditions are consistent with the ensemble FRET studies and show that Mtf1 is needed to bend the promoter [49]. The advantage of single-molecule studies is separation of populations and measurement of dynamics of bending/unbending of single molecules in real time. Single-molecule time traces showed that Rpo41 alone can bend the promoter DNA to the same degree as Rpo41–Mtf1, however, with a very low frequency. Thus, the low E_{FRET} of the Rpo41-promoter complex observed in ensemble studies is not due to a small bending angle, but a consequence of the rare bending events [49, 17]. The single-molecule FRET time traces measured the DNA bending rates; k_{bending} by Rpo41 alone is 0.0052 s^{-1} and by Rpo41–Mtf1 is 2.8 s^{-1}. The $k_{\text{unbending}}$ by Rpo41 is 3.0 s^{-1} and by Rpo41–Mtf1 is 0.066 s^{-1} [49].

The donor and acceptor fluorophores were placed in the 31-bp T7 promoter at −22 and +9 positions. Due to the greater separation between the donor and acceptor, the E_{FRET} of free DNA is much lower (0.02). However, in the presence of T7 RNAP, the E_{FRET} increases to 0.06 and further to 0.12 in the presence of the initiating GTP, which indicates that GTP aids in promoter bending [22]. Similarly, the E_{FRET} of free pre-melted promoter (bubble of 6 bp) is low (0.02), and upon addition of T7 RNAP, E_{FRET} increases to 0.12 and addition of GTP has little effect. This indicates that GTP binding drives the T7 RNAP complex toward the bent open state that resembles the pre-melted promoter.

6.3.3.2 Measurement of the Donor–Acceptor Distance

The end-to-end distance of a linear doubly labeled DNA can be calculated from the observed E_{FRET} (Eq. 6.9) assuming helical geometry of the B-type duplex DNA in aqueous solution [52]. FRET efficiency is inversely proportional to the sixth power of the distance between the donor and the acceptor:

$$R = R_0 \left[\frac{1 - E_{\text{FRET}}}{E_{\text{FRET}}} \right]^{1/6} \tag{6.9}$$

where R_0 is the Förster radius of the donor–acceptor pair (D–A) at which 50 % of E_{FRET} occurs and R is the actual D–A distance measured from E_{FRET}. As indicated above, when Rpo41 binds to promoter or non-promoter DNA, the E_{FRET} of the DNA increases from 0.3 to 0.45, which indicates that the D–A distance decreases by ~8 Å [19]. On the other hand, when Mtf1 is added to the complex of Rpo41 and promoter DNA, the E_{FRET} increases from 0.45 to 0.74, which indicates that the D–A distance decreases by an additional 13 Å, resulting in a total shortening of ~21 Å compared to free DNA. The E_{FRET} of T7 promoter with fluorophores at −22 and +9 indicate D–A distance of 113 Å, which decreases to 99 Å upon binding T7 RNAP and 86 Å upon further addition of GTP.

6.3.3.3 Measurement of the DNA Bending Angle

Increase in E_{FRET} in the end-labeled promoter DNA upon RNAP binding is indicative of DNA bending. The bending angle is calculated from the measured distance between the donor–acceptor pair using simple trigonometry. If the DNA is relatively small, one can assume the simplest "single-kink" model, where RNAP binds the promoter and bends it at a single point. In such cases, the bending angle α can be calculated as shown in Eq. (6.10):

$$R^2 = {L_1}^2 + {L_2}^2 + 2L_1L_2 \cos\alpha \tag{6.10}$$

where R is the measured D–A distance, and L_1 and L_2 are the DNA lengths upstream and downstream of the bending center. Bending angles can vary depending on what model one assumes, for instance, single-kink, two-kink, or smooth bend [53], as well as the positions of these kinks, which can be distinguished only through additional experiments such as crystallography.

Based on the calculated end-to-end distances and assuming a bend centered at the −1 position in the promoter DNA (+1 transcription start site), it was calculated that Rpo41 bends the DNA by ~52° [19]. On the other hand, Rpo41–Mtf1 bends the promoter DNA by ~90°. Similarly, T7 RNAP bends the promoter by 80–90° in the presence of GTP [22, 32]. These values are consistent and close to ~55–80° bending angle of the *Escherichia coli* RNAP initiation open complex [54, 55].

6.4 Promoter Opening

Melting of the duplex promoter is a key step wherein a specific region of the promoter is unwound by the RNAP, so the template strand becomes accessible for binding to the incoming NTPs. Many footprinting methods have been developed over the years to map the open complex [56–59] using DNase I, potassium permanganate, diethylpyrocarbonate, or hydroxyl radicals, each with its own set of merits and demerits. Fluorescence footprinting using 2-AP has gained popularity in recent times, being less time consuming and relatively easy to carry out compared to other gel-based techniques, and has been successfully used to map open complexes with single nucleotide resolution, as well as to study opening–closing dynamics in real-time [23, 60, 61, 49, 43, 62]. In both T7 RNAP and yeast mt RNAP complexes, the open DNA region is mapped to −4 to +2 region using 2-AP fluorescence assays described below [22, 19, 63].

6.4.1 2-Aminopurine Fluorescence

2-AP acts as a fluorescent adenine analogue that can form Watson−Crick-type base pairs with thymine (Fig. 6.4a). Since promoters contain many A:T base pairs in the region that is melted by RNAPs, promoters can be modified with 2-AP to spectroscopically monitor open complex formation [43, 23, 64]. The fluorescence intensity of 2-AP is highly sensitive to base-stacking interactions in the DNA helix [65, 66]. The 2-AP fluorescence emission intensity varies depending on the identity of the neighboring bases, as well as external factors like pH and temperature [67, 68, 66]. Compared to the fluorescence of free 2-AP nucleoside, the fluorescence of 2-AP (excitation at 315 nm and emission at 370 nm) in the duplex DNA is quenched and partly recovered when the DNA is melted on addition of RNAP and 2-AP is in the ssDNA region (Fig. 6.4b). If the base is flipped and unstacked, then there is a much larger 2-AP fluorescence increase.

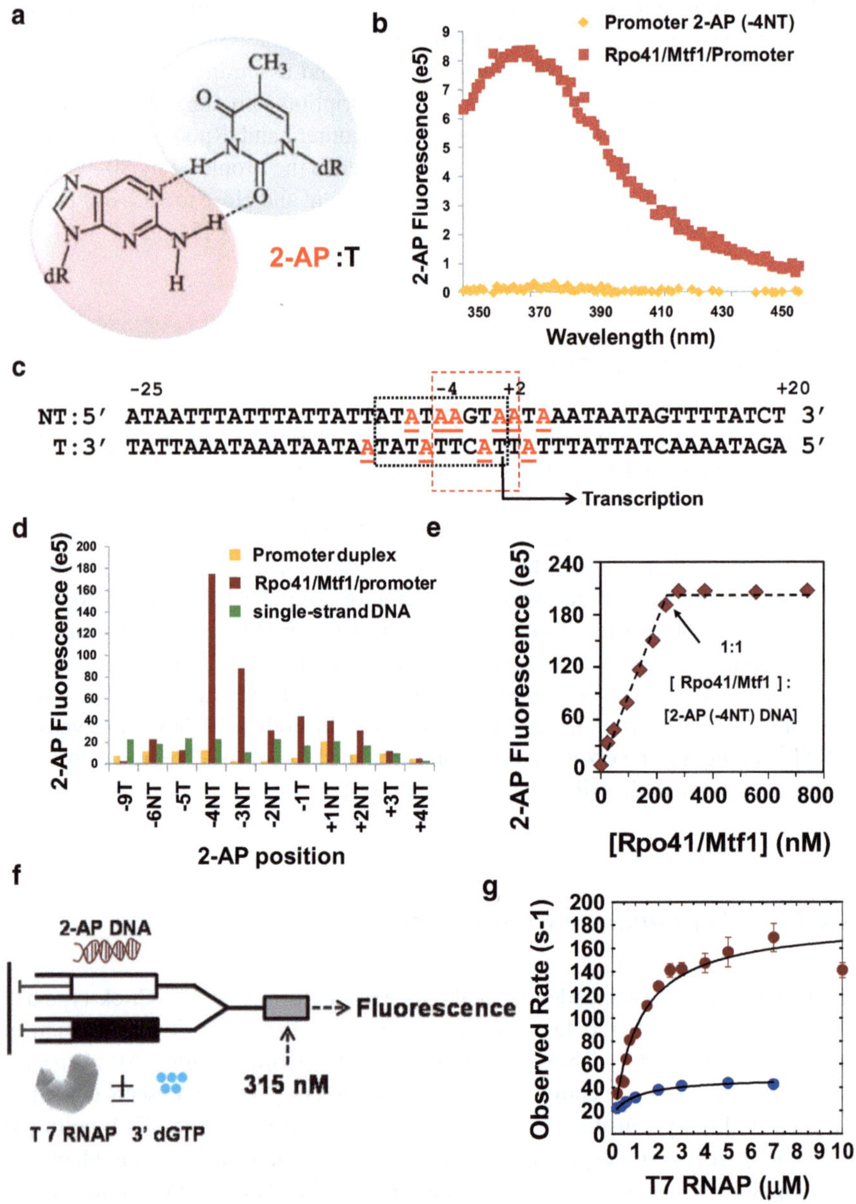

Fig. 6.4 Promoter opening and initial bubble collapse using 2-AP modified promoters. (**a**) The adenine analogue 2-AP base pairs with thymine in DNA. (**b**) The 2-AP fluorescence is quenched when 2-AP base is stacked in the duplex DNA and increases when the DNA is melted by RNAP. (**c**) Positions of the 2-AP (*red underlined*) in the template and non-template strands relative to the −8 to +1 consensus region (*dotted black box*) of the yeast mt 15S rRNA promoter. (**d**) Fluorescence mapping indicates that the −4 to +2 region is melted in the open complex [*dotted red box* in (**c**)]. (**e**) 2-AP(−4NT) mt promoter (200 nM) was titrated with increasing concentrations of a 1:1 mixture of Rpo41 and Mtf1. The 2-AP fluorescence increases and saturates at ~270 nM Rpo41–Mtf1 demonstrating the 1:1:1 stoichiometry of Rpo41–Mtf1–DNA. (**f**) Stopped-flow assay to

6.4.2 *Equilibrium Studies of Open Complex Formation*

6.4.2.1 Mapping of the Open Complex

To use fluorescence footprinting to map the melted DNA region, chemically synthesized promoter DNA is prepared where the adenines are systematically substituted with 2-AP, one at a time (Fig. 6.4c). The duplex promoter DNA contains only one 2-AP at a known position. The steady-state fluorescence measurements are carried out on a spectrofluorometer. The 2-AP modified promoter and RNAP are thoroughly mixed and incubated at desired temperature and for a predetermined time interval to reach equilibrium, before exciting at 315 nm. A control experiment using RNAP with unmodified DNA is necessary, because of the spectral overlap between RNAP and 2-AP fluorescence. After subtracting fluorescence intensity contributions from the RNAP, the corrected 2-AP fluorescence intensities at 370 nm are compared for different positions. Such fluorescence footprinting showed that the promoter DNA in the open complexes of T7 RNAP and yeast mt RNAP are melted from −4 to +2, with respect to the start site at +1 [61, 43, 69]. In T7 RNAP, this result was confirmed by crystallography studies of the open complex [69]. The greatest increase in fluorescence was observed at the −4 position indicative of flipping and unstacking of the −4 base from the neighboring base (Fig. 6.4d). This was confirmed by crystal structures of T7 RNAP that showed that the promoter template is bent between positions −4 and −5, and the −4 template base is unstacked from −3 base in the open complex. Even with mt promoter, highest fluorescence is observed at the −4 position, indicating a similar bend between positions −4 and −5 in the template. In case of mt RNAP, both Rpo41 and Mtf1 are required to observe 2-AP fluorescence increase in the promoter. Thus, the inability of Rpo41 to bend the promoter correlates with its inability to melt the promoter. This temporal coupling of bending and opening steps during open complex formation was verified using triply labeled promoter with FRET probes and 2-AP for the T7 RNAP complex [22, 30].

6.4.2.2 Equilibrium Binding of Promoter by 2-AP Fluorescence Changes

One can also use 2-AP fluorescence change as a signal to measure the K_d and stoichiometry of the promoter-RNAP complex. Typically, RNAP is titrated with

Fig. 6.4 (continued) measure the kinetics of promoter opening. The 2-AP modified T7 promoter from one syringe is mixed with T7 RNAP (at various concentrations) with and without the initiating NTPs. (**g**) The observed rates in the absence (*red*) or the presence (*blue*) of 3′dGTP plotted as a function of T7 RNAP concentration provide estimates of promoter opening and closing rates [43] [71] increasing concentrations of 2-AP modified promoter [23, 70]. Since 2-AP at −4 undergoes the largest increase in fluorescence, a single 2-AP at the −4 position can provide the signal for these experiments. To determine the stoichiometry of the complex, concentrations are kept well above their K_d values. To determine the K_d values, concentrations are kept close to the K_d values. Light absorption and emission by RNAP itself can be significant, because 2-AP excitation (315 nm) and emission (370 nm) are close to the tryptophan and tyrosine absorption (280–290 nm) and emission (340 nm). Therefore, it is critical to correct the observed 2-AP fluorescence for inner-filter effects in addition to volume changes using Eq. (6.11):

$$F_C = F_{obs} \times \left(\frac{v_f}{v_0}\right) \times 10^{0.5(Abs_{ex}+Abs_{em})} \tag{6.11}$$

where F_c is corrected fluorescence, F_{obs} is the observed fluorescence intensity, v_f is the final volume of the solution, v_0 is the initial volume, Abs_{ex} is the absorbance of the RNAP–DNA solution at the 2-AP excitation wavelength of 315 nm, and Abs_{em} is the absorbance of the same solution at the 2-AP emission wavelength of 370 nm. The concentration of the total RNAP and DNA at each point in titration can be determined by Eqs. (6.12) and (6.13):

$$[E]_i = \frac{v_0}{v_{if}}[E]_0 \tag{6.12}$$

$$[D]_i = \frac{v_0}{v_{if}}[D]_0 \tag{6.13}$$

where $[E]_i$ and $[D]_i$ are the concentrations of T7 RNAP and DNA, respectively, up to point i in titration, $[E]_0$ and $[D]_0$ are the initial concentrations, $v_{i,f}$ is the volume of the solution up to point i in the titration, and v_0 is the initial volume.

By conducting multiple titrations, where Mtf1 or 2-AP modified promoter or Rpo41 were titrated against each other (Fig. 6.4e), it was determined that a 20 bp mt RNAP promoter forms a 1:1:1 complex with Rpo41 and Mtf1 [43]. Similar experiments with longer promoters showed multiple Rpo41–Mtf1 per DNA.

6.4.3 Kinetics of Open Complex Formation

Transcription can be regulated at the promoter opening step, where the extent and/or the rate of opening is modulated by transcription factors and small ligands. To determine the rate of promoter opening, the 2-AP fluorescence change is monitored as a function of time after mixing the promoter with RNAP in a stopped-flow instrument (Fig. 6.4f). The experiment is conducted with increasing concentrations of RNAP under pseudo-first-order conditions, and multiple time traces (at least 7–8 shots) are averaged at each RNAP concentration, and the data are fit to a single exponential equation:

$$y = A \times \left(1 - \mathrm{e}^{-k_{obs}t}\right) + C \tag{6.14}$$

where y is observed fluorescence, A is fluorescence change, t is time, k_{obs} is the observed rate, and C is y-intercept.

In T7 RNAP, the observed rate of 2-AP fluorescence increase was not dependent on the position of the probe within the −4 to −1 region, indicating that the entire DNA region is melted in a coordinated fashion [61]. The observed rates increased in a hyperbolic manner with increasing T7 RNAP concentration (Fig. 6.4g) [71], which supported a two-step mechanism of promoter binding (Scheme 6.2) with an initial closed complex (ED_c) followed by open complex (ED_o):

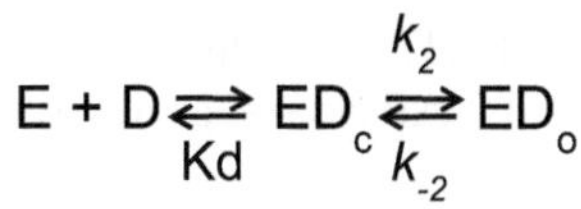

Scheme 6.2 Two-step mechanism of RNAP binding to DNA

Only one kinetic phase was observed in the 2-AP stopped-flow studies, which indicates that the closed complex intermediate is in rapid equilibrium with free RNAP and promoter, and the observed rate is a function of the equilibrium constant of the closed complex (K_d) and the sum of promoter opening and closing (k_{max}) (Eq. 6.15):

$$k_{obs} = C + \frac{k_{max}[\mathrm{E}]}{K_d + [\mathrm{E}]} \tag{6.15}$$

If E + D to ED_c is not in rapid equilibrium and one observes multiple phases in the binding kinetic trace, the data are computationally fit to the two-step model using KinTek Explorer Pro Software (KinTek Corp., Austin, TX). Data fitting showed that $k_{max} = 180\ \mathrm{s}^{-1}$ and K_d of the closed complex was 1 μM. In the presence of GTP, k_{max} decreased to 32 s^{-1}. Global fitting of the two sets of data, in absence and presence of GTP, yielded the promoter opening rate as ~30 s^{-1} and closing rate as ~150 s^{-1} [71].

6.5 Initial Transcription and Transition into Elongation

Once the open complex is formed, the initiating +1 and +2 NTPs bind at the RNAP's active site to base pair with the melted template bases and to catalytically react to form the first 2-mer RNA product. After RNA of certain length is synthesized, the RNAP transitions into the elongation mode by losing contacts with the promoter region [72, 73], where it couples processive RNA synthesis to downstream DNA unwinding. Along with transcription initiation, the transition from initiation to processive elongation is a critical step that regulates the synthesis of full-length RNA [21, 25, 26]. This transition from promoter-specific initiation to promoter-independent elongation is a multistep process, accompanied by conformational changes in both RNAP and DNA [20, 50, 33, 74]. It is extremely difficult to capture these intermediates in real time because of the transient and heterogeneous nature of the subpopulations. Hence, methods were developed to stall the RNAP at different positions by addition of limited NTPs and studying the associated DNA conformational changes using FRET and 2-AP assays, as described below.

6.5.1 FRET Studies to Monitor DNA Conformational Changes

FRET measurements on stalled complexes have provided snapshots of the conformational changes in RNAP–DNA complexes as reaction proceeds from IC2 to IC7 (IC: initiation complex) and as RNAP makes the transition from initiation to elongation (EC10: elongation complex at position 10) (Fig. 6.5a) [47]. Ensemble FRET studies of T7 RNAP carried out with T7 promoter labeled with TAMRA at −4, and Alexa647 at +18 positions ($R_0 \sim 67$ Å) showed a progressive increase in E_{FRET} from a low value with free DNA to a maximum value with RNAP-bound IC7 and then a decrease from IC7 to EC10. These results indicate progressive bending of the promoter DNA as the RNA:DNA hybrid grows from 2 to 7 bp and then DNA unbending upon transition into elongation. This conclusion was confirmed by single-molecule FRET measurements carried out with Cy3–Cy5-labeled T7 promoter (−4 and +17 positions) that showed increase in E_{FRET} with increasing lengths of RNA:DNA hybrid. Additionally, the single-molecule FRET studies were able to separate the IC population from the EC population that started appearing after IC7 [50].

The bending–unbending dynamics of DNA were monitored in real time by following the changes in E_{FRET} in stopped-flow assays. In these experiments, T7 RNAP was preincubated with T7 promoter labeled with TAMRA at −22 and Alexa647 at +18, and E_{FRET} change was measured as a function of time after adding a limited set of NTPs to walk RNAP to a particular position on the template DNA (Fig. 6.5b) [50]. At the start, a rapid increase in E_{FRET} in all the IC5–12 complexes was observed, indicative of DNA bending during initiation (Fig. 6.5c). However after IC7, there was a late phase of E_{FRET} decrease, consistent with DNA unbending. Both

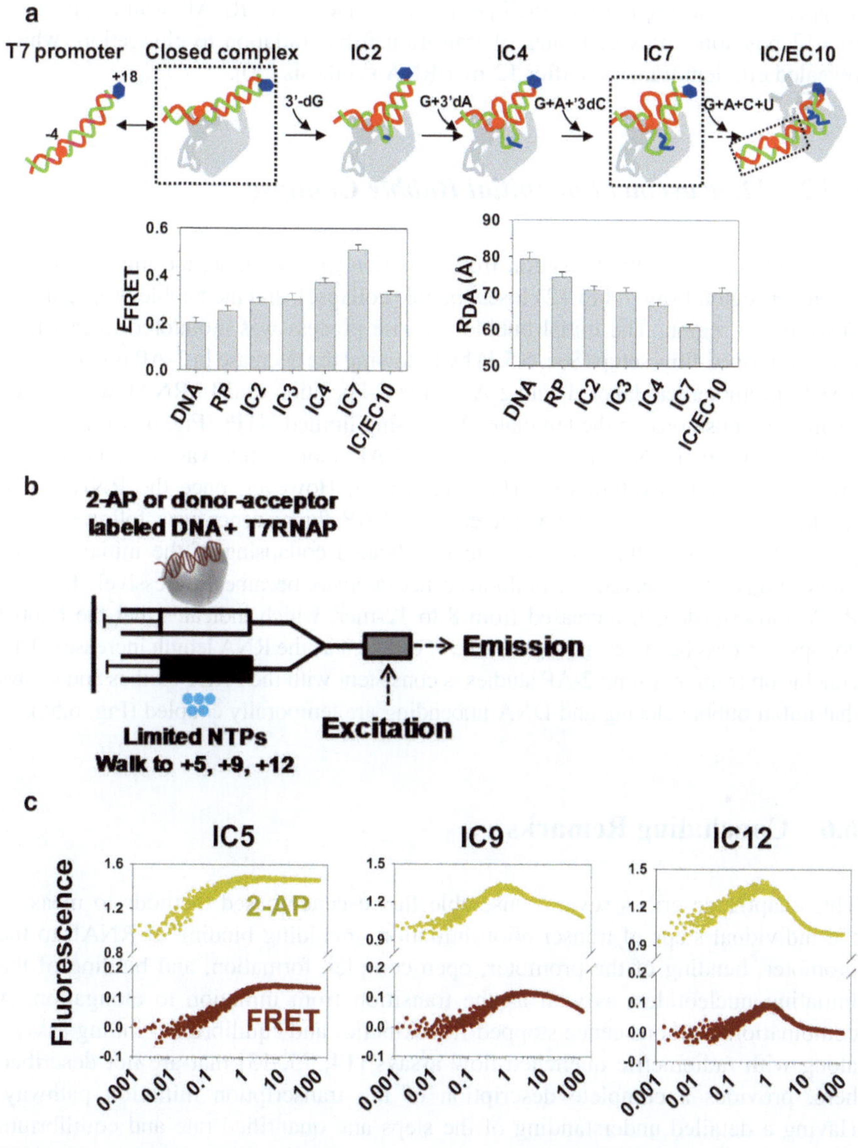

Fig. 6.5 FRET and 2-AP fluorescence assays to study conformational changes in the DNA during transition from initiation to elongation. (**a**) FRET between fluorophores attached to −4 and +18 bases of the T7 promoter measure DNA bending during transcription initiation. Complex RP (RNAP promoter) was formed by incubating T7 RNAP (120 nM) with dual labeled T7 promoter (100 nM). The IC2–7 and IC/EC10 stalled complexes were formed by addition of limited NTPs. The distances (R_{DA}) were calculated from E_{FRET} and R_0. (**b**) Stopped-flow assay to measure the kinetics of DNA unbending and initial bubble collapse. T7 RNAP was walked to +5, +9, and +12 positions on the template using limited NTPs, and acceptor fluorescence and 2-AP fluorescence were measured as a function of time after mixing. (**c**) Representative kinetic traces of FRET and 2-AP fluorescence changes for the IC5-12 complexes showing temporal coupling between DNA unbending and initial bubble collapse [47, 50, 72]

stopped-flow and single-molecule E_{FRET} time courses of T7 RNAP, walking from +8 to +12 positions, provided rates of transition from initiation to elongation, which revealed efficient transition after 12-mer RNA synthesis [50].

6.5.2 Measurement of Initial Bubble Collapse

When RNAP makes the transition from initiation to elongation, the initially melted promoter region from −4 to +2 closes (bubble collapse), and the bubble propagates to downstream region. The initial bubble collapse process was monitored in real time using a stopped-flow setup (Sec. 6.5.1) by following the decrease in 2-AP fluorescence. The T7 promoter was labeled with 2-AP at the −4 position, and T7 RNAP was walked to different positions on the template DNA using limited NTPs (Fig. 6.5b). Up to the synthesis of 7-mer RNA, only an increase in 2-AP fluorescence was observed as GTP drives open complex formation (Fig. 6.5c) [72]. However, once the RNA length reached 8-mer stage, the initial increase in 2-AP fluorescence was followed by a phase of decreasing fluorescence, which indicated collapsing of the initial bubble. Interestingly, the observed rate of fluorescence decrease became progressively faster as RNA transcript length increased from 8 to 12-mer, which indicates that the bubble collapse process becomes more probable or efficient as the RNA length increases. This conclusion from real-time 2-AP studies is consistent with the FRET studies and shows that initial bubble closing and DNA unbending are temporally coupled (Fig. 6.5c).

6.6 Concluding Remarks

This chapter describes several ensemble fluorescence-based methods to measure the individual steps of transcription initiation, including binding of RNAP to the promoter, bending of the promoter, open complex formation, and binding of the initiating nucleotides, as well as the transition from initiation to elongation. A combination of fluorescence stopped-flow kinetics and equilibrium binding assays, along with radiometric quenched-flow assays [14, 75, 13] that are not described here, provides a complete description of the transcription initiation pathway. Having a detailed understanding of the steps and quantified rate and equilibrium constants allows the investigator to ask how variations in promoter sequence, presence or absence of transcription factors, small ligands, or external conditions regulate transcription and which particular step is affected. Even if the individual steps are affected to a small extent, the cumulative effect on the final amount of RNA product can be significant. The findings from the ensemble studies provide the basis to conduct studies at the single-molecule level to tease out populations. These biochemical studies together with high-resolution structures are comprehensive means to obtain the fundamental insights that are required understand the mechanism of transcription.

Acknowledgments This work was supported by the NIH grants GM55310 and GM51966 to S.S.P.

References

1. Arnold JJ, Smidansky ED, Moustafa IM, Cameron CE (2012) Human mitochondrial RNA polymerase: structure-function, mechanism and inhibition. Biochim Biophys Acta 1819(9–10):948–960. doi:10.1016/j.bbagrm.2012.04.002, S1874-9399(12)00093-4 [pii]
2. Martin CT, Esposito EA, Theis K, Gong P (2005) Structure and function in promoter escape by T7 RNA polymerase. Prog Nucleic Acid Res Mol Biol 80:323–347. doi:10.1016/S0079-6603(05)80008-X, S0079-6603(05)80008-X [pii]
3. Tunitskaya VL, Kochetkov SN (2002) Structural-functional analysis of bacteriophage T7 RNA polymerase. Biochemistry (Mosc) 67(10):1124–1135, BCM67101360 [pii]
4. Steitz TA (2004) The structural basis of the transition from initiation to elongation phases of transcription, as well as translocation and strand separation, by T7 RNA polymerase. Curr Opin Struct Biol 14(1):4–9. doi:10.1016/j.sbi.2004.01.006, S0959440X04000077 [pii]
5. Cheetham GM, Steitz TA (2000) Insights into transcription: structure and function of single-subunit DNA-dependent RNA polymerases. Curr Opin Struct Biol 10(1):117–123, S0959-440X(99)00058-5 [pii]
6. Kochetkov SN, Rusakova EE, Tunitskaya VL (1998) Recent studies of T7 RNA polymerase mechanism. FEBS Lett 440(3):264–267, S0014-5793(98)01484-7 [pii]
7. McAllister WT (1993) Structure and function of the bacteriophage T7 RNA polymerase (or, the virtues of simplicity). Cell Mol Biol Res 39(4):385–391
8. Young BA, Gruber TM, Gross CA (2002) Views of transcription initiation. Cell 109(4):417–420, S0092867402007523 [pii]
9. Hirata A, Murakami KS (2009) Archaeal RNA polymerase. Curr Opin Struct Biol 19(6):724–731. doi:10.1016/j.sbi.2009.10.006, S0959-440X(09)00156-0 [pii]
10. Borukhov S, Lee J (2005) RNA polymerase structure and function at lac operon. C R Biol 328(6):576–587. doi:10.1016/j.crvi.2005.03.007, doi:S1631-0691(05)00066-1 [pii]
11. Cramer P (2002) Multisubunit RNA polymerases. Curr Opin Struct Biol 12(1):89–97, S0959440X02002944 [pii]
12. Hsu LM (2009) Monitoring abortive initiation. Methods 47(1):25–36. doi:10.1016/j.ymeth.2008.10.010
13. Patel SS, Bandwar RP (2003) Fluorescence methods for studying the kinetics and thermodynamics of transcription initiation. Methods Enzymol 370:668–686. doi:10.1016/S0076-6879(03)70055-X
14. Patel SS, Bandwar RP, Levin MK (2003) Transient-state kinetics and computational analysis of transcription initiation. In: Johnson KA (ed) Kinetic analysis of macromolecules a practical approach. Oxford University Press, New York, pp 87–129
15. Robinson A, van Oijen AM (2013) Bacterial replication, transcription and translation: mechanistic insights from single-molecule biochemical studies. Nat Rev Microbiol 11(5):303–315. doi:10.1038/nrmicro2994
16. Martin CT, Újvári A, Liu C (2003) Evaluation of fluorescence spectroscopy methods for mapping melted regions of DNA along the transcription pathway. In: Sankar LA, Susan G (eds) Methods in enzymology, vol 371. Academic Press, pp 13–33. doi:http://dx.doi.org/10.1016/S0076-6879(03)71002-7
17. Deshpande AP, Patel SS (2012) Mechanism of transcription initiation by the yeast mitochondrial RNA polymerase. Biochim Biophys Acta 1819(9–10):930–938. doi:10.1016/j.bbagrm.2012.02.003

18. Bandwar RP, Jia Y, Stano NM, Patel SS (2002) Kinetic and thermodynamic basis of promoter strength: multiple steps of transcription initiation by T7 RNA polymerase are modulated by the promoter sequence. Biochemistry 41(11):3586–3595
19. Tang GQ, Deshpande AP, Patel SS (2011) Transcription factor-dependent DNA bending governs promoter recognition by the mitochondrial RNA polymerase. J Biol Chem 286 (44):38805–38813. doi:10.1074/jbc.M111.261966
20. Bandwar RP, Ma N, Emanuel SA, Anikin M, Vassylyev DG, Patel SS, McAllister WT (2007) The transition to an elongation complex by T7 RNA polymerase is a multistep process. J Biol Chem 282(31):22879–22886. doi:10.1074/jbc.M702589200
21. Hsu LM (2002) Promoter clearance and escape in prokaryotes. Biochim Biophys Acta 1577 (2):191–207
22. Tang GQ, Patel SS (2006) T7 RNA polymerase-induced bending of promoter DNA is coupled to DNA opening. Biochemistry 45(15):4936–4946. doi:10.1021/bi0522910
23. Jia Y, Kumar A, Patel SS (1996) Equilibrium and stopped-flow kinetic studies of interaction between T7 RNA polymerase and its promoters measured by protein and 2-aminopurine fluorescence changes. J Biol Chem 271(48):30451–30458
24. LiCata VJ, Wowor AJ (2008) Applications of fluorescence anisotropy to the study of protein-DNA interactions. Methods Cell Biol 84:243–262. doi:10.1016/S0091-679X(07)84009-X, S0091-679X(07)84009-X [pii]
25. Reppas NB, Wade JT, Church GM, Struhl K (2006) The transition between transcriptional initiation and elongation in E. coli is highly variable and often rate limiting. Mol Cell 24 (5):747–757. doi:10.1016/j.molcel.2006.10.030
26. Panov KI, Friedrich JK, Zomerdijk JC (2001) A step subsequent to preinitiation complex assembly at the ribosomal RNA gene promoter is rate limiting for human RNA polymerase I-dependent transcription. Mol Cell Biol 21(8):2641–2649. doi:10.1128/MCB.21.8.2641-2649.2001
27. Anderson BJ, Larkin C, Guja K, Schildbach JF (2008) Using fluorophore-labeled oligonucleotides to measure affinities of protein-DNA interactions. Methods Enzymol 450:253–272. doi:10.1016/S0076-6879(08)03412-5
28. Wang ZX (1995) An exact mathematical expression for describing competitive binding of two different ligands to a protein molecule. FEBS Lett 360(2):111–114
29. Toptygin D, Brand L (1995) Spectrabind user's guide. The Johns Hopkins University, Baltimore, MD
30. Tang GQ, Patel SS (2006) Rapid binding of T7 RNA polymerase is followed by simultaneous bending and opening of the promoter DNA. Biochemistry 45(15):4947–4956. doi:10.1021/bi052292s
31. Schultz SC, Shields GC, Steitz TA (1991) Crystal structure of a CAP-DNA complex: the DNA is bent by 90 degrees. Science 253(5023):1001–1007
32. Yin YW, Steitz TA (2002) Structural basis for the transition from initiation to elongation transcription in T7 RNA polymerase. Science 298(5597):1387–1395. doi:10.1126/science.1077464
33. Durniak KJ, Bailey S, Steitz TA (2008) The structure of a transcribing T7 RNA polymerase in transition from initiation to elongation. Science 322(5901):553–557. doi:10.1126/science.1163433
34. Rettig M, Germann MW, Wang S, Wilson WD (2013) Molecular basis for sequence-dependent induced DNA bending. Chembiochem 14(3):323–331. doi:10.1002/cbic.201200706
35. Hudson BP, Quispe J, Lara-Gonzalez S, Kim Y, Berman HM, Arnold E, Ebright RH, Lawson CL (2009) Three-dimensional EM structure of an intact activator-dependent transcription initiation complex. Proc Natl Acad Sci U S A 106(47):19830–19835. doi:10.1073/pnas.0908782106
36. Hardin AH, Sarkar SK, Seol Y, Liou GF, Osheroff N, Neuman KC (2011) Direct measurement of DNA bending by type IIA topoisomerases: implications for non-equilibrium topology simplification. Nucleic Acids Res 39(13):5729–5743. doi:10.1093/nar/gkr109

37. Noah JW, Park S, Whitt JT, Perutka J, Frey W, Lambowitz AM (2006) Atomic force microscopy reveals DNA bending during group II intron ribonucleoprotein particle integration into double-stranded DNA. Biochemistry 45(41):12424–12435. doi:10.1021/bi060612h
38. Ujvari A, Martin CT (2000) Evidence for DNA bending at the T7 RNA polymerase promoter. J Mol Biol 295(5):1173–1184. doi:10.1006/jmbi.1999.3418, S0022-2836(99)93418-6 [pii]
39. Kahn JD (1999) Methods for analyzing DNA bending. Methods Mol Biol 94:109–123. doi:10.1385/1-59259-259-7:109
40. Akiyama T, Hogan ME (1997) Structural analysis of DNA bending induced by tethered triple helix forming oligonucleotides. Biochemistry 36(8):2307–2315. doi:10.1021/bi9624292
41. Dlakic M, Harrington RE (1998) Unconventional helical phasing of repetitive DNA motifs reveals their relative bending contributions. Nucleic Acids Res 26(18):4274–4279
42. Koo HS, Drak J, Rice JA, Crothers DM (1990) Determination of the extent of DNA bending by an adenine-thymine tract. Biochemistry 29(17):4227–4234
43. Tang GQ, Paratkar S, Patel SS (2009) Fluorescence mapping of the open complex of yeast mitochondrial RNA polymerase. J Biol Chem 284(9):5514–5522. doi:10.1074/jbc.M807880200
44. Mukhopadhyay J, Mekler V, Kortkhonjia E, Kapanidis AN, Ebright YW, Ebright RH (2003) Fluorescence resonance energy transfer (FRET) in analysis of transcription-complex structure and function. Methods Enzymol 371:144–159. doi:10.1016/S0076-6879(03)71010-6
45. Joo C, Balci H, Ishitsuka Y, Buranachai C, Ha T (2008) Advances in single-molecule fluorescence methods for molecular biology. Annu Rev Biochem 77:51–76. doi:10.1146/annurev.biochem.77.070606.101543
46. Sorokina M, Koh HR, Patel SS, Ha T (2009) Fluorescent lifetime trajectories of a single fluorophore reveal reaction intermediates during transcription initiation. J Am Chem Soc 131 (28):9630–9631. doi:10.1021/ja902861f
47. Tang GQ, Roy R, Ha T, Patel SS (2008) Transcription initiation in a single-subunit RNA polymerase proceeds through DNA scrunching and rotation of the N-terminal subdomains. Mol Cell 30(5):567–577. doi:10.1016/j.molcel.2008.04.003
48. Roy R, Hohng S, Ha T (2008) A practical guide to single-molecule FRET. Nat Methods 5 (6):507–516. doi:10.1038/nmeth.1208, nmeth.1208 [pii]
49. Kim H, Tang GQ, Patel SS, Ha T (2012) Opening-closing dynamics of the mitochondrial transcription pre-initiation complex. Nucleic Acids Res 40(1):371–380. doi:10.1093/nar/gkr736
50. Tang GQ, Roy R, Bandwar RP, Ha T, Patel SS (2009) Real-time observation of the transition from transcription initiation to elongation of the RNA polymerase. Proc Natl Acad Sci USA 106(52):22175–22180. doi:10.1073/pnas.0906979106
51. Dragan AI, Privalov PL (2008) Use of fluorescence resonance energy transfer (FRET) in studying protein-induced DNA bending. Methods Enzymol 450:185–199. doi:10.1016/S0076-6879(08)03409-5
52. Clegg RM, Murchie AI, Zechel A, Lilley DM (1993) Observing the helical geometry of double-stranded DNA in solution by fluorescence resonance energy transfer. Proc Natl Acad Sci USA 90(7):2994–2998
53. Wu J, Parkhurst KM, Powell RM, Brenowitz M, Parkhurst LJ (2001) DNA bends in TATA-binding protein-TATA complexes in solution are DNA sequence-dependent. J Biol Chem 276 (18):14614–14622. doi:10.1074/jbc.M004402200
54. Rippe K, Guthold M, von Hippel PH, Bustamante C (1997) Transcriptional activation via DNA-looping: visualization of intermediates in the activation pathway of E. coli RNA polymerase x sigma 54 holoenzyme by scanning force microscopy. J Mol Biol 270(2):125–138
55. Rivetti C, Guthold M, Bustamante C (1999) Wrapping of DNA around the E.coli RNA polymerase open promoter complex. EMBO J 18(16):4464–4475. doi:10.1093/emboj/18.16.4464
56. Saluz HP, Jost JP (1993) Approaches to characterize protein-DNA interactions in vivo. Crit Rev Eukaryot Gene Expr 3(1):1–29

57. Davis CA, Bingman CA, Landick R, Record MT Jr, Saecker RM (2007) Real-time footprinting of DNA in the first kinetically significant intermediate in open complex formation by Escherichia coli RNA polymerase. Proc Natl Acad Sci USA 104(19):7833–7838. doi:10.1073/pnas.0609888104
58. Holstege FC, Timmers HT (1997) Analysis of open complex formation during RNA polymerase II transcription initiation using heteroduplex templates and potassium permanganate probing. Methods 12(3):203–211. doi:10.1006/meth.1997.0472
59. Kahl BF, Paule MR (2009) The use of diethyl pyrocarbonate and potassium permanganate as probes for strand separation and structural distortions in DNA. Methods Mol Biol 543:73–85. doi:10.1007/978-1-60327-015-1_6
60. Liu C, Martin CT (2001) Fluorescence characterization of the transcription bubble in elongation complexes of T7 RNA polymerase. J Mol Biol 308(3):465–475. doi:10.1006/jmbi.2001.4601
61. Bandwar RP, Patel SS (2001) Peculiar 2-aminopurine fluorescence monitors the dynamics of open complex formation by bacteriophage T7 RNA polymerase. J Biol Chem 276(17):14075–14082. doi:10.1074/jbc.M011289200
62. Ujvari A, Martin CT (1996) Thermodynamic and kinetic measurements of promoter binding by T7 RNA polymerase. Biochemistry 35(46):14574–14582. doi:10.1021/bi961165g, bi961165g [pii]
63. Sastry SS, Ross BM (1996) A direct real-time spectroscopic investigation of the mechanism of open complex formation by T7 RNA polymerase. Biochemistry 35(49):15715–15725. doi:10.1021/bi960729d, bi960729d [pii]
64. Sullivan JJ, Bjornson KP, Sowers LC, deHaseth PL (1997) Spectroscopic determination of open complex formation at promoters for Escherichia coli RNA polymerase. Biochemistry 36 (26):8005–8012. doi:10.1021/bi970363k
65. Stivers JT (1998) 2-Aminopurine fluorescence studies of base stacking interactions at abasic sites in DNA: metal-ion and base sequence effects. Nucleic Acids Res 26(16):3837–3844
66. Jean JM, Hall KB (2001) 2-Aminopurine fluorescence quenching and lifetimes: role of base stacking. Proc Natl Acad Sci USA 98(1):37–41. doi:10.1073/pnas.011442198
67. Rachofsky EL, Osman R, Ross JB (2001) Probing structure and dynamics of DNA with 2-aminopurine: effects of local environment on fluorescence. Biochemistry 40(4):946–956
68. Ward DC, Reich E, Stryer L (1969) Fluorescence studies of nucleotides and polynucleotides. I. Formycin, 2-aminopurine riboside, 2,6-diaminopurine riboside, and their derivatives. J Biol Chem 244(5):1228–1237
69. Cheetham GM, Jeruzalmi D, Steitz TA (1999) Structural basis for initiation of transcription from an RNA polymerase-promoter complex. Nature 399(6731):80–83. doi:10.1038/19999
70. Paratkar S, Deshpande AP, Tang GQ, Patel SS (2011) The N-terminal domain of the yeast mitochondrial RNA polymerase regulates multiple steps of transcription. J Biol Chem 286 (18):16109–16120. doi:10.1074/jbc.M111.228023
71. Stano NM, Levin MK, Patel SS (2002) The +2 NTP binding drives open complex formation in T7 RNA polymerase. J Biol Chem 277(40):37292–37300. doi:10.1074/jbc.M201600200
72. Bandwar RP, Tang GQ, Patel SS (2006) Sequential release of promoter contacts during transcription initiation to elongation transition. J Mol Biol 360(2):466–483. doi:10.1016/j.jmb.2006.05.029
73. Guo Q, Nayak D, Brieba LG, Sousa R (2005) Major conformational changes during T7RNAP transcription initiation coincide with, and are required for, promoter release. J Mol Biol 353 (2):256–270. doi:10.1016/j.jmb.2005.08.016
74. Spitalny P, Thomm M (2003) Analysis of the open region and of DNA-protein contacts of archaeal RNA polymerase transcription complexes during transition from initiation to elongation. J Biol Chem 278(33):30497–30505. doi:10.1074/jbc.M303633200
75. Kuzmine I, Martin CT (2001) Pre-steady-state kinetics of initiation of transcription by T7 RNA polymerase: a new kinetic model. J Mol Biol 305(3):559–566. doi:10.1006/jmbi.2000.4316

Chapter 7
Single-Molecule and Single-Particle Imaging of Molecular Motors In Vitro and In Vivo

Natalia Fili

Abstract Motor proteins are multi-potent molecular machines, whose localisation, function and regulation are achieved through tightly controlled processes involving conformational changes and interactions with their tracks, cargos and binding partners. Understanding how these complex machines work requires dissection of these processes both in space and time. Complementing the traditional ensemble measurements, single-molecule assays enable the detection of rare or short-lived intermediates and molecular heterogeneities, and the measurements of subpopulation dynamics. This chapter is focusing on the fluorescence imaging of single motors and their cargo. It discusses what is required in order to achieve single-molecule imaging with high temporal and spatial resolution and how these requirements are met both in vitro and in vivo. It also presents a general overview and applied examples of the major single-molecule imaging techniques and experimental assays which have been used to study motor proteins.

Keywords Single molecule • Motors • Fluorescence • Surface passivation • Fluorophore photostability • Tracking • Dynamics • TIRFM • Confocal microscopy • In vivo • In vitro • Super-resolution imaging

Abbreviations

AFM	Atomic force microscopy
ALEX	Alternating-laser excitation
APD	Avalanche photodiodes
CCD	Charge-coupled device
EM-CCD	Electron-multiplying CCD

N. Fili (✉)
Department of Cellular Physiology, Ludwig-Maximilians-Universität München, Schillerstrasse 44, 80336 Munich, Germany
e-mail: natali.fili@gmail.com; natali.fili@med.lmu.de

C.P. Toseland and N. Fili (eds.), *Fluorescent Methods for Molecular Motors*, Experientia Supplementum 105, DOI 10.1007/978-3-0348-0856-9_7,

FIONA	Fluorescence imaging with one nanometre accuracy
FCS	Fluorescence correlation spectroscopy
FPALM	Fluorescence photoactivated localisation microscopy
HILO	Highly inclined and laminated optical sheet microscopy
HSM	High-speed hyperspectral microscope
ICCD	Intensified charge-couple devices
LSCM	Laser scanning confocal microscopy
MUM	Multifocal plane microscopy
NDC	Nipkow disk confocal
PhADE	Photoactivation, diffusion and excitation
PMT	Photomultiplier tube
PSF	Point spread function
PEG	Polyethylene glycol
QD	Quantum dot
RLS	Reflected light-sheet microscopy
ROXS	Reducing and oxidising system
SNR	Signal-to-noise ratio
smFRET	Single-molecule Fluorescence/Förster resonance energy transfer
SPEED	Single-point edge-excitation subdiffraction microscopy
SPIM	Selective plane illumination microscopy
STORM	Stochastic optical reconstruction microscopy
sptPALM	Single-particle tracking PALM
TIRFM	Total internal reflection microscopy
ZMWs	Zero-mode waveguides

7.1 Introduction

The term "molecular motors" refers to a vast group of functionally distinct proteins which share a common feature: the conversion of chemical energy from ATP hydrolysis into mechanical work. The group includes rotary motors, such as the mitochondrial F_0F_1-ATPase synthase and the flagellar motor proteins; cytoskeletal motors, such as myosins, kinesins and dyneins; and nucleic acid motors, such as helicases, nucleases and polymerases. Molecular motors are multi-potent molecular machines that play a key role in vital cellular processes including vesicular transport, cellular and organelle morphology, cell motility, cell division, DNA replication, etc. Their function, conformational changes, regulation and dynamics are characterised by an inherent complexity and heterogeneity at the single-molecule level. This is further aggravated by their non-homogeneous interaction with cellular tracks, their interplay with multiple cellular components and, in general, the spatially and temporally stochastic nature of cellular events, especially in the crowded intracellular environment.

The conventional ensemble measurements provide a wealth of mechanistic information on motor proteins. However, they are intrinsically based on the averaging of molecular heterogeneity, which can mask interesting features of subpopulations both in space and time. Complementary to the ensemble assays, the rapidly expanding single-molecule approach, being free of ensemble averaging, allows us to track with high resolution and in real time, individual motors in relation to their tracks, cargos and other associated proteins. In this way, it can reveal key aspects of motor dynamics and regulation and can provide a rounded understanding of these complex biomolecules.

During the past decades, various single-molecule techniques have been developed to achieve either force measurements, such as optical trapping, magnetic tweezers and atomic force microscopy, or optical imaging of motors [1–3]. Together, they revealed complementary aspects of the mechanistic function and regulation of various motors, such as applied forces, step size, processivity, directionality, coordination between force and movement, interaction with binding partners, associated networks and cargo, oligomerisation and coordination between different motors [3–6].

The visualisation of single molecules by fluorescence detection has been growing increasingly popular. However, it is subject to various requirements which, if not met, can severely compromise the spatial and temporal resolution of the measurements. The first requirement is that, within the sampling volume, the individual molecules are spatially separated to a distance greater than the diffraction limit. This can be particularly challenging to achieve inside cells by controlling expression levels. The second requirement is to use a bright and photostable fluorescent probe, which during its life cycle emits enough photons for the accurate detection and tracking of the molecule against the fluorescence background. The fluorescent signal can be further enhanced by the addition of components which increase the fluorophore's photostability and minimise the background noise. The optical technique used for visualisation can then provide further signal enhancement by rejecting background fluorescence through optical sectioning. Together with a highly sensitive detection device, a higher spatial and temporal resolution can be achieved. Finally, signal analysis algorithms could improve even further the accuracy of single-molecule localisation.

This chapter will discuss in detail how to image motor proteins and their cargo at a single-particle level both in vitro and in vivo. More specifically, the chapter will provide a detailed overview on how to increase the photostability of fluorescent probes, how to minimise the fluorescence background and which optical techniques ensure a high signal-to-noise ratio (SNR). This chapter will also describe how these imaging techniques can be incorporated into single-molecule assays and present various experimental schemes available in the toolbox for single motor studies.

7.2 How to Achieve Signal Enhancement

The spatial and temporal resolution of single-molecule imaging is limited by the number of photons emitted by the fluorescent label. Therefore, the first step in setting up this type of experiments is selecting the most suitable fluorescent probe. A detailed overview of the types of fluorescent labels available and their suitability for single-molecule imaging is presented in Chap. 1 and in [7]. Briefly, photostable and bright fluorophores, with both high extinction coefficient and quantum yield, are the most suitable. Moreover, the choice of the fluorescent probe also depends on the type of the single-molecule assay and the type of motor protein under investigation. Even in the case of bright and photostable probes, single fluorophores have intrinsically weak fluorescent signals. In order to achieve imaging with high spatial and temporal resolution, the signal needs to be maintained over time under conditions of minimal fluorescent noise. By combining high-quality optics, high numerical aperture objectives and highly sensitive detectors, with a suitable optical sectioning technique which ensures high contrast levels, single molecules can be imaged and localised with high accuracy. This section presents in detail the consecutive steps that one can follow to achieve optimal single-molecule imaging.

7.2.1 Increase of Fluorophore Photostability

During their lifetime, fluorophores undergo multiple rounds of transition between the singlet ground (S_0) and singlet excited (S_1) state before their irreversible conversion into a dark state, a process called photobleaching. Occasionally, about once every 1,000 excitations, the fluorophore populates the triplet excited state (T_1) (triplet blinking) which corresponds to long-lived dark species [8]. Due to its lifetime, which is in the order of few microseconds in aqueous solutions, T_1 is considered to be the most photochemically reactive state and it does lead to photobleaching.

Molecular oxygen, which in normal experimental conditions has a concentration of ~0.5 mM, is the main cause of photobleaching, either through its direct interaction with the fluorophores or through photo-oxidation by the reactive oxygen radicals [8]. In addition, a few motor proteins can be deactivated through these processes. Oxygen removal is therefore the first step required for fluorescent imaging of motor proteins. This can be achieved first by "degassing" the experimental buffers immediately prior to use, whereby the oxygen gets physically removed by applying reduced pressure. Given that such removal is not sufficient, oxygen can be continuously "pumped out" the reaction volume during imaging by including into the buffers enzymatic oxygen scavenger systems. The most popular one is the combination of glucose, glucose oxidase and catalase [9], which however has been shown to increase the solution's acidity over time [10]. An alternative system which is more effective and does not have this limitation is the combination of protocatechuic acid and protocatechuate-3,4-dioxygenase [11].

Fluorophore photostability can also be enhanced by quenching the T_1 state. This is particularly important following oxygen removal, because oxygen is a T_1 quencher and its removal increases T_1 lifetimes to milliseconds, seriously compromising the imaging. Reducing agents such β-mercaptoethanol (BME) and the most efficient Trolox [12], an analogue of vitamin E, are being used for T_1 quenching. An alternative approach which achieves quenching of the T_1, as well as of the photo-induced free radicals, is to combine oxygen removal with a reducing and oxidising system (ROXS), such methylviologen and ascorbic acid [13]. Through electron transfer reactions, the ROXS forces the T_1 species into the ground state and thus prevents photobleaching.

All the above strategies are applicable only in vitro, since oxygen removal from cells would be detrimental. For in vivo imaging, enhanced signal can only be achieved through the selection of the brightest fluorescent probe and by minimising the cell autofluorescence (Sect. 7.2.2).

7.2.2 Minimise Fluorescence Background

In vitro single-molecule assays are based on the surface immobilisation of the motor protein of interest; its interacting track, such as nucleic acids, actin filaments or microtubules; or its binding partners. Immobilisation provides high flexibility in terms of the components used, high accessibility for manipulation and a wide observation field. The most commonly used immobilisation scheme is the surface tethering through biotin–streptavidin or biotin–neutravidin interaction. However, non-specific interactions of the components of the assay, especially the fluorescently labelled ones, with the surface can induce artefacts and high background, which can seriously compromise the quality of single-molecule data. In addition, contact with the surface could also affect the activity of the proteins. Therefore, surface passivation is crucial for both preserving the functionality of the components and ensuring a high SNR. The choice of passivation approach depends on the components of the assay and has to be optimised depending on the application.

The most commonly used blocking agents are proteins, such as bovine serum albumin (BSA) and casein, which passivate the surface by getting non-specifically absorbed (Fig. 7.1a). Following specific immobilisation of selected components of the assay, passivation is achieved by incubating the surface with the blocking protein. In addition, by maintaining the blocking agent in all experimental buffers, one ensures surface passivation during the experiment. However, proteins like BSA are negatively charged. As a result, BSA coating can be very adhesive and therefore unsuitable for many applications, especially the ones involving positively charged DNA-binding proteins such as helicases [16, 17].

An alternative and significantly more effective method to generate inert surfaces is the use of polyethylene glycol (PEG) [16, 18] (Fig. 7.1b). Linear PEG is the most

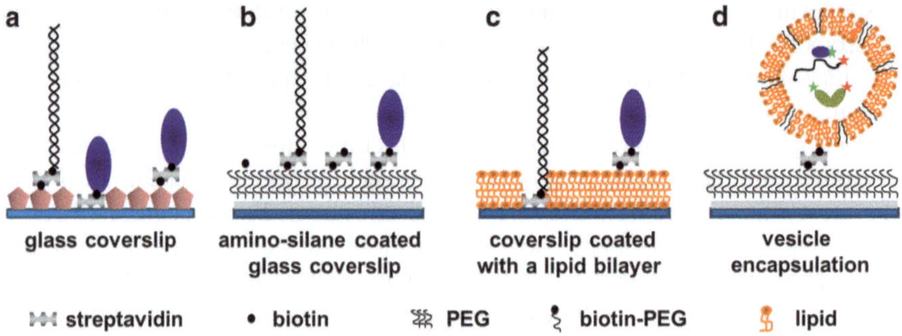

Fig. 7.1 Strategies to generate inert surface substrates for single-molecule assays. (**a**) Surface passivation by the non-specific absorption of blocking proteins, such as BSA and casein. Specific immobilisation of the molecules of interest can be achieved through biotin–streptavidin interactions with either streptavidin or a biotin-modified version of the blocking protein, non-specifically absorbed at low density. (**b**) Surface passivation by the formation of a dense brush of linear amino-reactive PEG polymers covalently attached on an aminosilane-treated glass coverslip. The use of a small fraction of biotinylated-PEG allows the specific surface-tethering of the molecules of interest. (**c**) Generating an inert surface by coating it with a lipid bilayer. The molecules of interest can be specifically immobilised on lipids functionalised with biotin and present at low density (adapted from Visnapuu et al. [14]). (**d**) Encapsulation of the molecules of interest within a lipid vesicle which is tethered on PEGylated surface through biotin–streptavidin interactions (adapted from Cisse et al. [15])

commonly used, and its capacity to resist non-specific absorption depends on the molecular weight of the polymer and its surface density. Linear PEGs of various molecular weights are commercially available as mono- or bifunctional chains, containing a single or two reactive groups, respectively. The most commonly used PEGylation method involves the covalent coupling of amino-reactive PEG, such as *N*-hydroxysuccinimidyl PEG, on an aminosilane (Vectabond)-coated surface [16, 19]. PEGylation results in a dense brush of PEG polymers which blocks non-specific absorption of proteins or fluorophores to the underlying surface. By mixing monofunctional PEG chains with bifunctional ones, such as biotinylated PEGs, at an optimised ratio, one can simultaneously passivate the surface and functionalise it, in order to specifically immobilise the molecules of interest. An alternative PEG version, which is suitable for rendering inert highly adhesive surfaces, is the end-functionalised star-shaped PEGs [20]. Through intermolecular cross-linking, these polymers form a very dense coating which is more resistant to non-specific absorption than linear PEG.

Another way to generate inert surfaces for single-molecule studies is to coat the surface with a lipid bilayer [21, 22] (Fig. 7.1c). Given their natural occurrence in the cell, lipid bilayers are compatible with many proteins and macromolecules like nucleic acids [14, 21]. There are also perfect substrates for investigating the lipid–protein interaction of motor proteins which naturally interact through their tail with intracellular membranes, such as myosins, kinesins and dyneins. Lipid bilayers can be very easily prepared by allowing lipid vesicles, made by sonication or extrusion, to fuse together in a silica flow cell, until a maximum size after which they burst and coat the surface as a bilayer [14]. For specific attachment, the bilayer can be functionalised by incorporating lipids whose head groups are modified with reactive moieties, such as biotin. In order to achieve solid immobilisation points, streptavidin or neutravidin can be non-specifically absorbed at low density on the surface prior to the formation of the bilayer, generating single-molecule anchoring points surrounded by lipids [21].

Due to their inert nature, lipids can also be used to encapsulate the molecules of interest into small unilamellar vesicles which are tethered to the surface [23] (Fig. 7.1d). Vesicle encapsulation enables monitoring of single-molecule dynamics in the absence of perturbations caused by surface immobilisation. By introducing pores into the vesicles, the enclosed molecules can also be accessible for external manipulation [15]. This approach is suitable for setting up single-molecule fluorescence/Förster resonance energy transfer (smFRET) assays to monitor the dynamics of conformational transitions within proteins or complexes and the interactions between partners [15, 24].

Unlike in vitro studies, in vivo imaging allows minimal external manipulations and is therefore limited by the cellular scattering and the inherent fluorescent background of the cell, called the autofluorescence. Autofluorescence is generated by internal organelles, such as mitochondria and lysosomes, aromatic amino acids and cellular metabolites, mainly flavin and nicotinamide adenine dinucleotide (NADH) [25, 26]. With excitation in the ultraviolet (UV)/blue spectral region, autofluorescence has a very broad emission spectrum, extending from the blue to the yellow region. This overlaps with the emission spectra of many dyes and fluorescent proteins (FPs) and can therefore significantly compromise the SNR. In order to minimise the contribution of autofluorescence, it is critical to choose fluorophores with strong emission in the yellow–red band of the spectrum, such as the yellow-green EYFP [27], the orange mKOκ [28] and the red mRuby [29] (Chap. 1). If however there is no flexibility in the selection of fluorophore, cell background can be minimised by limiting the excitation within a thin section close to the cell membrane by total internal fluorescence microscopy (TIRFM) (Sect. 7.2.3.1) or by using other optical sectioning techniques (Sects. 7.2.3.2 and 7.2.3.3). Alternatively, one could first photobleach autofluorescence with an intense excitation pulse, before imaging the fluorescent probe [30]. However, this requires fluorophores with lifetime longer than cell autofluorescence.

7.2.3 Optical Sectioning

7.2.3.1 Total Internal Reflection Illumination

In TIRFM, optical sectioning is achieved with a prism or a high numerical aperture objective which directs the excitation light to strike the interface between the glass substrate and the aqueous sample at an angle higher than the critical angle [31,

32]. Total internal reflection of the incident beam generates an evanescent wave whose intensity exponentially decays with the distance from the interface. In this way, the excitation volume and therefore the fluorescence emission are confined to a thin vertical section of about 100–150 nm, depending on the refractive indices and the incidence angle of the excitation light. This sectioning effect eliminates the fluorescence background coming from emitters within the above volume and therefore allows imaging with high SNR.

TIRFM has been extensively used for single-molecule imaging because of its simplicity, versatility and affordability. There are two TIRFM configurations, the objective-type and the prism-type, each with its advantages and downsides [31–33]. In the objective-based set-up, both the excitation and the emission light are collected by the same oil-immersion objective which has a high numerical aperture (>1.45). This configuration is therefore simple to use, requires minimal alignment, has higher fluorescence collection efficiency, offers high accessibility to the sample and therefore allows quick and easy external manipulations. It is therefore the configuration of choice especially for in vivo imaging. However, it suffers from higher background light and light scattering, as well as a narrow range of incidence angles and thus penetration depths. In contrast, the prism-type configuration, in which the excitation and emission paths are uncoupled, has the benefit of lower background, less scattered light, wider field of view, precise control over the incidence angle within a wide range and therefore variable penetration depths. However, it is less simple to use due to the required alignments and it is limited by the types of objectives used and the restricted accessibility to the sample. It is therefore mostly used for in vitro studies.

Although custom-made set-ups offer higher flexibility, affordable commercial systems of objective-type TIRFM are currently available. A TIRFM suitable for single-molecule imaging should combine high excitation power and highly sensitive detectors. A high illumination power density (~0.5 kW/cm^2) is achieved by diode-pumped solid state lasers of at least 50 mW [34]. Single-molecule imaging of slow events (500 ms) can be achieved with charge-coupled device (CCD) cameras by integrating the fluorescent signal over long periods of times [35]. However, CCDs have limited photon sensitivity due to their read-out noise. Also, most biological processes are very fast and require imaging at video rate or higher. This can be achieved with highly sensitive cameras, such as intensified charge-coupled devices (ICCDs) or deep-cooled electron-multiplying CCDs (EM-CCDs). ICCDs achieve their high sensitivity through a multichannel plate intensifier coupled to the CCD chip. However, their use is limited by image artefacts due to misalignment between the plate and CCD and the narrow linear range between photon number and pixel intensity. EM-CCDs combine high sensitivity without the limitations of ICCDs. Signal amplification occurs on the CCD chip by reading the pixel array through a gain register. EM-CCDs currently available have high quantum efficiencies (almost reaching 95 %) over a wide range of wavelengths, high dynamic range and significantly reduced read-out noise due to their deep cooling.

Multi-colour single-molecule TIRF imaging can be achieved by selecting fluorophores with no overlapping emission spectra and combining various excitation lines coupled to the microscope through fibre optics with suitable dichroic mirrors, excitation and emission filters. Various additional optical components might be required to ensure independent adjustment of laser intensities and wavelength-specific focusing of the beams. Alternatively, excitation of multiple fluorophores using a single wavelength or supercontinuum source has been also used for single-molecule imaging [36, 37]. Detection can be achieved either by using two EM-CCD cameras or by splitting the sensor in two regions, one for every colour, using a commercial image splitting device (such as Photometrics Dual Viewer or Cairn Research Optisplit).

TIRFM is suitable for both in vitro and in vivo imaging. In addition to its increased SNR and higher time resolution, TIRFM minimises both photobleaching and cellular photodamage and ensures a significantly higher vertical resolution compared to the 1- and 2-photon confocal systems (500–800 nm). However, the excitation, and therefore, the fluorescence intensity in TIRFM vary with the distance of the fluorophore from the interface. This causes limitations in measurements based on signal intensity, especially in cells, where the thickness of the plasma membrane is variable [38]. One way to circumvent this is to in situ calibrate the evanescent wave using tilted, in vitro reconstituted fluorescently labelled microtubules as nanoscale rulers [39].

Another downside of TIRF imaging is the narrow imaging depth, which limits its application for studies using surface immobilising molecules or studying events on the cell plasma membrane. For live imaging of cellular events occurring away from the plasma membrane, the penetration depth can be increased by tuning the angle of the TIRF illumination to values slightly lower than the critical angle [40, 41]. In such an approach, described as highly inclined and laminated optical sheet (HILO) microscopy, a highly inclined illumination generates a few microns thick optical layer within the cell, which ensures higher SNR than epi-illumination and enables visualisation of single molecules [42]. However, HILO suffers from a reduced SNR due to autofluorescence and out-of-focus background, and from a limited imaging depth only few microns away from the interface.

7.2.3.2 Confocal Illumination

Laser scanning confocal microscopy (LSCM) achieves optical sectioning by scanning a diffraction-limited excitation volume across the specimen and collecting its fluorescence using spatial filtering. Pinholes placed at conjugated (confocal) planes in the excitation and emission paths ensure a point-by-point illumination and then detection of fluorescence only from the focal plane, while rejecting secondary signals from out-of-focus areas. The diameter of the pinhole determines the thickness of the optical section which is free from out-of-focus noise. Typically, in LSCM, excitation occurs via multiple laser lines, whereas fluorescence is collected by point detection systems such as photomultiplier tubes (PMTs) or avalanche photodiodes (APDs).

Confocal illumination has been extensively used for live and fixed cell imaging. By optimising the pinhole size, LSCM can achieve submicron optical sectioning and significantly higher SNR compared to wide-field microscopy, allowing high-contrast three-dimensional (3D) imaging. However, due the high intensity laser illumination, photobleaching and sample photodamage can be significant. In addition, due to its point scanning nature, LSCM suffers from low time resolution (only few frames per second (fps) frame rate). Even in a high-speed scanning mode, where the laser dwell time per pixel is decreased, the number of detected photons is too low for single-molecule detection. Therefore, its use as a single-molecule imaging tool is limited. A version of confocal illumination that allows faster imaging speed and higher detection efficiency is the spinning-disk or the Nipkow disk confocal (NDC). This is achieved by combining CCD detection with two coupled disks, one with an array of microlenses and one with an array of pinholes, which spin at high frequency and illuminate simultaneously every point of the region of interest. However, the use of NDC in single-molecule imaging is limited due to low SNR.

Recently, various labs have modified the conventional confocal illumination to render it suitable for single-molecule detection in vivo. By combining fast scanning speeds with detection by an APD, which has higher sensitivity than the commonly used PMTs, LSCM has been integrated with fluorescence correlation spectroscopy (FCS) and was able to provide quantitative imaging of single molecules in vitro and in vivo [43]. Another variation of LSCM is the single-point edge-excitation sub-diffraction (SPEED) microscopy, whereby the PMT was replaced by a CCD camera, the detection pinhole was removed to increase the detected signal and the laser beam was shifted at the edge of the objective and focused on the focal plane in a 45° angle to reduce the axial illumination volume [44]. SPEED was successfully used to visualise 3D density maps of transient interactions within single nuclear pore complexes with a high spatio-temporal resolution. Confocal illumination has been also modified to achieve the 3D tracking of single quantum dot (QD)-labelled proteins in live cells with a Z tracking range of 10 μm [45]. By using as spatial filters four fibre optics to create four overlapping confocal volumes, which were simultaneously monitored, and an active feedback of a fast XYZ piezo stage, Wells et al. were able to track molecules with localisation accuracy of ~50 nm in X, Y and ~80 nm in the Z direction. A video-rate confocal microscope with high imaging rate (up to 33 fps) and 1 μm optical sectioning capability has been developed on the basis of a line-scan confocal microscope [46]. By simultaneously scanning the illumination line on the sample and the fluorescence image on the CCD camera with independent galvanometric scanners, this microscope allows visualisation of single molecules at any depth within cells with a higher SNR compared to HILO imaging (Sect. 7.2.3.1). A line scanning confocal microscope coupled to an EM-CCD camera has also been used as a base for the development of a high-speed hyperspectral microscope (HSM) which allows single-particle tracking of eight or more spectrally distinct fluorescent probes at video rate [47].

7.2.3.3 Alternative Optical Sectioning Approaches

Observing single-molecule events within the cell requires optical sectioning with penetration depth of several microns above the coverslip. This is particularly crucial for studying motor proteins within the nucleus, where the high density of molecules generates high fluorescent background. One approach for single-molecule imaging as deep as the cell nucleus is narrow-field epifluorescence microscopy [48, 49]. By incorporating in the excitation path, in an image conjugate plane, a 400 μm pinhole, the illuminated area at the sample plane is reduced to 8 μm in diameter. This generates an effective focal plane thickness of 300 nm and a ~2-fold increase in the SNR compared to wide-field imaging. In this way, molecular interactions within the nuclear pore complex were monitored with 2 ms and 15 nm resolution [48, 49].

Another approach is sheet illumination microscopy or selective plane illumination microscopy (SPIM). In SPIM, the optical sectioning is achieved by a thin light sheet generated by an independent objective lens placed perpendicular to the imaging lens [50, 51]. By scanning the light sheet simultaneously with the imaging objective, single molecules within the working range of the objective can be visualised in the nucleus. However, the application of this technique is limited by steric constraints imposed by the set-up. Such limitation is overcome by the reflected light-sheet (RLS) microscopy, which uses a miniature mirror positioned next to the cell to reflect the light sheet and to thus generate a thin optical section within the cell nucleus [52]. This ensures higher SNR and therefore enhanced resolution. RLS has been combined with super-resolution microscopy to visualise the special organisation of RNA polymerase II in cell nuclei [53].

7.3 How to Probe the Properties, Function and Regulation of Single Motors

Molecular motors move in nanometre-scale steps and for micrometre-scale run-lengths. They also undergo multiple and fast conformational changes during nucleotide binding and upon their interaction with their track, cargo and binding partners. By monitoring the dynamics and movement of single-motor proteins along their track, we can obtain information about many aspects of their function and regulation: their velocity; their directionality; their processivity (which describes how far they move before dissociating from their track); their step size per ATP hydrolysed; their mechanism of translocation (hand-over-hand versus inchworm, active versus passive); their mode of movement (directed, diffusive or confined); the coordination between monomers in a dimer or oligomer; their conformational fluctuations during the ATP cycle and their interaction with their track, partners or cargo; the coordination between different motors in cargo transport; and the mechanism underlying their intracellular localisation and track selectivity. This section gives a general overview of the techniques currently available to extract such information in vitro and in vivo.

7.3.1 Tracking Single Motors In Vitro

Tracking single motors involves labelling the motor with a bright and photostable fluorescent probe (Chap. 1). A single-labelled motor appears as a diffraction-limited spot, whose point spread function (PSF) is defined by the microscope's optics. The trajectories of the molecules are built by tracking algorithms which localise the centre of the spots in every frame and then link them across consecutive frames. The most commonly used algorithms achieve that by fitting the intensity profile of the PSF with a two dimensional (2D) Gaussian function [54], with accuracy only limited by the SNR [55]. Single-molecule/particle tracking can be done with various freely available software, such as ImageJ with its suitable plug-ins (http://rsbweb.nih.gov/ij/plugins/index.html) and GMimPro (http://www.nimr.mrc.ac.uk/gmimpro/) [56].

A high-resolution tracking technique which is based on the 2D Gaussian fitting of the PSF is fluorescence imaging with one nanometre accuracy (FIONA). FIONA, discussed in detail in Chap. 10, is suitable for tracking single motors both in vitro and in vivo and can achieve localisation with a spatial resolution of 1.5 nm [35] and a temporal resolution of 1–500 ms [57]. This approach has been used to measure the step size of various motors, such as myosin V [35], myosin VI [58, 59] and kinesin-1 [60]. High-resolution tracking has been also able to dissect the head-to-head coordination of dimeric motors and discriminate between the "hand-over-hand" stepping shown by myosins and kinesins [35, 58–60] and the uncoordinated stepping of dynein [61, 62]. High-resolution tracking can be achieved in two colours, which provides large amount of data from tracking and distance measurements (Chap. 10). This approach has been used to dissect the stepping mechanism of myosin V [63], VI [64] and dynein [61].

In vitro single-molecule tracking is restricted to the use of low concentration of fluorescent molecules, which could vary between 1 nM and 100 nM depending on the imaging technique [65–67]. These concentrations are below the dissociation constant of many molecular interactions. Therefore, using fluorescently labelled ATP, substrates or binding partners to assess motor activity is limited to very slow kinetics, if any. One approach that allows single-molecule imaging at concentrations near to physiological is the use of linear zero-mode waveguides (ZMWs) [68]. ZMWs are sub-wavelength wells created by lithography on an aluminium film, which prevent the propagation of light and generate an evanescent wave which is significantly shallower than in TIRFM. In this way, ZMWs result in a considerably reduced background which allows motor activity to be studied at fluorophore concentrations up to 10 μM, almost 100-fold higher than with conventional TIRFM. This approach has been used to visualise the activity of DNA polymerase [68, 69] and translating ribosomes [70] at micromolar concentrations of fluorescent nucleotides and tRNAs, respectively. Recently, ZWGs have been modified to allow imaging of cytoskeletal motors moving along long filamentous tracks at high ATP concentrations [71]. Another approach which allows imaging at high protein concentrations is photoactivation, diffusion and excitation (PhADE) [72]. This is a TIRF-based method, which relies on the labelling of the motor of interest with a photoactivatable probe (Chap. 1), such as the protein mKikGR. Photoactivation of mKikGR near the surface where the motor substrates are immobilised and then diffusion of the unbound proteins allows the selective imaging of only the motors bound to their substrates. PhADE has been used to image the activity of the eukaryotic DNA replication protein flap endonuclease 1 at micromolar concentrations [72].

7.3.2 Tracking Single Motors and Their Cargo In Vivo

Tracking single molecules in cells is challenging not only because of the inherent autofluorescence and light scattering (Sect. 7.2.2), but also because cellular environment is very crowded and cellular processes involve many players and multiple transient interactions. Using bright and photostable labels and low background imaging techniques, various groups have tracked single motors in vivo. Tracking single QD-labelled kinesin-1 [73] and myosin V [74, 75] with high spatial and temporal resolution revealed key properties of these motors in vivo, such as their velocity, processivity and step size. In addition, analysis of the trajectories of single motors has been able to discriminate between different modes of movement, such as directed, diffusive or confined motion [73]. Also, in vivo imaging of single kinesin-1, -2 and -3 molecules labelled with a three-tandem monomeric citrine has highlighted that different members of the kinesin family have differential selectivity for microtubules subpopulations [40, 41].

Another challenge in tracking single motors in vivo is obtaining their 3D trajectories which provide a more complete description of their movement. Various strategies for 3D tracking in cells have been reported. One approach incorporates in the detection path a cylindrical lens which induces asymmetry in the defocused image of single fluorophores [76]. By analysing changes in the image shape, Kao and Verkman were able to extract information about the Z position. Another strategy, which is based on bifocal imaging, the simultaneous imaging of focused and defocused planes, allowed the 3D tracking of organelles with 2–5 nm spatial and 2–50 ms time resolution [77]. Recently, Ram et al., using multifocal plane microscopy (MUM), have achieved 3D single-molecule tracking across a depth of 10 μm [78]. Although these approaches have not been applied to motor proteins, but rather to their intracellular cargos, such an application would be an interesting aim for the future.

Similar to in vitro tracking (Sect. 7.3.2), single-motor imaging inside cells requires that the intracellular concentration of the labelled molecules is low enough to allow visualisation of single molecules. This is achieved either by controlling the expression levels of genetically tagged motors [79] or by ensuring that the cytoplasmic delivery of pre-tagged molecules occurs at low concentrations. One way to image single molecules, which are naturally present in cells at high levels, is to use the stochastic photoactivation or photoswitching super-resolution techniques, such as fluorescence photoactivated localisation microscopy (FPALM) [80] and stochastic optical reconstruction microscopy (STORM) [81]. Using photoactivatable or photoswitchable fluorophores as labels (Chap. 1), these methods separate in time labelled molecules located closer than the diffraction limit through their stochastic activation at low spatial density. By fitting the PSF of individual molecules to a 2D Gaussian function, their centroid position can be localised with sub-diffraction accuracy. Through multiple cycles of stochastic excitation–localisation–photobleaching, the trajectories of a large number of single molecules can be extracted. FPALM imaging in live cells, named single-particle tracking PALM (sptPALM) [82], has been shown to obtain dynamic, high-density maps of transmembrane proteins in one and two colours [82, 83]. Also, STORM has been recently shown to achieve 2D and 3D single-molecule imaging of membrane receptors and clathrin-coated pits in live cells with high temporal resolution [84]. Although very powerful, these techniques are limited by the inherent trade-off between their spatial and temporal resolution, which stems from the large number of localisations required. So far, the 2D spatial resolution ranges between 20 and70 nm with a 0.5–60 s time resolution [84–86] and in 3D, 50 nm and 1–2 s [84], respectively. For this reason, these techniques have not been widely applied in the motor field. However, they offer a promising prospect for tracking large numbers of motors in vivo. For instance, PALM tracking has been recently used to visualise the activity of single DNA polymerase I molecules during the DNA repair process in live *Escherichia coli* [87].

Another promising application of super-resolution imaging on motors has been demonstrated by Lee and colleagues [88]. They used PALM imaging, not for tracking, but for counting the number of motors in vivo. Using Dendra2 as the optimum photoactivatable label and a novel photoactivation scheme and analysis algorithm, they were able to count the number of the bacterial flagellar motor protein FliM in live cells. Such a protein quantification method, which allows counting of large numbers of a protein species, is an important addition in the single-molecule toolbox for studying molecular motors and can address pending questions about the number of motors present on intracellular organelles and structures.

7.3.3 Motor Dynamics: Conformational Changes and Interactions with Partners and Substrates

In addition to tracking the movement of single motors, understanding these complex molecular machines requires dissection of their intramolecular dynamics and also the dynamics of their interaction with their tracks, cargos and regulatory partners. Acquiring such information requires measurements at single-molecule level over sub-nanometre distances and in millisecond time scales. These requirements are fulfilled by a technique, widely used in the motor field, termed single-molecule fluorescence resonance energy transfer (smFRET) [89, 90]. The principle, methodology and data analysis for smFRET measurements are described in detail in Chaps. 6, 8 and 12. Briefly, FRET refers to the energy transfer from a donor to an acceptor fluorophore when they are in close proximity, in a range of 2–8 nm [90], and they share a spectral overlap between the emission of the donor and the absorption of the acceptor. Depending on the position of the fluorescent labels, smFRET measures the relative distance changes between the two molecules, or two positions within the same molecule, in a way which is uncoupled from the Brownian motion of the molecule or the complex studied. As a result, smFRET can report local fluctuations in real time and with a high spatial resolution. Such fluctuations can be visualised by TIRFM, whereby the track of the motor is immobilised on a passivated surface, or by confocal imaging of individual molecules diffusing through the confocal volume (Fig. 7.2a, b).

Widely used in the field of nucleic acid motors, smFRET was able to reveal many functional aspects of these proteins. For instance, smFRET has been used to characterise the mechanism of initiation of unwinding by the DNA helicase Rep [18], to define its DNA-binding orientation [16] and was able to detect a "repetitive shuttling" mechanism of movement [91]. Also, smFRET between labels at different positions on the DNA substrate was able to determine the step size of the hepatitis C virus helicase NS3, to discriminate between an active and passive unwinding mechanism and to reveal hidden steps which indicated a spring loading mechanism of unwinding [92]. smFRET has also revealed that Rrp44 exoribonuclease stores elastic energy and unwinds 4 bps of its RNA substrate in one burst [97].

In addition, smFRET can probe and measure the conformational transitions that motors and their substrates undergo. This is exemplified in Chap. 6, whereby combination of single-molecule and ensemble FRET measurements was able to monitor in real time the conformational changes induced by RNA polymerase and its effectors to its promoter during the initiation of transcription [24] and during the transition from initiation to elongation [98]. Also, smFRET data can reveal the kinetics of conformational transitions of the core of DEAD-box helicases during their catalytic cycle and of their RNA substrates (Chap. 8).

A powerful single-molecule FRET-based method which allows simultaneous measurement of intramolecular distances and molecular interactions is the alternating-laser excitation (ALEX), which uses rapid switching between donor and acceptor excitation [99]. In this way, the donor excitation provides structural information, whereas the simultaneous donor and acceptor excitation yields interaction information. For complex systems, ALEX can also be used in a three-colour format and measure three intermolecular distances and interaction stoichiometries [100].

Finally, smFRET can be combined with force spectroscopy in order to investigate tension-dependent conformational fluctuations with nanometre resolution and

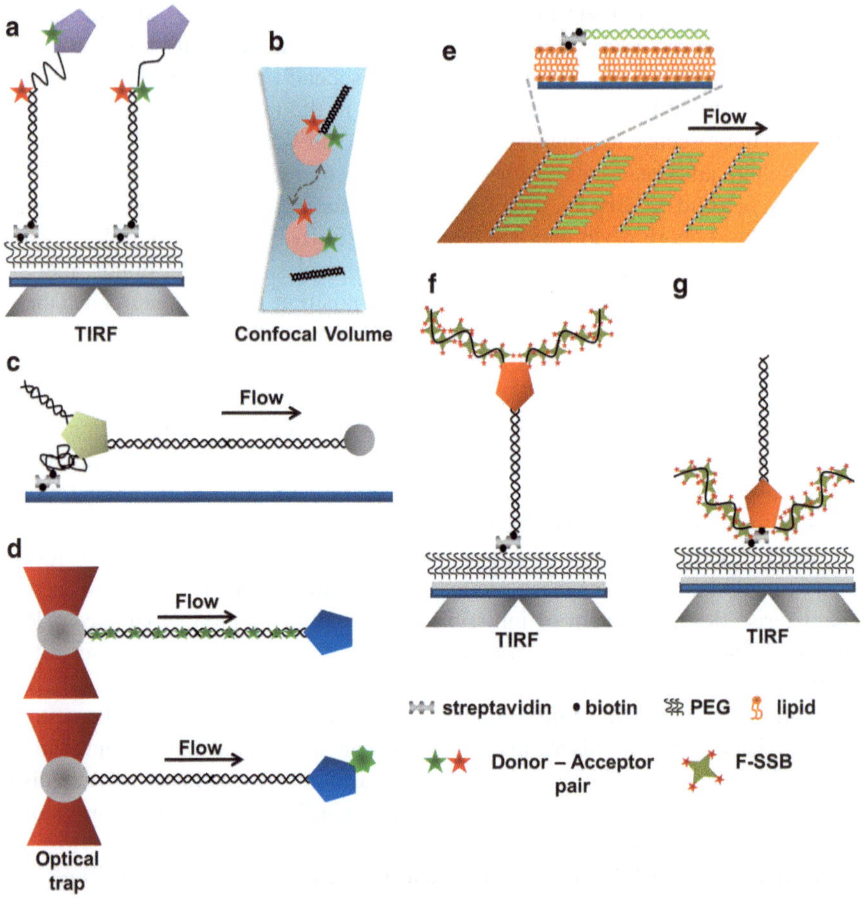

Fig. 7.2 Experimental schemes for studying nucleic acid motors in vitro. (**a**) smFRET assays on DNA substrates immobilised on a PEGylated surface. The binding of the motor of interest on its substrate, its traslocation and/or unwinding activity is visualised by TIRFM (adapted from Myong et al. [91, 92]). (**b**) smFRET by confocal microscopy. The conformational fluctuations of single motors, double-labelled with a donor–acceptor pair, are visualised by confocal microscopy as they diffuse across the confocal volume (adapted from Yodh et al. [3]). (**c**) Tethered-bead assay. The DNA substrate of the helicase is surface-immobilised on one end, whereas attached to a microsphere on its other end. Hydrodynamic flow is used to induce stretching of the DNA parallel to the surface. Given the differential elastic properties between ssDNA and dsDNA, the trajectory of the bead, monitored by bright-field microscopy, reports the unwinding activity of the helicase or, in the case of reconstituted replisome, the replication kinetics (adapted from Lee et al. [93]). (**d**) Optical trap-based hydrodynamic stretching. The DNA substrate is attached to one end on an optically trapped bead and is stretched parallel to the surface by laminar flow. Its free end is allowed to interact with the helicase, and unwinding activity can be monitored either as a fluorescence decrease due to the displacement of a dsDNA-specific intercalating dye or by following the fluorescent nanoparticle bound to the helicase (adapted from Bianco et al. [94] and Handa et al. [95]). (**e**) DNA curtains. Hundreds of lipid-tethered DNA molecules are aligned through laminar flow due to the inability of the lipids to cross microscale diffusion barriers etched on the surface. TIRFM can be used to visualise changes on fluorescently labelled DNA or to follow fluorescently labelled DNA motor proteins (adapted from Visnapuu et al. [14]). (**f, g**) SSB-based helicase assay. The DNA substrate (**f**) or the helicase of interest (**g**) is immobilised on an inert

at single-molecule level. A hybrid system which combines smFRET and optical tweezers has been used to study the effect of tension on the conformation of Holliday junctions (HJs) [101] and investigate how force modulates the dissociation and diffusion of single-stranded DNA-binding protein (SSB) along single-stranded DNA (ssDNA) [102]. Another hybrid scheme is smFRET coupled to magnetic tweezers. In addition to their simplicity, magnetic tweezers are advantageous over optical trapping, since they allow the simultaneous monitoring of multiple individual molecules and do not induce overheating nor photodamage of the sample. The later advantage is important in the hybrid system, because it ensures longer lifetime for the fluorophores. A detailed description of the experimental set-up, sample preparation and data analysis for such experiments, as well as of the application of this hybrid system on HJs and supercoiling-induced cruciform formation, are presented in detail in Chap. 12.

Finally, another powerful approach to probe the structural dynamics of motor proteins in three dimensions is by single-molecule fluorescence polarisation [103]. By monitoring the changes in orientation of a motor domain in real time with a temporal resolution of 20–40 ms, this technique was able to demonstrate the rotation of the lever arm of myosin V. Recently, by combining high-speed polarised TIRFM with time-correlated single-photon counting and change-point analysis, Beausang et al. were able to improve the temporal resolution by ~50-fold and were able to detect short-lived substeps during myosin V stepping, probably corresponding to rotational wobble of the motor domain as it searches for its next binding site [104].

7.4 Experimental Schemes for Visualising the Activity of Single Motors

In addition to single-molecule force spectroscopy assays which are based on atomic force microscopy (AFM) and optical and magnetic tweezers, the past few decades have witnessed the development of various fluorescence-based experimental assays [1, 3]. Complementary to force-inducing schemes, these assays enabled visualisation of the activity of single motors in relation to their substrates, tracks, interacting partners and cargos. Depending on the type of motor studied and the biological question of interest, each assay was built around the most suitable imaging and detection technique among the ones described above. This section gives an overview of the most commonly used single-motor assays reported to date.

Fig. 7.2 (continued) PEGylated surface through biotin–streptavidin interactions. The DNA unwinding can be monitored by TIRFM, in real time, by means of the ssDNA-specific fluorescent biosensor F-SSB which gradually accumulates on the ssDNA product of unwinding (adapted from Fili et al. [96])

7.4.1 Assays for Nucleic Acid Motors In Vitro

One approach frequently used to study DNA replication and helicase activity is flow-induced DNA stretching, whereby immobilised DNA is extended by means of laminar flow (Fig. 7.2c, d). This can be achieved in two experimental set-ups: (a) the tethered-bead set-up, in which the one end of the DNA substrate is surface immobilised through biotin–streptavidin interactions, whereas the other is specifically tethered to a microsphere (Fig. 7.2c). Laminar flow acts on the bead and induces hydrodynamic stretching of the DNA. At the low forces applied (below 5 pN), ssDNA is more compact than double-stranded DNA (dsDNA), and therefore dsDNA unwinding results in the gradual movement of the bead towards the anchoring point [3]. By monitoring the movement by bright-field microscopy, one can follow processive DNA unwinding with a spatial resolution of 100 bp and temporal resolution of 200–500 ms. This approach has been used to dissect DNA replication [93, 105]. (b) The optical trap-based hydrodynamic stretching (Fig. 7.2d), whereby the one end of the DNA substrate is specifically tethered via biotin–streptavidin interactions to a 1 μm polystyrene bead which is optically trapped, whereas the other end remains free and is stretched by laminar flow [94]. DNA unwinding can be visualised either indirectly through the displacement of a dsDNA-specific interacting dye, such YOYO [94, 106], or directly by attaching to the helicase of interest a 40 nm fluorescent nanoparticle [95]. This approach has been able to measure the unwinding rate and processivity of the helicase RecBCD [94] and detect pauses and changes of activity induced by hotspot sequences [95, 106]. With a spatial resolution of ~800 bp and a temporal resolution of ~300 ms, this assay is only suitable for studying highly processive motors. Also, this assay allows monitoring of only one molecule at a time, which limits the statistical analysis.

One methodology that allows simultaneous visualisation of hundreds of individual DNA molecules, aligned in a defined orientation, in a single field of view is the "DNA curtains" [21, 107]. The DNA molecules are tethered through biotin–neutravidin interaction to lipids which form a lipid bilayer on a glass surface, etched to contain microscale diffusion barriers. Since lipids are unable to transverse the barriers, application of flow forces the lipid-bound DNA molecules to stretch and accumulate at the edge of the barriers (Fig. 7.2e). DNA curtains can also be generated with double tethers. This format is suitable for following diffusive processes which get disrupted by flow [108, 109]. The near-surface stretching of DNA molecules makes this set-up compatible with TIRFM and therefore allows high signal-to-noise real-time imaging with millisecond temporal resolution and tens-of-nanometres spatial resolution [110]. Based on the use of the lipid bilayers, DNA curtains ensure minimal non-specific interactions and are therefore suitable for investigating a wide range of DNA-processing enzymes. This scheme also allows the use of very long DNA substrates, like the 48 kb long phage λDNA, which can be visualised by intercalating dyes, such as YOYO-1. So far, DNA curtains have been used to investigate how specific sequences control the positioning of DNA-interacting proteins, such as the nucleosomes [111]; to dissect the mechanism by which the two different damage-sensing mismatch repair protein complexes Msh2-Msh6 [112] and Mlh1-Pms1 [109] travel along DNA and how they react to roadblocks during DNA repair; to define the translocation mechanism of the chromatin remodelling motor Rdh54 which is involved in DNA recombination [113] and to unravel how the bacterial helicase–nuclease RecBCD evicts DNA-bound proteins during its translocation [114].

Another way to monitor the activity of single nucleic acid motors is to probe the products of motor activity using fluorescent biosensors (Chap. 2). Such an approach has been followed by Fili and co-workers who developed a generic single-molecule DNA helicase assay [96]. The assay uses a fluorescent version of the single-stranded DNA-binding protein (F-SSB) to monitor in real time the unwinding of multiple linear DNA molecules simultaneously by TIRFM. By immobilising either the DNA substrate or the helicase, unwinding can be visualised as a localised increase in fluorescence which reports the accumulation of F-SSB on the progressively grown ssDNA products (Fig 7.2f, g). Analysis of the intensity traces of multiple events allows one to determine the processivity and unwinding rate of the helicase, assess the population's heterogeneity and also detect hidden phases of unwinding, such as pausing. This assay has been recently modified to enable visualisation of plasmid unwinding and was used to assess the oligomeric state of the bacterial helicase PcrA [115]. With a resolution limited by the binding mode of F-SSB to ssDNA (~65 bases at high salt conditions), this assay is suitable for studying helicases of medium to high processivity. Moreover, the conditions of the assay have to be adjusted so that the unwinding rate of the helicase is significantly higher than the photobleaching rate of F-SSB. Also, due to the exponential decaying evanescent field, the maximum length of DNA substrates, which ensures linear relationship between intensity and DNA length, is 2 kb. For longer DNA substrates, the assay can be performed under hydrodynamic flow. Future development of fluorescent biosensors which are suitable for single-molecule imaging would enable the development of similar assays applicable to other types of nucleic acid motors.

7.4.2 Assays for Cytoskeletal Motors In Vitro

Some of the single-molecule or single-cargo assays for cytoskeletal motors are described in detail in dedicated chapters of this book. For instance, Chap. 9 gives a detailed overview of motility assays that are widely used to track and dissect the function and regulation of myosins. Briefly, myosin motility assays are performed in two experimental formats: (**a**) the actin gliding assay, whereby surface-immobilised motors are moving fluorescently labelled actin filaments, and (**b**) the single-molecule motility assay, whereby single fluorescently labelled myosins are tracked along immobilised actin filaments.

Similar motility assays have been developed for microtubule (MT) interacting motors. More specifically, in a MT gliding assay, kinesin molecules are specifically or non-specifically immobilised on the surface and their activity on fluorescently labelled, in vitro assembled, MTs can be monitored with high accuracy [116]. Although this geometry reports the activity of multiple kinesins, it presents the advantage of an easy determination of motor directionality, enabling monitoring of non-processive motors and addressing questions about the cooperativity between multiple motors [117]. In addition, various strategies are available to define the number of motors involved in the transport. The gliding assay can be modified into a single-molecule scheme by diluting down the surface density of immobilised kinesins, so that each MT can interact with an individual motor [118]. Alternatively, the single-molecule format is based on tracking individual fluorescently labelled kinesins on surface-immobilised MTs [119–121]. These assays have been widely used to determine the processivity, step size, mechanism of translocation and MT depolymerisation of many members of the kinesin family [60, 122, 123], as well as to study their activity and how their translocation is affected by roadblocks or other MT interacting proteins [124–126]. Single-molecule MT-based assays have also provided insights into the properties and translocation mechanism of the complex motor dynein [61, 62, 127].

Although rich in information, in vitro motility assays lack the structural complexity of the intracellular cytoskeleton, a dense 3D network where different filamentous species acquire multiple spatial arrangements and are decorated with various interacting proteins. This complex architecture is not only expected to affect the motility of motor proteins but also the preference of different motors for different subpopulations of filaments and to therefore regulate their spatial localisation and intracellular role. One way to address this question, at the interface of in vitro and in vivo measurements, is to perform single-molecule or single-cargo motility assays on extracted cytoskeletal networks. This approach was followed by Brawley and Rock who extracted the actin cytoskeleton from three different cell lines by gently solubilising the plasma membrane with detergent and simultaneously stabilising it with rhodamine–phalloidin. This "ex vivo" motility assay showed that different myosins have differential preference for different cytoskeletal species and different intracellular regions [128]. Similarly, Sivaramakrishnan and Spudich have used detergent-extracted keratocyte lamellipodium to assess the ability of multiple non-processive monomeric myosin VI molecules to coordinate in order to transport cargo [129]. Another strategy to isolate the native cytoskeleton is by exposing the cytosplasmic side of the plasma membrane, termed cell "unroofing". This is achieved by exposing adherent cells, attached to poly-L-lysine-coated coverslips, to a brief sonication pulse which shears the cell body leaving behind plasma membrane sheets [130, 131]. Unlike the detergent extraction method, this approach preserves intact both the natural lipid bilayer and its associated protein structures (such clathrin-coated pits, caveolae etc) and the membrane-associated cortical actin cytoskeleton and possibly microtubules [130, 132, 133]. Therefore, plasma membrane sheets combine an exposed natural environment with the ease and accessibility for in vitro manipulations. So far, cell "unroofing" has been used to study endocytosis [134–137]. However, it also offers an excellent 3D natural platform for semi-in vivo single-molecule motility assays.

The next step in gaining insight into the function of cytoskeletal motors in cells is to understand the mechanism underlying the interplay of multiple motors. One way to achieve this is the in vitro reconstitution of intracellular cargo transport (Chap. 11). Briefly, multiple motors, of the same or different species, are non-specifically absorbed or specifically attached to artificial cargos, such as beads, QDs and DNA- or protein-based scaffolds. Then, the motility of the reconstituted cargo on surface-immobilised cytoskeletal filaments is visualised. By using the minimum number of components in a tightly controlled manner and gradually increasing the complexity of the system, this approach can dissect the highly complex mechanisms of intracellular transport and can provide a unique insight into how many different motors coordinate in vivo in order to achieve cargo motility. An increased complexity can be achieved by moving from an in vitro fully reconstituted cargo to purified intracellular cargos and studying their motility in vitro. In Chap. 11, Conway and Ross describe how these reconstitution experiments can be designed and controlled.

7.5 Conclusions

During the past few decades, single-molecule fluorescence imaging has gone through a continuous and incredibly fast progress. Novel bright and photostable fluorescent probes and powerful imaging techniques which ensure high spatial and temporal resolution have been added to the toolbox of researchers. In particular, the field of molecular motors has greatly benefited from this development. High-end imaging combined with ingenious experimental schemes, both in vitro, in vivo and at their interface, enabled researchers to study these complex and multi-potent molecular machines in great depth and detail and gain invaluable insight into their mechanistic properties, regulation and cellular function. However, still many aspects of the biology of motor proteins remain elusive. Therefore, the future would welcome new additions in our current toolbox which will allow us to advance our knowledge on these proteins even further.

References

1. Kim H, Ha T (2013) Single-molecule nanometry for biological physics. Rep Progr Phys Soc 76(1):016601. doi:10.1088/0034-4885/76/1/016601
2. Walter NG, Huang CY, Manzo AJ, Sobhy MA (2008) Do-it-yourself guide: how to use the modern single-molecule toolkit. Nat Methods 5(6):475–489. doi:10.1038/nmeth.1215
3. Yodh JG, Schlierf M, Ha T (2010) Insight into helicase mechanism and function revealed through single-molecule approaches. Q Rev Biophys 43(2):185–217. doi:10.1017/S0033583510000107

4. Cai D, Kaul N, Lionberger TA, Wiener DM, Verhey KJ, Meyhofer E (2010) Recording single motor proteins in the cytoplasm of mammalian cells. Methods Enzymol 475:81–107. doi:10.1016/S0076-6879(10)75004-7
5. Hilario J, Kowalczykowski SC (2010) Visualizing protein-DNA interactions at the single-molecule level. Curr Opin Chem Biol 14(1):15–22. doi:10.1016/j.cbpa.2009.10.035
6. Park H, Toprak E, Selvin PR (2007) Single-molecule fluorescence to study molecular motors. Q Rev Biophys 40(1):87–111. doi:10.1017/S0033583507004611
7. Toseland CP (2013) Fluorescent labeling and modification of proteins. J Chem Biol 6 (3):85–95. doi:10.1007/s12154-013-0094-5
8. Ha T, Tinnefeld P (2012) Photophysics of fluorescent probes for single-molecule biophysics and super-resolution imaging. Annu Rev Phys Chem 63:595–617. doi:10.1146/annurev-physchem-032210-103340
9. Harada Y, Sakurada K, Aoki T, Thomas DD, Yanagida T (1990) Mechanochemical coupling in actomyosin energy transduction studied by in vitro movement assay. J Mol Biol 216 (1):49–68. doi:10.1016/S0022-2836(05)80060-9
10. Shi X, Lim J, Ha T (2010) Acidification of the oxygen scavenging system in single-molecule fluorescence studies: in situ sensing with a ratiometric dual-emission probe. Anal Chem 82 (14):6132–6138. doi:10.1021/ac1008749
11. Aitken CE, Marshall RA, Puglisi JD (2008) An oxygen scavenging system for improvement of dye stability in single-molecule fluorescence experiments. Biophys J 94(5):1826–1835. doi:10.1529/biophysj.107.117689
12. Rasnik I, McKinney SA, Ha T (2006) Nonblinking and long-lasting single-molecule fluorescence imaging. Nat Methods 3(11):891–893. doi:10.1038/nmeth934
13. Vogelsang J, Kasper R, Steinhauer C, Person B, Heilemann M, Sauer M, Tinnefeld P (2008) A reducing and oxidizing system minimizes photobleaching and blinking of fluorescent dyes. Angew Chem 47(29):5465–5469. doi:10.1002/anie.200801518
14. Visnapuu ML, Duzdevich D, Greene EC (2008) The importance of surfaces in single-molecule bioscience. Mol Biosyst 4(5):394–403. doi:10.1039/b800444g
15. Cisse I, Okumus B, Joo C, Ha T (2007) Fueling protein DNA interactions inside porous nanocontainers. Proc Natl Acad Sci USA 104(31):12646–12650. doi:10.1073/pnas.0610673104
16. Rasnik I, Myong S, Cheng W, Lohman TM, Ha T (2004) DNA-binding orientation and domain conformation of the E. coli rep helicase monomer bound to a partial duplex junction: single-molecule studies of fluorescently labeled enzymes. J Mol Biol 336(2):395–408
17. Rasnik I, McKinney SA, Ha T (2005) Surfaces and orientations: much to FRET about? Acc Chem Res 38(7):542–548. doi:10.1021/ar040138c
18. Ha T, Rasnik I, Cheng W, Babcock HP, Gauss GH, Lohman TM, Chu S (2002) Initiation and re-initiation of DNA unwinding by the Escherichia coli Rep helicase. Nature 419 (6907):638–641. doi:10.1038/nature01083
19. Fili N, Toseland CP, Dillingham MS, Webb MR, Molloy JE (2011) A single-molecule approach to visualize the unwinding activity of DNA helicases. Methods Mol Biol 778:193–214. doi:10.1007/978-1-61779-261-8_13
20. Heyes CD, Groll J, Moller M, Nienhaus GU (2007) Synthesis, patterning and applications of star-shaped poly(ethylene glycol) biofunctionalized surfaces. Mol Biosyst 3(6):419–430. doi:10.1039/b700055n
21. Graneli A, Yeykal CC, Prasad TK, Greene EC (2006) Organized arrays of individual DNA molecules tethered to supported lipid bilayers. Langmuir 22(1):292–299. doi:10.1021/la051944a
22. Sackmann E (1996) Supported membranes: scientific and practical applications. Science 271 (5245):43–48
23. Rhoades E, Gussakovsky E, Haran G (2003) Watching proteins fold one molecule at a time. Proc Natl Acad Sci USA 100(6):3197–3202. doi:10.1073/pnas.2628068100

24. Kim H, Tang GQ, Patel SS, Ha T (2012) Opening-closing dynamics of the mitochondrial transcription pre-initiation complex. Nucleic Acids Res 40(1):371–380. doi:10.1093/nar/gkr736
25. Aubin JE (1979) Autofluorescence of viable cultured mammalian cells. J Histochem Cytochem 27(1):36–43
26. Benson RC, Meyer RA, Zaruba ME, McKhann GM (1979) Cellular autofluorescence–is it due to flavins? J Histochem Cytochem 27(1):44–48
27. Harms GS, Cognet L, Lommerse PH, Blab GA, Schmidt T (2001) Autofluorescent proteins in single-molecule research: applications to live cell imaging microscopy. Biophys J 80 (5):2396–2408. doi:10.1016/S0006-3495(01)76209-1
28. Tsutsui H, Karasawa S, Okamura Y, Miyawaki A (2008) Improving membrane voltage measurements using FRET with new fluorescent proteins. Nat Methods 5(8):683–685. doi:10.1038/nmeth.1235
29. Kredel S, Oswald F, Nienhaus K, Deuschle K, Rocker C, Wolff M, Heilker R, Nienhaus GU, Wiedenmann J (2009) mRuby, a bright monomeric red fluorescent protein for labeling of subcellular structures. PLoS One 4(2):e4391. doi:10.1371/journal.pone.0004391
30. Harms GS, Cognet L, Lommerse PH, Blab GA, Kahr H, Gamsjager R, Spaink HP, Soldatov NM, Romanin C, Schmidt T (2001) Single-molecule imaging of l-type Ca(2+) channels in live cells. Biophys J 81(5):2639–2646. doi:10.1016/S0006-3495(01)75907-3
31. Steyer JA, Almers W (2001) A real-time view of life within 100 nm of the plasma membrane. Nat Rev Mol Cell Biol 2(4):268–275. doi:10.1038/35067069
32. Toomre D, Manstein DJ (2001) Lighting up the cell surface with evanescent wave microscopy. Trends Cell Biol 11(7):298–303
33. Mattheyses AL, Simon SM, Rappoport JZ (2010) Imaging with total internal reflection fluorescence microscopy for the cell biologist. J Cell Sci 123(Pt 21):3621–3628. doi:10.1242/jcs.056218
34. Selvin PR, Ha T (2008) Single-molecule techniques: a laboratory manual. CSHL Press, New York, NY
35. Yildiz A, Forkey JN, McKinney SA, Ha T, Goldman YE, Selvin PR (2003) Myosin V walks hand-over-hand: single fluorophore imaging with 1.5-nm localization. Science 300 (5628):2061–2065. doi:10.1126/science.1084398
36. Testa I, Wurm CA, Medda R, Rothermel E, von Middendorf C, Folling J, Jakobs S, Schonle A, Hell SW, Eggeling C (2010) Multicolor fluorescence nanoscopy in fixed and living cells by exciting conventional fluorophores with a single wavelength. Biophys J 99 (8):2686–2694. doi:10.1016/j.bpj.2010.08.012
37. Webb SE, Zanetti-Domingues L, Coles BC, Rolfe DJ, Wareham RJ, Martin-Fernandez ML (2012) Multicolour single molecule imaging on cells using a supercontinuum source. Biomed Opt Exp 3(3):400–406. doi:10.1364/BOE.3.000400
38. Giebel K, Bechinger C, Herminghaus S, Riedel M, Leiderer P, Weiland U, Bastmeyer M (1999) Imaging of cell/substrate contacts of living cells with surface plasmon resonance microscopy. Biophys J 76(1 Pt 1):509–516
39. Gell C, Berndt M, Enderlein J, Diez S (2009) TIRF microscopy evanescent field calibration using tilted fluorescent microtubules. J Microsc 234(1):38–46. doi:10.1111/j.1365-2818.2009.03147.x
40. Cai D, McEwen DP, Martens JR, Meyhofer E, Verhey KJ (2009) Single molecule imaging reveals differences in microtubule track selection between Kinesin motors. PLoS Biol 7(10): e1000216. doi:10.1371/journal.pbio.1000216
41. Cai D, Verhey KJ, Meyhofer E (2007) Tracking single Kinesin molecules in the cytoplasm of mammalian cells. Biophys J 92(12):4137–4144. doi:10.1529/biophysj.106.100206
42. Tokunaga M, Imamoto N, Sakata-Sogawa K (2008) Highly inclined thin illumination enables clear single-molecule imaging in cells. Nat Methods 5(2):159–161. doi:10.1038/nmeth1171

43. Vukojevic V, Heidkamp M, Ming Y, Johansson B, Terenius L, Rigler R (2008) Quantitative single-molecule imaging by confocal laser scanning microscopy. Proc Natl Acad Sci USA 105(47):18176–18181. doi:10.1073/pnas.0809250105
44. Ma J, Yang W (2010) Three-dimensional distribution of transient interactions in the nuclear pore complex obtained from single-molecule snapshots. Proc Natl Acad Sci USA 107 (16):7305–7310. doi:10.1073/pnas.0908269107
45. Wells NP, Lessard GA, Goodwin PM, Phipps ME, Cutler PJ, Lidke DS, Wilson BS, Werner JH (2010) Time-resolved three-dimensional molecular tracking in live cells. Nano Lett 10 (11):4732–4737. doi:10.1021/nl103247v
46. Lee J, Miyanaga Y, Ueda M, Hohng S (2012) Video-rate confocal microscopy for single-molecule imaging in live cells and superresolution fluorescence imaging. Biophys J 103 (8):1691–1697. doi:10.1016/j.bpj.2012.09.014
47. Cutler PJ, Malik MD, Liu S, Byars JM, Lidke DS, Lidke KA (2013) Multi-color quantum dot tracking using a high-speed hyperspectral line-scanning microscope. PLoS One 8(5):e64320. doi:10.1371/journal.pone.0064320
48. Yang W, Gelles J, Musser SM (2004) Imaging of single-molecule translocation through nuclear pore complexes. Proc Natl Acad Sci USA 101(35):12887–12892. doi:10.1073/pnas.0403675101
49. Yang W, Musser SM (2006) Visualizing single molecules interacting with nuclear pore complexes by narrow-field epifluorescence microscopy. Methods 39(4):316–328. doi:10.1016/j.ymeth.2006.06.002
50. Cella Zanacchi F, Lavagnino Z, Perrone Donnorso M, Del Bue A, Furia L, Faretta M, Diaspro A (2011) Live-cell 3D super-resolution imaging in thick biological samples. Nat Methods 8 (12):1047–1049. doi:10.1038/nmeth.1744
51. Ritter JG, Veith R, Veenendaal A, Siebrasse JP, Kubitscheck U (2010) Light sheet microscopy for single molecule tracking in living tissue. PLoS One 5(7):e11639. doi:10.1371/journal.pone.0011639
52. Gebhardt JC, Suter DM, Roy R, Zhao ZW, Chapman AR, Basu S, Maniatis T, Xie XS (2013) Single-molecule imaging of transcription factor binding to DNA in live mammalian cells. Nat Methods 10(5):421–426. doi:10.1038/nmeth.2411
53. Zhao ZW, Roy R, Gebhardt JC, Suter DM, Chapman AR, Xie XS (2014) Spatial organization of RNA polymerase II inside a mammalian cell nucleus revealed by reflected light-sheet superresolution microscopy. Proc Natl Acad Sci USA 111(2):681–686. doi:10.1073/pnas.1318496111
54. Cheezum MK, Walker WF, Guilford WH (2001) Quantitative comparison of algorithms for tracking single fluorescent particles. Biophys J 81(4):2378–2388. doi:10.1016/S0006-3495(01)75884-5
55. Thompson RE, Larson DR, Webb WW (2002) Precise nanometer localization analysis for individual fluorescent probes. Biophys J 82(5):2775–2783. doi:10.1016/S0006-3495(02)75618-X
56. Mashanov GI, Molloy JE (2007) Automatic detection of single fluorophores in live cells. Biophys J 92(6):2199–2211. doi:10.1529/biophysj.106.081117
57. Kural C, Kim H, Syed S, Goshima G, Gelfand VI, Selvin PR (2005) Kinesin and dynein move a peroxisome in vivo: a tug-of-war or coordinated movement? Science 308 (5727):1469–1472. doi:10.1126/science.1108408
58. Okten Z, Churchman LS, Rock RS, Spudich JA (2004) Myosin VI walks hand-over-hand along actin. Nat Struct Mol Biol 11(9):884–887. doi:10.1038/nsmb815
59. Yildiz A, Park H, Safer D, Yang Z, Chen LQ, Selvin PR, Sweeney HL (2004) Myosin VI steps via a hand-over-hand mechanism with its lever arm undergoing fluctuations when attached to actin. J Biol Chem 279(36):37223–37226. doi:10.1074/jbc.C400252200
60. Yildiz A, Tomishige M, Vale RD, Selvin PR (2004) Kinesin walks hand-over-hand. Science 303(5658):676–678. doi:10.1126/science.1093753

61. DeWitt MA, Chang AY, Combs PA, Yildiz A (2012) Cytoplasmic dynein moves through uncoordinated stepping of the AAA + ring domains. Science 335(6065):221–225. doi:10.1126/science.1215804
62. Qiu W, Derr ND, Goodman BS, Villa E, Wu D, Shih W, Reck-Peterson SL (2012) Dynein achieves processive motion using both stochastic and coordinated stepping. Nat Struct Mol Biol 19(2):193–200. doi:10.1038/nsmb.2205
63. Warshaw DM, Kennedy GG, Work SS, Krementsova EB, Beck S, Trybus KM (2005) Differential labeling of myosin V heads with quantum dots allows direct visualization of hand-over-hand processivity. Biophys J 88(5):L30–32. doi:10.1529/biophysj.105.061903
64. Nishikawa S, Arimoto I, Ikezaki K, Sugawa M, Ueno H, Komori T, Iwane AH, Yanagida T (2010) Switch between large hand-over-hand and small inchworm-like steps in myosin VI. Cell 142(6):879–888. doi:10.1016/j.cell.2010.08.033
65. Adachi K, Oiwa K, Nishizaka T, Furuike S, Noji H, Itoh H, Yoshida M, Kinosita K Jr (2007) Coupling of rotation and catalysis in F(1)-ATPase revealed by single-molecule imaging and manipulation. Cell 130(2):309–321. doi:10.1016/j.cell.2007.05.020
66. Komori T, Nishikawa S, Ariga T, Iwane AH, Yanagida T (2009) Simultaneous measurement of nucleotide occupancy and mechanical displacement in myosin-V, a processive molecular motor. Biophys J 96(1):L04–06. doi:10.1016/j.bpj.2008.09.031
67. van Oijen AM (2011) Single-molecule approaches to characterizing kinetics of biomolecular interactions. Curr Opin Biotechnol 22(1):75–80. doi:10.1016/j.copbio.2010.10.002
68. Levene MJ, Korlach J, Turner SW, Foquet M, Craighead HG, Webb WW (2003) Zero-mode waveguides for single-molecule analysis at high concentrations. Science 299(5607):682–686. doi:10.1126/science.1079700
69. Eid J, Fehr A, Gray J, Luong K, Lyle J, Otto G, Peluso P, Rank D, Baybayan P, Bettman B, Bibillo A, Bjornson K, Chaudhuri B, Christians F, Cicero R, Clark S, Dalal R, Dewinter A, Dixon J, Foquet M, Gaertner A, Hardenbol P, Heiner C, Hester K, Holden D, Kearns G, Kong X, Kuse R, Lacroix Y, Lin S, Lundquist P, Ma C, Marks P, Maxham M, Murphy D, Park I, Pham T, Phillips M, Roy J, Sebra R, Shen G, Sorenson J, Tomaney A, Travers K, Trulson M, Vieceli J, Wegener J, Wu D, Yang A, Zaccarin D, Zhao P, Zhong F, Korlach J, Turner S (2009) Real-time DNA sequencing from single polymerase molecules. Science 323 (5910):133–138. doi:10.1126/science.1162986
70. Uemura S, Aitken CE, Korlach J, Flusberg BA, Turner SW, Puglisi JD (2010) Real-time tRNA transit on single translating ribosomes at codon resolution. Nature 464 (7291):1012–1017. doi:10.1038/nature08925
71. Elting MW, Leslie SR, Churchman LS, Korlach J, McFaul CM, Leith JS, Levene MJ, Cohen AE, Spudich JA (2013) Single-molecule fluorescence imaging of processive myosin with enhanced background suppression using linear zero-mode waveguides (ZMWs) and convex lens induced confinement (CLIC). Opt Express 21(1):1189–1202. doi:10.1364/OE.21.001189
72. Loveland AB, Habuchi S, Walter JC, van Oijen AM (2012) A general approach to break the concentration barrier in single-molecule imaging. Nat Methods 9(10):987–992. doi:10.1038/nmeth.2174
73. Courty S, Luccardini C, Bellaiche Y, Cappello G, Dahan M (2006) Tracking individual kinesin motors in living cells using single quantum-dot imaging. Nano Lett 6(7):1491–1495. doi:10.1021/nl060921t
74. Nelson SR, Ali MY, Trybus KM, Warshaw DM (2009) Random walk of processive, quantum dot-labeled myosin Va molecules within the actin cortex of COS-7 cells. Biophys J 97 (2):509–518. doi:10.1016/j.bpj.2009.04.052
75. Pierobon P, Achouri S, Courty S, Dunn AR, Spudich JA, Dahan M, Cappello G (2009) Velocity, processivity, and individual steps of single myosin V molecules in live cells. Biophys J 96(10):4268–4275. doi:10.1016/j.bpj.2009.02.045

76. Kao HP, Verkman AS (1994) Tracking of single fluorescent particles in three dimensions: use of cylindrical optics to encode particle position. Biophys J 67(3):1291–1300. doi:10.1016/S0006-3495(94)80601-0
77. Toprak E, Balci H, Blehm BH, Selvin PR (2007) Three-dimensional particle tracking via bifocal imaging. Nano Lett 7(7):2043–2045. doi:10.1021/nl0709120
78. Ram S, Kim D, Ober RJ, Ward ES (2012) 3D single molecule tracking with multifocal plane microscopy reveals rapid intercellular transferrin transport at epithelial cell barriers. Biophys J 103(7):1594–1603. doi:10.1016/j.bpj.2012.08.054
79. Douglass AD, Vale RD (2008) Single-molecule imaging of fluorescent proteins. Methods Cell Biol 85:113–125. doi:10.1016/S0091-679X(08)85006-6
80. Betzig E, Patterson GH, Sougrat R, Lindwasser OW, Olenych S, Bonifacino JS, Davidson MW, Lippincott-Schwartz J, Hess HF (2006) Imaging intracellular fluorescent proteins at nanometer resolution. Science 313(5793):1642–1645. doi:10.1126/science.1127344
81. Rust MJ, Bates M, Zhuang X (2006) Sub-diffraction-limit imaging by stochastic optical reconstruction microscopy (STORM). Nat Methods 3(10):793–795. doi:10.1038/nmeth929
82. Manley S, Gillette JM, Patterson GH, Shroff H, Hess HF, Betzig E, Lippincott-Schwartz J (2008) High-density mapping of single-molecule trajectories with photoactivated localization microscopy. Nat Methods 5(2):155–157. doi:10.1038/nmeth.1176
83. Subach FV, Patterson GH, Renz M, Lippincott-Schwartz J, Verkhusha VV (2010) Bright monomeric photoactivatable red fluorescent protein for two-color super-resolution sptPALM of live cells. J Am Chem Soc 132(18):6481–6491. doi:10.1021/ja100906g
84. Jones SA, Shim SH, He J, Zhuang X (2011) Fast, three-dimensional super-resolution imaging of live cells. Nat Methods 8(6):499–508. doi:10.1038/nmeth.1605
85. Biteen JS, Thompson MA, Tselentis NK, Bowman GR, Shapiro L, Moerner WE (2008) Super-resolution imaging in live Caulobacter crescentus cells using photoswitchable EYFP. Nat Methods 5(11):947–949. doi:10.1038/nmeth.1258
86. Shroff H, Galbraith CG, Galbraith JA, Betzig E (2008) Live-cell photoactivated localization microscopy of nanoscale adhesion dynamics. Nat Methods 5(5):417–423. doi:10.1038/nmeth.1202
87. Uphoff S, Reyes-Lamothe R, Garza de Leon F, Sherratt DJ, Kapanidis AN (2013) Single-molecule DNA repair in live bacteria. Proc Natl Acad Sci USA 110(20):8063–8068. doi:10.1073/pnas.1301804110
88. Lee SH, Shin JY, Lee A, Bustamante C (2012) Counting single photoactivatable fluorescent molecules by photoactivated localization microscopy (PALM). Proc Natl Acad Sci USA 109 (43):17436–17441. doi:10.1073/pnas.1215175109
89. Ha T, Enderle T, Ogletree DF, Chemla DS, Selvin PR, Weiss S (1996) Probing the interaction between two single molecules: fluorescence resonance energy transfer between a single donor and a single acceptor. Proc Natl Acad Sci USA 93(13):6264–6268
90. Roy R, Hohng S, Ha T (2008) A practical guide to single-molecule FRET. Nat Methods 5 (6):507–516. doi:10.1038/nmeth.1208
91. Myong S, Rasnik I, Joo C, Lohman TM, Ha T (2005) Repetitive shuttling of a motor protein on DNA. Nature 437(7063):1321–1325. doi:10.1038/nature04049
92. Myong S, Bruno MM, Pyle AM, Ha T (2007) Spring-loaded mechanism of DNA unwinding by hepatitis C virus NS3 helicase. Science 317(5837):513–516. doi:10.1126/science.1144130
93. Lee JB, Hite RK, Hamdan SM, Xie XS, Richardson CC, van Oijen AM (2006) DNA primase acts as a molecular brake in DNA replication. Nature 439(7076):621–624. doi:10.1038/nature04317
94. Bianco PR, Brewer LR, Corzett M, Balhorn R, Yeh Y, Kowalczykowski SC, Baskin RJ (2001) Processive translocation and DNA unwinding by individual RecBCD enzyme molecules. Nature 409(6818):374–378. doi:10.1038/35053131
95. Handa N, Bianco PR, Baskin RJ, Kowalczykowski SC (2005) Direct visualization of RecBCD movement reveals cotranslocation of the RecD motor after chi recognition. Mol Cell 17(5):745–750. doi:10.1016/j.molcel.2005.02.011

96. Fili N, Mashanov GI, Toseland CP, Batters C, Wallace MI, Yeeles JT, Dillingham MS, Webb MR, Molloy JE (2010) Visualizing helicases unwinding DNA at the single molecule level. Nucleic Acids Res 38(13):4448–4457. doi:10.1093/nar/gkq173
97. Lee G, Bratkowski MA, Ding F, Ke A, Ha T (2012) Elastic coupling between RNA degradation and unwinding by an exoribonuclease. Science 336(6089):1726–1729. doi:10.1126/science.1216848
98. Tang GQ, Roy R, Bandwar RP, Ha T, Patel SS (2009) Real-time observation of the transition from transcription initiation to elongation of the RNA polymerase. Proc Natl Acad Sci USA 106(52):22175–22180. doi:10.1073/pnas.0906979106
99. Kapanidis AN, Laurence TA, Lee NK, Margeat E, Kong X, Weiss S (2005) Alternating-laser excitation of single molecules. Acc Chem Res 38(7):523–533. doi:10.1021/ar0401348
100. Lee NK, Kapanidis AN, Koh HR, Korlann Y, Ho SO, Kim Y, Gassman N, Kim SK, Weiss S (2007) Three-color alternating-laser excitation of single molecules: monitoring multiple interactions and distances. Biophys J 92(1):303–312. doi:10.1529/biophysj.106.093211
101. Hohng S, Zhou R, Nahas MK, Yu J, Schulten K, Lilley DM, Ha T (2007) Fluorescence-force spectroscopy maps two-dimensional reaction landscape of the holliday junction. Science 318 (5848):279–283. doi:10.1126/science.1146113
102. Zhou R, Kozlov AG, Roy R, Zhang J, Korolev S, Lohman TM, Ha T (2011) SSB functions as a sliding platform that migrates on DNA via reptation. Cell 146(2):222–232. doi:10.1016/j.cell.2011.06.036
103. Forkey JN, Quinlan ME, Shaw MA, Corrie JE, Goldman YE (2003) Three-dimensional structural dynamics of myosin V by single-molecule fluorescence polarization. Nature 422 (6930):399–404. doi:10.1038/nature01529
104. Beausang JF, Shroder DY, Nelson PC, Goldman YE (2013) Tilting and wobble of myosin V by high-speed single-molecule polarized fluorescence microscopy. Biophys J 104 (6):1263–1273. doi:10.1016/j.bpj.2013.01.057
105. Hamdan SM, Loparo JJ, Takahashi M, Richardson CC, van Oijen AM (2009) Dynamics of DNA replication loops reveal temporal control of lagging-strand synthesis. Nature 457 (7227):336–339. doi:10.1038/nature07512
106. Spies M, Bianco PR, Dillingham MS, Handa N, Baskin RJ, Kowalczykowski SC (2003) A molecular throttle: the recombination hotspot chi controls DNA translocation by the RecBCD helicase. Cell 114(5):647–654
107. Greene EC, Wind S, Fazio T, Gorman J, Visnapuu ML (2010) DNA curtains for high-throughput single-molecule optical imaging. Methods Enzymol 472:293–315. doi:10.1016/S0076-6879(10)72006-1
108. Gorman J, Fazio T, Wang F, Wind S, Greene EC (2010) Nanofabricated racks of aligned and anchored DNA substrates for single-molecule imaging. Langmuir 26(2):1372–1379. doi:10.1021/la902443e
109. Gorman J, Plys AJ, Visnapuu ML, Alani E, Greene EC (2010) Visualizing one-dimensional diffusion of eukaryotic DNA repair factors along a chromatin lattice. Nat Struct Mol Biol 17 (8):932–938. doi:10.1038/nsmb.1858
110. Duzdevich D, Greene EC (2013) Towards physiological complexity with in vitro single-molecule biophysics. Philos Trans R Soc Lond Ser B Biol Sci 368(1611):20120271. doi:10.1098/rstb.2012.0271
111. Visnapuu ML, Greene EC (2009) Single-molecule imaging of DNA curtains reveals intrinsic energy landscapes for nucleosome deposition. Nat Struct Mol Biol 16(10):1056–1062. doi:10.1038/nsmb.1655
112. Gorman J, Chowdhury A, Surtees JA, Shimada J, Reichman DR, Alani E, Greene EC (2007) Dynamic basis for one-dimensional DNA scanning by the mismatch repair complex Msh2-Msh6. Mol Cell 28(3):359–370. doi:10.1016/j.molcel.2007.09.008
113. Prasad TK, Robertson RB, Visnapuu ML, Chi P, Sung P, Greene EC (2007) A DNA-translocating Snf2 molecular motor: Saccharomyces cerevisiae Rdh54 displays

processive translocation and extrudes DNA loops. J Mol Biol 369(4):940–953. doi:10.1016/j.jmb.2007.04.005
114. Finkelstein IJ, Visnapuu ML, Greene EC (2010) Single-molecule imaging reveals mechanisms of protein disruption by a DNA translocase. Nature 468(7326):983–987. doi:10.1038/nature09561
115. Chisty LT, Toseland CP, Fili N, Mashanov GI, Dillingham MS, Molloy JE, Webb MR (2013) Monomeric PcrA helicase processively unwinds plasmid lengths of DNA in the presence of the initiator protein RepD. Nucleic Acids Res 41(9):5010–5023. doi:10.1093/nar/gkt194
116. Nitzsche B, Bormuth V, Brauer C, Howard J, Ionov L, Kerssemakers J, Korten T, Leduc C, Ruhnow F, Diez S (2010) Studying kinesin motors by optical 3D-nanometry in gliding motility assays. Methods Cell Biol 95:247–271. doi:10.1016/S0091-679X(10)95014-0
117. Leduc C, Ruhnow F, Howard J, Diez S (2007) Detection of fractional steps in cargo movement by the collective operation of kinesin-1 motors. Proc Natl Acad Sci USA 104 (26):10847–10852. doi:10.1073/pnas.0701864104
118. Howard J, Hudspeth AJ, Vale RD (1989) Movement of microtubules by single kinesin molecules. Nature 342(6246):154–158. doi:10.1038/342154a0
119. Korten T, Nitzsche B, Gell C, Ruhnow F, Leduc C, Diez S (2011) Fluorescence imaging of single Kinesin motors on immobilized microtubules. Methods Mol Biol 783:121–137. doi:10.1007/978-1-61779-282-3_7
120. Pierce DW, Hom-Booher N, Vale RD (1997) Imaging individual green fluorescent proteins. Nature 388(6640):338. doi:10.1038/41009
121. Vale RD, Funatsu T, Pierce DW, Romberg L, Harada Y, Yanagida T (1996) Direct observation of single kinesin molecules moving along microtubules. Nature 380(6573):451–453. doi:10.1038/380451a0
122. Helenius J, Brouhard G, Kalaidzidis Y, Diez S, Howard J (2006) The depolymerising kinesin MCAK uses lattice diffusion to rapidly target microtubule ends. Nature 441(7089):115–119. doi:10.1038/nature04736
123. Varga V, Helenius J, Tanaka K, Hyman AA, Tanaka TU, Howard J (2006) Yeast kinesin-8 depolymerizes microtubules in a length-dependent manner. Nat Cell Biol 8(9):957–962. doi:10.1038/ncb1462
124. Dixit R, Ross JL, Goldman YE, Holzbaur EL (2008) Differential regulation of dynein and kinesin motor proteins by tau. Science 319(5866):1086–1089. doi:10.1126/science.1152993
125. Korten T, Diez S (2008) Setting up roadblocks for kinesin-1: mechanism for the selective speed control of cargo carrying microtubules. Lab Chip 8(9):1441–1447. doi:10.1039/b803585g
126. Telley IA, Bieling P, Surrey T (2009) Obstacles on the microtubule reduce the processivity of Kinesin-1 in a minimal in vitro system and in cell extract. Biophys J 96(8):3341–3353. doi:10.1016/j.bpj.2009.01.015
127. Reck-Peterson SL, Yildiz A, Carter AP, Gennerich A, Zhang N, Vale RD (2006) Single-molecule analysis of dynein processivity and stepping behavior. Cell 126(2):335–348. doi:10.1016/j.cell.2006.05.046
128. Brawley CM, Rock RS (2009) Unconventional myosin traffic in cells reveals a selective actin cytoskeleton. Proc Natl Acad Sci USA 106(24):9685–9690. doi:10.1073/pnas.0810451106
129. Sivaramakrishnan S, Spudich JA (2009) Coupled myosin VI motors facilitate unidirectional movement on an F-actin network. J Cell Biol 187(1):53–60. doi:10.1083/jcb.200906133
130. Heuser J (2000) The production of 'cell cortices' for light and electron microscopy. Traffic 1 (7):545–552
131. Moore MS, Mahaffey DT, Brodsky FM, Anderson RG (1987) Assembly of clathrin-coated pits onto purified plasma membranes. Science 236(4801):558–563
132. Clarke M, Schatten G, Mazia D, Spudich JA (1975) Visualization of actin fibers associated with the cell membrane in amoebae of Dictyostelium discoideum. Proc Natl Acad Sci USA 72(5):1758–1762

133. Heuser JE, Anderson RG (1989) Hypertonic media inhibit receptor-mediated endocytosis by blocking clathrin-coated pit formation. J Cell Biol 108(2):389–400
134. Mahaffey DT, Moore MS, Brodsky FM, Anderson RG (1989) Coat proteins isolated from clathrin coated vesicles can assemble into coated pits. J Cell Biol 108(5):1615–1624
135. Mahaffey DT, Peeler JS, Brodsky FM, Anderson RG (1990) Clathrin-coated pits contain an integral membrane protein that binds the AP-2 subunit with high affinity. J Biol Chem 265 (27):16514–16520
136. Wu M, De Camilli P (2012) Supported native plasma membranes as platforms for the reconstitution and visualisation of endocytic membrane budding. Methods Cell Biol 108:3–18. doi:10.1016/B978-0-12-386487-1.00001-8
137. Wu M, Huang B, Graham M, Raimondi A, Heuser JE, Zhuang X, De Camilli P (2010) Coupling between clathrin-dependent endocytic budding and F-BAR-dependent tubulation in a cell-free system. Nat Cell Biol 12(9):902–908. doi:10.1038/ncb2094

Chapter 8
Fluorescence Methods in the Investigation of the DEAD-Box Helicase Mechanism

Alexandra Z. Andreou and Dagmar Klostermeier

Abstract DEAD-box proteins catalyze the ATP-dependent unwinding of RNA duplexes and accompany RNA molecules throughout their cellular life. Conformational changes in the helicase core of DEAD-box proteins are intimately linked to duplex unwinding. In the absence of ligands, the two RecA domains of the helicase core are separated. ATP and RNA binding induces a closure of the cleft between the RecA domains that is coupled to the distortion of bound RNA, leading to duplex destabilization and dissociation of one RNA strand. Reopening of the helicase core occurs after ATP hydrolysis and is coupled to phosphate release and dissociation of the second RNA strand.

Fluorescence spectroscopy provides an array of approaches to study intermolecular interactions, local structural rearrangements, or large conformational changes of biomolecules. The fluorescence intensity of a fluorophore reports on its environment, and fluorescence anisotropy reflects the size of the molecular entity the fluorophore is part of. Fluorescence intensity and anisotropy are therefore sensitive probes to report on binding and dissociation events. Fluorescence resonance energy transfer (FRET) reports on the distance between two fluorophores and thus on conformational changes. Single-molecule FRET experiments reveal the distribution of conformational states and the kinetics of their interconversion. This chapter summarizes fluorescence approaches for monitoring individual aspects of DEAD-box protein activity, from nucleotide and RNA binding and RNA unwinding to protein and RNA conformational changes in the catalytic cycle, and illustrates exemplarily how fluorescence-based methods have contributed to understanding the mechanism of DEAD-box helicase-catalyzed RNA unwinding.

Keywords DEAD-box • Helicase • DNA • FRET • RNA

A.Z. Andreou • D. Klostermeier (✉)
University of Muenster, Institute for Physical Chemistry, Corrensstrasse 30, 48149 Muenster, Germany
e-mail: aandr_01@uni-muenster.de; dagmar.klostermeier@uni-muenster.de

C.P. Toseland and N. Fili (eds.), *Fluorescent Methods for Molecular Motors*, Experientia Supplementum 105, DOI 10.1007/978-3-0348-0856-9_8,

Abbreviations

2-AP	2-Aminopurine
BSA	Bovine serum albumin
CCD	Charge-coupled device
FCS	Fluorescence correlation spectroscopy
FRET	Fluorescence resonance energy transfer
Mant	2′(3′)-O-(*N*-Methylanthraniloyl)-
PEG	Polyethylene glycol
RBD	RNA binding domain
RRM	RNA recognition motif
TIRF	Total internal reflection fluorescence
TNP	Trinitrophenyl
MDCC	*N*-[2-(1-Maleimidyl)ethyl]-7-diethylaminocoumarin-3-carboxamide
smFRET	Single-molecule fluorescence resonance energy transfer

8.1 ATP-Dependent RNA Unwinding by DEAD-box Helicases

RNA is a key player in a multitude of cellular processes. Large RNA molecules form double-stranded regions and fold into stable three-dimensional structures, leading to a propensity to misfold [1]. RNA helicases unwind RNA duplex regions in an ATP-dependent reaction (reviewed in [2–4]). They accompany RNA molecules in all processes throughout their cellular life and are involved in RNA transcription, splicing, translation, and degradation (reviewed in [5]). DEAD-box proteins are the largest RNA helicase family (reviewed in [6, 7]). Members of the DEAD-box family share a set of helicase signature motifs that confer ATP binding and hydrolysis activity and RNA binding and unwinding [8] and derive their name from one of these motifs, a conserved Asp-Glu-Ala-Asp sequence (DEAD-box [9]). The helicase motifs are arranged on a ~400-amino-acid helicase core formed by two RecA-like domains. In the absence of nucleotide and RNA substrates, these two domains are separated [10–15] (Fig. 8.1). In contrast, they extensively interact in the closed conformation of the ATP- and RNA-bound state [11, 15–23] (Fig. 8.1), aligning the conserved motifs from both domains to form the ATPase and the RNA binding sites (reviewed in [24, 25]). The conformational transitions of the helicase core have been mapped in single-molecule FRET (smFRET) experiments [11, 26, 27] and are intimately linked to the local duplex unwinding activity of DEAD-box proteins (reviewed in [28]). The transition of the DEAD-box helicase core to the closed conformation distorts the backbone of a bound RNA duplex [20], and the resulting destabilization facilitates release of one RNA strand. Reopening of the helicase core after ATP hydrolysis, upon phosphate release, then disrupts the

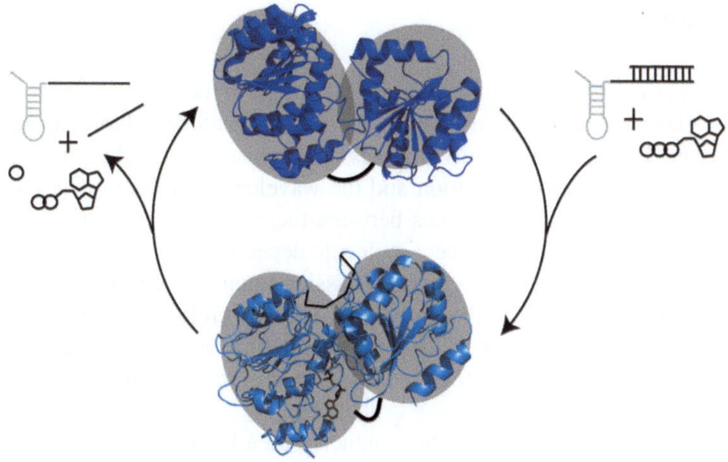

Fig. 8.1 Conformational changes in the catalytic cycle of DEAD-box proteins. The helicase core of DEAD-box proteins consists of two RecA-like domains (*ovals*), connected by a flexible linker. In the absence of nucleotide and RNA, the RecA domains are separated, and the helicase core is in an open conformation (*top*). Cooperative binding of ATP and RNA (*cartoons*) promotes a conformational change to a closed state (*bottom*). In this compact conformation, the two RecA-like domains interact, and the conserved motifs are aligned to form the nucleotide and RNA binding sites. The RNA bound to the closed helicase core is kinked, leading to a local destabilization of the duplex that ultimately leads to its dissociation. Reopening of the helicase core occurs after ATP hydrolysis, upon phosphate release. The open state is represented by the crystal structure of *M. jannaschii* DeaD ([13], PDB-ID 1hv8), the closed state by the crystal structure of Dbp5 ([19], PDB-ID 3pew). A 32/9mer RNA derived from 23S rRNA, which has been used as minimal unwinding substrate for DEAD-box helicases YxiN, Hera, and DbpA, is depicted as a representative RNA substrate, with the duplex in *black* and the hairpin that is bound by the C-terminal domains of these helicases in *gray*

bipartite RNA binding site formed by both RecA domains. Dissociation of the second RNA strand completes duplex separation and resets the enzyme for subsequent catalytic turnovers (reviewed in [3, 29]).

Fluorescence-based approaches have been instrumental in unraveling the mechanism of RNA unwinding by DEAD-box proteins. Fluorescently labeled nucleotides and RNAs are readily commercially available, and advances in instrumentation have made fluorescence methods widely accessible. This chapter summarizes fluorescence approaches that have been used in our own and other laboratories to follow individual events during RNA unwinding by DEAD-box proteins, ranging from nucleotide and RNA substrate binding to duplex separation, nucleotide-dependent conformational changes in the catalytic cycle, and structural changes in their RNA substrates.

8.2 Fluorescence

Upon irradiation with light, molecules are excited from their electronic ground state S_0 to the electronically excited state S_1. Fluorescent moieties (fluorophores) return to the ground state in a radiative process, by emission of light (Fig. 8.2). The wavelengths required for excitation and the wavelength of fluorescence emission depend on the energetic differences between the S_0 and S_1 states (Fig. 8.2). The excitation efficiency of a fluorescent molecule depends on its extinction coefficient ε at the excitation wavelength λ_{ex}. The emission intensity depends on the fluorescence quantum yield ϕ, i.e., the number of photons emitted per excitation event ($0 < \phi < 1$), and the lifetime τ, i.e., the mean time a molecule spends in the electronically excited state before returning to the ground state (typically a few nanoseconds).

Fluorescence experiments can be conducted on a large ensemble of molecules, leading to averaged fluorescence intensities, or at the single-molecule level, revealing fluorescence intensities from single fluorophores and intensity distributions over a large number of individually observed molecules. Ensemble fluorescence experiments are performed with commercial spectrofluorimeters. A xenon lamp is used as light source for excitation, monochromators allow for the selection of the excitation and emission wavelengths, and fluorescence is detected by a photomultiplier. Fluorescence emission spectra are measured by exciting fluorescence at a constant excitation wavelength λ_{ex} and scanning the emission wavelength λ_{em}. For excitation spectra, the emission is measured at a fixed wavelength length λ_{em}, and the excitation wavelength λ_{ex} is varied. Fluorescence lifetimes can be determined in time-resolved fluorescence experiments using pulsed excitation sources and a microchannel plate detector. Fluorescence decays are constructed from histograms of photon arrival times (time-correlated single-photon counting [30]). Decay curves can be described by an exponential function, where the time constant is the fluorescence lifetime. Quantum yields can be determined relative to standard molecules with known quantum yields [31, 32].

Single-molecule experiments are performed using fluorescence microscopes with laser excitation (Sect. 8.2.3.2). In confocal microscopy, the fluorescence of freely diffusing molecules in solution is detected by avalanche photodiodes. By total internal reflection microscopy, fluorescence of surface-immobilized molecules is detected by imaging using a charge-coupled device (CCD) camera.

8.2.1 *Fluorescence Intensity as a Probe of Fluorophore Environment*

The fluorescence intensity is determined by the extinction coefficient ε and by the fluorescence quantum yield ϕ and lifetime τ. While the extinction coefficient shows little dependence on the fluorophore surroundings, quantum yield and lifetime are

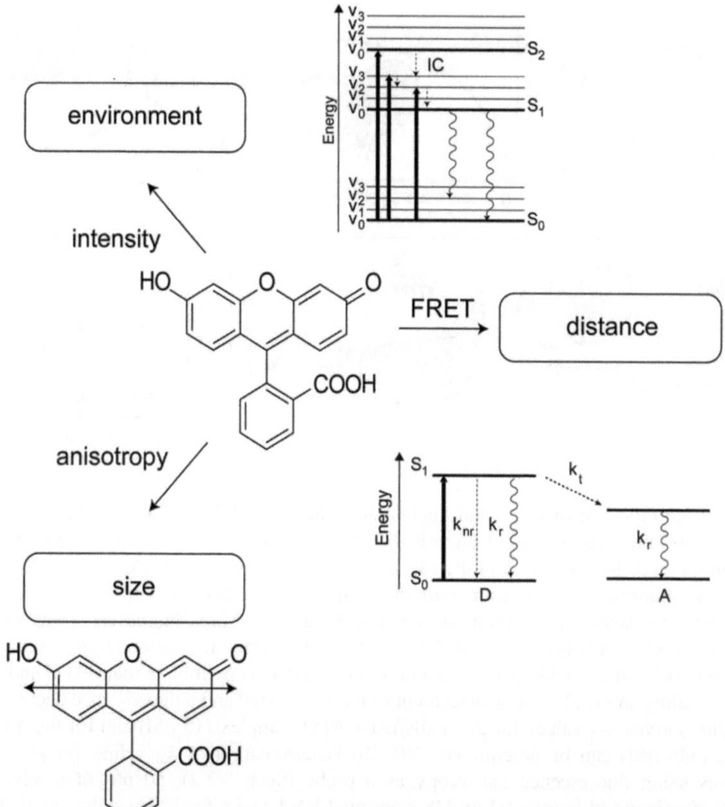

Fig. 8.2 Different fluorescence read-outs reporting on molecular events. *Center*: The fluorescence of a reporter fluorophore, here fluorescein as an example, reports on the fluorophore environment (fluorescence intensity), on its distance to a second fluorophore (FRET), or on the size of the molecular entity (fluorescence anisotropy). *Top right*: Jablonski diagram summarizing the transitions between energy levels that are relevant for fluorescence. Absorption of light by a molecule in the electronic ground state S_0 induces a transition to a higher electronic state S_1 or S_2 (*thick arrows*). V_0–V_3 (*thin lines*) depict vibrational sub-states of each electronic state (*thick line*). An excited molecule rapidly relaxes to S_1 via non-radiative processes (internal conversion, IC) and to the lowest vibrational level of S_1 (*broken arrows*) before it returns to the ground state with emission of fluorescent light (*undulating arrows*). The nanosecond fluorescence lifetime of the excited state renders fluorescence a sensitive reporter on the fluorophore environment. *Bottom right*: Jablonski diagram for fluorescence resonance energy transfer (FRET) between a donor (D) and an acceptor fluorophore (A). Upon excitation (*thick arrow*), the donor fluorophore can relax from the excited state to the ground state by non-radiative processes (k_{nr}, *broken arrow*) or by fluorescence (k_r, *undulating arrow*). Alternatively, it can transfer its excitation energy to a suitable second fluorophore via dipole–dipole interactions (k_t, *broken arrow*). The acceptor fluorophores will then emit fluorescence at a wavelength red-shifted compared to the excitation wavelength (*undulating arrow*). The FRET efficiency depends on the distance between the two fluorophores. FRET measurements can thus serve as a molecular ruler. *Bottom left*: Upon excitation with linearly polarized light, fluorophores whose transition dipole (*arrow*) is parallel to the polarization direction are excited preferentially. Fixed fluorophores then emit polarized fluorescence light.

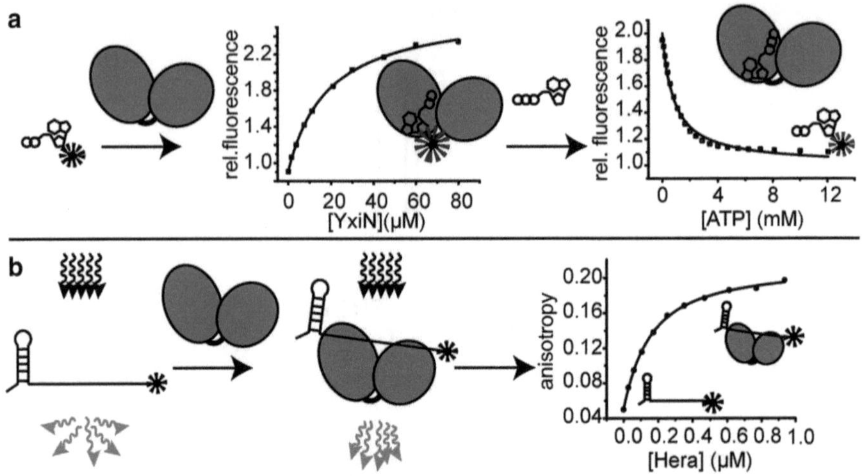

Fig. 8.3 Determination of RNA and nucleotide affinities of DEAD-box proteins from fluorescence equilibrium titrations. (**a**) Determination of K_d values for protein/nucleotide complexes using fluorescence intensity as a probe (Sect. 8.3.1). In a titration of 1 μM mant-ADP (*cartoon, black sphere*: fluorescent mant-group) with increasing concentrations of YxiN (*cartoon*), the mant-fluorescence increases until saturation is reached (*center*). Mant-fluorescence was excited at 365 nm and emission was detected at 440 nm. In a displacement titration (*right*), a mant-ADP/YxiN complex is titrated with increasing amounts of ATP. Competition of mant-ADP and ATP for the same binding site leads to a displacement of the mant-ADP and a fluorescence decrease. From the titration curves, K_d values for the YxiN/mant-ADP complex (15 μM) and for the YxiN/ATP complex (346 μM) can be determined [59]. (**b**) Determination of K_d values for protein/RNA complexes using fluorescence anisotropy as a probe (Sect. 8.3.2). 50 nM of a 32mer RNA substrate, consisting of hairpin 92 of 23S ribosomal RNA and a flanking single-stranded region, and labeled with fluorescein at the 5′-end (*black sphere*), was titrated with Hera [73]. The fluorescein anisotropy increases upon complex formation until saturation is reached. Fluorescein was excited with linearly polarized light (indicated by the *black undulated arrows*) at 495 nm and fluorescence anisotropy (*gray undulated arrows*) was detected at 530 nm. The K_d value of the Hera/RNA complex determined from the titration curve is 0.13 μM [73]

strongly dependent on the local environment. Fluorescence intensity thus serves as a sensitive spectroscopic probe for changes in the fluorophore environment upon ligand binding, local structural rearrangements, or large conformational changes (Figs. 8.2, 8.3a, 8.4b).

The fluorescence intensity can be measured in standard fluorescence spectrometers, either integrated over time at a fixed excitation and emission wavelength or with spectral resolution. In the first case, the optimal excitation and emission

---

Fig. 8.2 (continued) However, during the fluorescence lifetime the orientation of these fluorophores will change due to rotational diffusion. The degree of randomization and the resulting anisotropy of the emitted light depends on the tumbling time and thus on the size of the molecule carrying the fluorophore. Fluorescence anisotropy serves as a probe for binding events

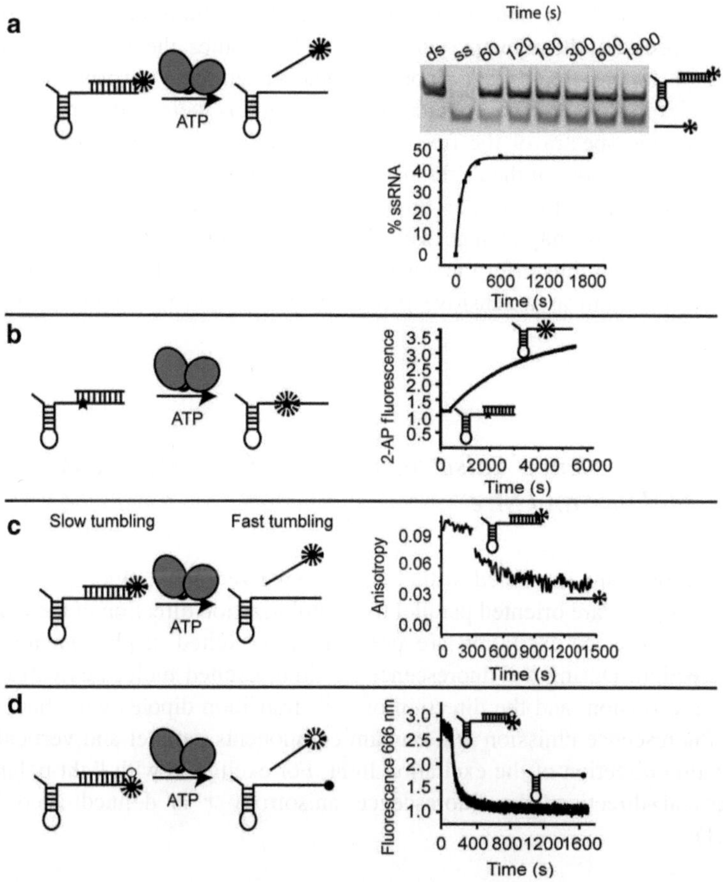

Fig. 8.4 Fluorescence-based methods to monitor RNA unwinding. (**a**) Discontinuous RNA helicase assay (Sect. 8.3.3.1). A 32/9mer duplex RNA (5 μM, *cartoon*), labeled with fluorescein (*black sphere*) at the 3′-end of the 9mer, was incubated with 10 μM of a motif III mutant (SAT to AAA) of YxiN in the presence of ATP. Unwinding reactions were stopped at different time points, and duplex and single-stranded RNA were separated by native gel electrophoresis (*top right*). The fraction of released single strand was determined at each time point (*bottom right*), and a rate constant of unwinding of $k_{\text{unwind}} = 12 \times 10^{-3}\ \text{s}^{-1}$ was determined [26]. (**b**) RNA unwinding monitored by 2-AP fluorescence (Sect. 8.3.3.2). 2-AP fluorescence (*black star*) is quenched when it is part of a 32mer/9mer duplex RNA. In the presence of ATP, eIF4A unwinds the RNA duplex, leading to an increase in 2-AP emission. The intrinsic RNA helicase activity of eIF4A is very low, and the rate constant for unwinding is $k_{\text{unwind}} = 0.2 \times 10^{-3}\ \text{s}^{-1}$ [92]. (**c**) RNA unwinding monitored by fluorescence anisotropy (Sect. 8.3.3.3). The fluorescence anisotropy of fluorescein attached to the 3′-end of a 9mer RNA (50 nM) is low (fast tumbling). The slower tumbling of the fluorophores upon duplex formation with a complementary 32mer (100 nM) leads to an increase in fluorescein anisotropy. Upon addition of ATP to a YxiN/RNA complex, duplex unwinding and release of the 9mer can be followed by a decrease in anisotropy. The rate constant for unwinding is $k_{\text{unwind}} = 4 \times 10^{-3}\ \text{s}^{-1}$ [59]. (**d**) RNA unwinding monitored by FRET (Sect. 8.3.3.4). A duplex RNA substrate formed by annealing of a Cy3-labeled (*white*) 9mer and a Cy5-labeled (*black*) 32mer exhibits high acceptor fluorescence due to FRET. In the presence of Hera (5 μM) and ATP (2 mM), the duplex is unwound, leading to a decrease in acceptor fluorescence and FRET. The rate constant for unwinding k_{unwind} is $17 \times 10^{-3}\ \text{s}^{-1}$ (plus a slow phase of $3 \times 10^{-3}\ \text{s}^{-1}$) [73]

wavelengths should be determined beforehand from fluorescence excitation and emission spectra of the relevant states. For binding studies, the maximum excitation wavelength for the fluorescent reporter is determined from absorption or fluorescence excitation spectra. This excitation wavelength is then used to record fluorescence emission spectra of the free and bound states of the fluorescently labeled ligand. The maximum of the difference between these reference spectra defines the optimal wavelength to monitor binding. In addition to intensity changes, the emission maximum may change with the local environment. Measurements with spectral resolution thus offer the advantage to monitor reactions using the shift in emission maximum as a sensitive probe for changes in the environment of the reporter.

8.2.2 *Fluorescence Anisotropy as a Probe of Molecular Mobility and Size*

When samples are irradiated with linearly polarized light, fluorophores whose transition dipoles are oriented parallel to the polarization direction of the excitation light at the time of excitation are preferentially excited, a phenomenon called photoselection. During the fluorescence lifetime, excited molecules will undergo rotational diffusion, and the direction of their transition dipoles will change. As a result, fluorescence emission will contain components parallel and vertical to the polarization direction of the excitation light. For excitation with light polarized in the vertical direction, the fluorescence anisotropy r is defined according to Eq. (8.1):

$$r = \frac{I_{VV} - I_{VH}}{I_{VV} + 2I_{VH}} \tag{8.1}$$

with the vertical and horizontal components of the emitted fluorescent light, I_{VV} and I_{VH}, respectively. To measure fluorescence anisotropy, instruments with emission and excitation polarizers are required to select for the vertical polarization of the excitation light and to measure horizontally and vertically polarized components of the emitted light. In the so-called L-format, the intensities I_{VV} and I_{VH} are measured sequentially on a single detector, alternating the orientation of the emission polarizer. In the T-format, the two components are measured in parallel using two (fixed) emission polarizers and two detectors.

A fixed fluorophore (with parallel emission and excitation transition dipoles) that does not undergo rotational diffusion during the excited state lifetime has a fluorescence anisotropy value of 0.4. Rotational diffusion of the fluorophore, or of the molecule it is attached to, leads to a decrease of the anisotropy. Hence, fluorescence anisotropy is a measure for the size of the molecule or complex (Fig. 8.2). Molecules binding to the fluorophore, or to the molecule it is attached to, decrease the rotation rate of the fluorophore and increase the fluorescence anisotropy. Anisotropy is thus a convenient probe for monitoring binding and dissociation events. To maximize the anisotropy increase upon binding, the smaller interaction partner should carry the fluorescent reporter.

8.2.3 Fluorescence Resonance Energy Transfer as a Probe of Inter-dye Distance

A single excited fluorophore returns to the ground state by fluorescence emission. In the presence of a suitable second fluorophore, however, fluorescence resonance energy transfer (FRET) from the first (donor) to the second (acceptor) fluorophore can occur through coupling of their transition dipoles (Fig. 8.2). The efficiency of this process depends on the inter-fluorophore distance [33]. The distance dependence allows the FRET efficiency to be used as a molecular ruler. With commonly used fluorophores, distances in the range of 2–8 nm can be measured.

The rate constant k_t of the energy transfer is given by Eq. (8.2) [33]:

$$k_t = \frac{1}{\tau_D}\left(\frac{R_0}{r_{DA}}\right)^6 \tag{8.2}$$

where r_{DA} is the distance between donor and acceptor, τ_D is the donor fluorescence lifetime in absence of the acceptor, and R_0 is the so-called Förster distance that depends on the spectral properties of the donor and acceptor fluorophores and represents the donor/acceptor distance at which the FRET efficiency is 50 %.

The energy transfer efficiency E results from the competition of FRET with all other processes that depopulate the excited state of the donor (Fig. 8.2). The FRET efficiency can be expressed as the ratio of the energy transfer rate constant k_t and the sum of the rate constants for deactivation of the excited state by energy transfer, by non-radiative (k_{nr}) and radiative processes (k_r) [Eq. (8.3)]:

$$E = \frac{k_t}{k_t + k_r + k_{nr}} \tag{8.3}$$

By combining Eqs. (8.2) and (8.3), and with $k_r + k_{nr} = \tau_d^{-1}$, the distance dependence of the FRET efficiency is obtained as

$$E = \frac{R_0^6}{R_0^6 + r_{DA}^6} \tag{8.4}$$

Experimentally, the transfer efficiency E can be determined from the donor and acceptor fluorescence intensities, I_D and I_A, upon excitation of the donor after Eq. (8.5):

$$E = \frac{I_A}{I_D + I_A} \tag{8.5}$$

This experimentally determined FRET efficiency can be converted into a donor/acceptor distance using Eq. (8.4) and the Förster distance for the donor/acceptor pair. A quantitative interpretation of FRET efficiencies as a molecular ruler requires correction of the measured intensities for instrument non-idealities (reviewed in [28], Sects. 8.2.3.3 and 8.3.6).The FRET efficiency also serves as a valuable qualitative probe for processes that alter the distances between donor and acceptor dyes. Fluorophores can be attached to two different molecules (intermolecular FRET) to monitor binding or to two sites within the same molecule (intramolecular FRET) to follow conformational changes (Sects. 8.3.4 and 8.3.5).

8.2.3.1 Ensemble FRET

For ensembles of molecules, the FRET efficiency can be determined from fluorescence spectra of molecules that carry a donor and an acceptor fluorophore. The fractional contributions f_D and f_A of donor and acceptor fluorescence to the measured spectrum $I_{DA}(\lambda)$ can be determined by describing the spectra as a linear combination of the reference spectra from molecules carrying only a donor, $I_D(\lambda)$, and carrying only the acceptor, $I_A(\lambda)$ [Eq. (8.6)]

$$I_{DA}(\lambda) = f_D \cdot I_D(\lambda) + f_A \cdot I_A(\lambda) \tag{8.6}$$

The FRET efficiency can be calculated according to Eq. (8.5), using the obtained coefficients f_D and f_A as a measure for I_D and I_A. Alternatively, FRET efficiencies can be calculated from donor quenching or from sensitized emission of the acceptor. For the calculation from donor quenching, the donor fluorescence in the absence of the acceptor, I_D, is determined from the emission spectrum of the molecules that carry only the donor fluorophore. The fluorescence of the donor in the presence of the acceptor, I_{DA}, is determined from the fluorescence spectrum of the molecules carrying both donor and acceptor. The FRET efficiency E is then calculated after Eq. (8.7):

$$E = 1 - \frac{I_{DA}}{I_D} \tag{8.7}$$

For the calculation of the FRET efficiency from sensitized acceptor emission, the emission spectra of acceptor-labeled molecules upon excitation of the donor have to be measured in the absence and presence of the donor. The FRET efficiency E is calculated according to Eq. (8.8):

$$E = \frac{1}{f_D} \cdot \frac{\varepsilon_A}{\varepsilon_D} \left(\frac{I_{AD}}{I_A} - 1 \right) \tag{8.8}$$

Here, f_D denotes the fraction of molecules carrying only a donor moiety, ε_A and ε_D are the extinction coefficients of the acceptor and donor fluorophores at the excitation wavelength of the donor, and I_{AD} and I_A are the fluorescence intensities of the acceptor in the presence and absence of the donor fluorophore, respectively. The acceptor emission has to be corrected for contributions from the fluorescence of the donor, using a reference spectrum from molecules carrying only a donor.

The FRET efficiency determined from fluorescence spectra represents a population-weighted average over all molecules in the sample. For heterogeneous samples, it will thus not represent a true intramolecular distance. This limitation can be overcome by determining FRET efficiencies from fluorescence lifetimes [34] and by performing FRET experiments at the single-molecule level.

8.2.3.2 smFRET Microscopy

Single-molecule fluorescence experiments offer the advantage to determine the donor and acceptor fluorescence intensities for individual molecules without averaging over different populations. They directly reveal the distributions of FRET efficiencies over all molecules in the sample that reflect the different populations present.

Fluorescence Correlation Spectroscopy and Confocal Microscopy

In confocal microscopy, a laser beam is focused with a high numerical aperture objective into a very small spot, the confocal volume (~1 femtoliter) (Fig. 8.5a). At low concentrations of the fluorescent species, only a few molecules will be present within this excitation volume at any given time. Diffusion of individual molecules in and out of the confocal volume will lead to fluctuations in the number of molecules present. Fluorescence from this volume, collected by the same objective and detected by avalanche photodiodes, will therefore also fluctuate. From the fluorescence as a function of time, it is possible to calculate the intensity fluctuations at times t and $t+\tau$ from the mean value F(t), $\delta F(t)$ and $\delta F(t+\tau)$. The autocorrelation function $G(\tau)$ can then be calculated after Eq. (8.9):

$$G(\tau) = \frac{\langle \delta F(t) \cdot \delta F(t+\tau) \rangle}{\langle F(t) \rangle^2} \tag{8.9}$$

The amplitude of the autocorrelation function depends on the inverse of the average number of fluorescent molecules within the confocal volume and thus on their concentration. The inflection point corresponds to the diffusion time τ_{diff}, i.e.,

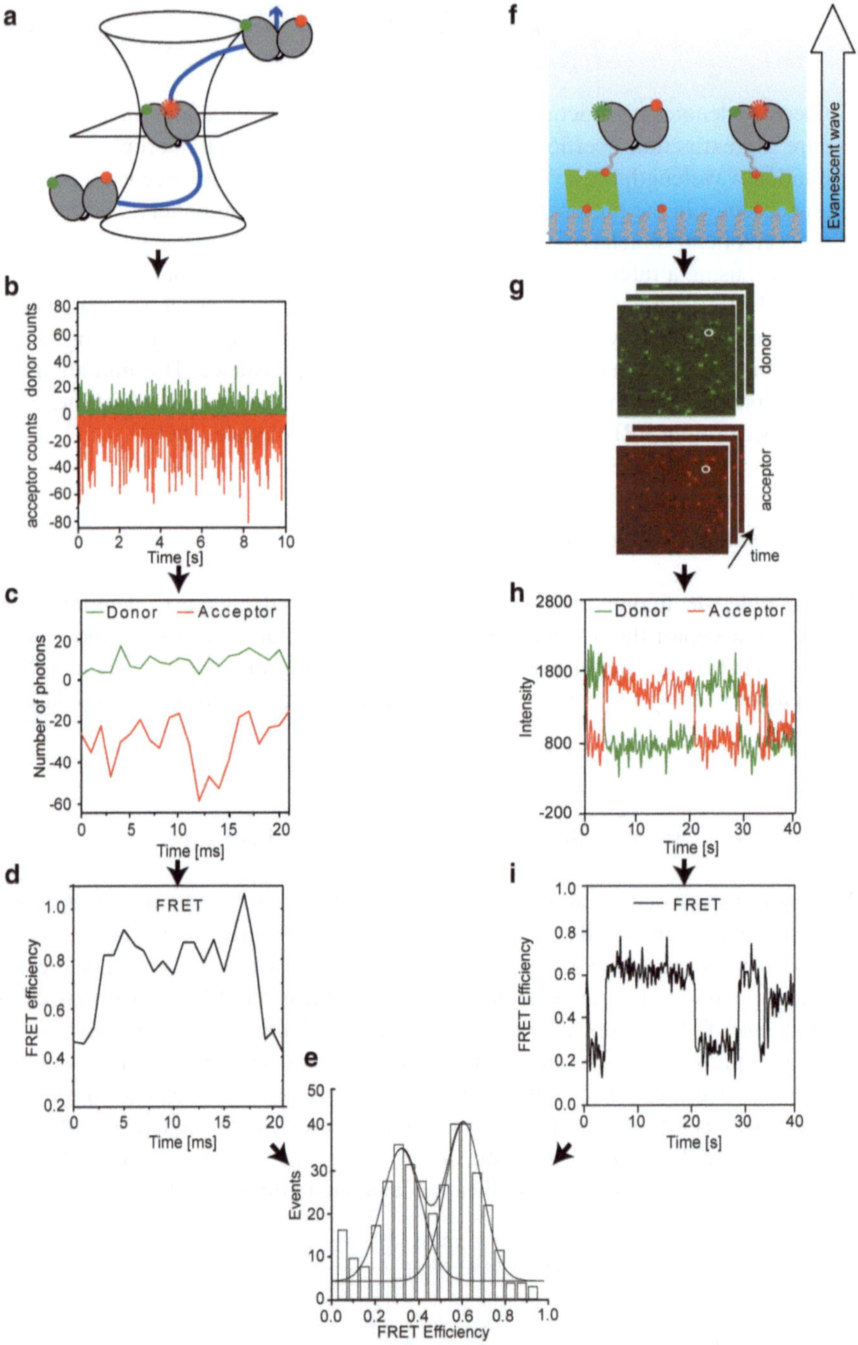

Fig. 8.5 Monitoring conformational changes of DEAD-box proteins by single-molecule FRET. (**a–e**) Single-molecule FRET on freely diffusing molecules using confocal microscopy. (**a**) A DEAD-box protein (*cartoon*) labeled with a donor (*green*) and acceptor fluorophore (*red*) is diffusing through the confocal volume (*double cone*). If the two domains are in close proximity,

the mean time a molecule spends in the confocal volume. The diffusion time τ_{diff} depends on the size of the species carrying the fluorophore and can thus be used as a probe for binding (Sect. 8.3.2).

An even lower concentration of the fluorescently labeled molecule (~50–100 pM) minimizes the probability that more than one molecule is present in the confocal volume at any given time. Freely diffusing single molecules traversing the confocal volume lead to bursts of fluorescence that are collected by the objective, split into donor and acceptor spectral regions, and detected by two avalanche photodiodes (Fig. 8.5a). The FRET efficiency for individual molecules is calculated from donor and acceptor intensities integrated over each burst (Fig. 8.5b, c, d). Histograms of FRET efficiencies, observed molecule by molecule, directly reveal the FRET efficiency distributions and the populations of individual molecules with different FRET efficiencies and thus allow for the determination of the relative populations of these species (Fig. 8.5e). In principle, FRET can be calculated as function of time during fluorescence bursts. The observation time is limited by the millisecond diffusion time of molecules through the confocal volume, however, and large-scale conformational changes are unlikely to occur during the duration of a fluorescence burst.

Total Internal Reflection Microscopy

Longer observation times of individual molecules can be achieved by immobilizing donor-/acceptor-labeled molecules on a functionalized slide and imaging of the surface with confocal scanning, wide-field or total internal reflection fluorescence

---

Fig. 8.5 (continued) FRET will occur and acceptor fluorescence (*red*) will be detected. (**b**) Raw data obtained are donor (*green*) and acceptor (*red*) fluorescence time traces from single molecules traversing through the confocal volume. For clarity, the acceptor signal has been inverted. Each burst of fluorescence is caused by a single molecule traversing the confocal volume. (**c**) Time trace of donor and acceptor intensities during a fluorescence burst. (**d**) Time-dependent changes in FRET efficiency during a fluorescence burst. (**e**) The histogram of FRET efficiencies, calculated burst-wise for individual molecules from donor and acceptor fluorescence intensities, or from images of surface-immobilized molecules (see **f–i**), unveils different conformations that are present in equilibrium. (**f–i**) Single-molecule FRET on surface-immobilized molecules using TIRF microscopy. (**f**) Schematic representation of a biotinylated (*orange sphere*), donor-/acceptor-labeled (*green*/*red*) DEAD-box protein, immobilized on a streptavidin-functionalized (*light green*), PEG-passivated (*gray*) glass surface. An evanescent field (*light blue gradient*) is generated upon total internal reflection of the excitation light at the glass–water interface. Fluorescence excitation is selective for molecules that are close to the surface and within the evanescent field. (**g**) Images of molecules on the surface in terms of donor (*green*) and acceptor (*red*) fluorescence. Each single molecule is detected as a spot, with a brightness that corresponds to the donor and acceptor fluorescence intensity. (**h**) Donor (*green*) and acceptor (*red*) fluorescence time traces of an immobilized single molecule, obtained from a series of images taken at different time points (*arrow* in **g**). (**i**) Time trace of the FRET efficiency, showing switching of the molecule between a low-FRET state ($E = 0.2$) and a high-FRET state ($E = 0.6$)

(TIRF) microscopy. TIRF microscopy allows for selective excitation of molecules near the surface (Fig. 8.5f). The laser excitation beam, either focused off axis into the back-focal plane of a high numerical aperture objective lens (objective-type TIRF) or passed through a prism (prism-type TIRF), impinges on the glass–water interface at an incident angle larger than the critical angle θ_C. Under these conditions, the beam undergoes total internal reflection, with the concomitant generation of an evanescent wave on the distant side of the interface within the sample. The intensity of the evanescent wave decreases exponentially with increasing distance from the surface, allowing for selective excitation of surface-immobilized molecules. Donor and acceptor fluorescence from these molecules is collected by an objective and imaged on adjacent regions of a CCD camera.

The donor and acceptor intensities for each molecule can be determined from the brightness of the detected spots and are used to calculate FRET efficiencies and to construct FRET histograms (Fig. 8.5e, f, g). From a series of images determined at different time points, the FRET efficiency for individual molecules can be determined as a function of time, with observation times of seconds to minutes that are only limited by photobleaching of the dyes. FRET time traces directly reveal the switching between conformational states with different FRET efficiencies and the sequence in which these states occur. The analysis of dwell times of individual states allows for the determination of the kinetic constants for conformational transitions (Sect. 8.2.5).

Surface immobilization frequently takes advantage of the high-affinity biotin-streptavidin interaction. Quartz surfaces are functionalized with biotin, and streptavidin is coupled to the biotin moieties. Biotinylated biomolecules can then be immobilized on the streptavidin-functionalized surface. To avoid nonspecific interactions, the surface is passivated with BSA or PEG (reviewed in [35]).

8.2.3.3 Extracting Distance Information from FRET Efficiencies

FRET efficiencies can be used as a qualitative probe for conformational changes that change the inter-dye distances. Distances calculated from FRET efficiencies can also serve as distance restraints to model global conformations of large molecules that are not amenable to high-resolution structure determination or of transient species that may elude detection by other methods (Sect. 8.3.6). This section describes the procedure to convert corrected FRET efficiencies to inter-fluorophore distances. To account for instrument non-idealities, measured fluorescence intensities need to be corrected for cross talk between the donor and acceptor channels (α, β), different quantum yields and detection efficiencies (γ), and direct acceptor excitation (δ). A detailed description of the determination of these correction parameters is found in [28, 34]. The corrected FRET efficiency that reports on true inter-dye distances can be calculated after Eq. (8.10) [11]:

$$E_{\mathrm{FRET}} = \frac{(1+\beta\gamma\delta)\cdot\left(I_{\mathrm{A}} - \frac{\alpha+\gamma\delta}{1+\beta\gamma\delta}\cdot I_{\mathrm{D}}\right)}{(1+\beta\gamma\delta)\cdot\left(I_{\mathrm{A}} - \frac{\alpha+\gamma\delta}{1+\beta\gamma\delta}\cdot I_{\mathrm{D}}\right) + (\gamma+\gamma\delta)\cdot(I_{\mathrm{D}} - \beta I_{\mathrm{A}})} \tag{8.10}$$

To convert the corrected FRET efficiency into a distance, the Förster distance for the respective donor/acceptor pair has to be determined. The Förster distance depends on the spectral overlap of the donor emission and the acceptor absorbance, quantified in the overlap integral J [in M^{-1} cm^{-1} nm^4; see Eq. (8.12)], the relative orientation of their transition dipoles, expressed by the orientation factor κ^2, the fluorescence quantum yield of the donor ϕ_{D}, and the refractive index n of the medium [Eq. (8.11)]:

$$R_0^6 = 8.785 \times 10^{-11} \frac{\kappa^2 \phi_{\mathrm{D}} J}{n^4} \tag{8.11}$$

The Förster distance for a donor/acceptor pair depends on the local environment of the dyes. Therefore, it may differ for each construct used and may change in the presence of ligands.

The refractive index n in aqueous solution is usually approximated as 1.33 for water. The donor quantum yield is determined experimentally in comparison to standard fluorophores [31, 32]. The overlap integral is calculated from the normalized emission spectrum of the donor, $F_{\mathrm{D}}(\lambda)$, and the normalized acceptor emission spectrum $\varepsilon_{\mathrm{A}}(\lambda)$, by Eq. (8.12):

$$J = \int_0^\infty F_{\mathrm{D}}(\lambda)\cdot\varepsilon_{\mathrm{A}}(\lambda)\cdot\lambda^4 \mathrm{d}\lambda \approx \sum_i F_{\mathrm{D}}(\lambda_i)\cdot\varepsilon_{\mathrm{A}}(\lambda_i)\cdot\lambda_i^4 \tag{8.12}$$

The orientation factor κ^2 describes the relative orientation of the donor and acceptor transition dipoles and can be calculated from the angle between donor and acceptor dipole, θ_{DA}, and the angles between the vector connecting the two fluorophores and the donor and acceptor dipole, θ_{D} and θ_{A}, according to Eq. (8.13):

$$\kappa^2 = (\cos\theta_{\mathrm{DA}} - 3\cos\theta_{\mathrm{D}}\cdot\cos\theta_{\mathrm{A}})^2 \tag{8.13}$$

The orientation factor can adopt values between 0 and 4. The direction of transition dipoles is difficult to determine experimentally and therefore often unknown. If the dyes rotate freely, κ^2 is 2/3. To ensure their free rotation, donor and acceptor are usually coupled to the molecule of interest by long, flexible linkers. The rotational freedom of the donor and acceptor can be probed experimentally in anisotropy decays of fluorescently labeled molecules [11, 28]. The fluorophore is excited with pulsed, linearly polarized light, and the anisotropy decay is constructed from the vertically and horizontally polarized components of the emission, measured by time-correlated single-photon counting (Sect. 8.2.3.1).

Anisotropy decay curves of fluorophores attached to a biomolecule can often be described by a double exponential function:

$$r(t) = \beta_1 e^{-t/\tau_{c,1}} + \beta_2 e^{-t/\tau_{c,2}} \tag{8.14}$$

with the rotational correlation times $\tau_{c,1}$ and $\tau_{c,2}$ and the corresponding amplitudes β_1 and β_2. The two components of the anisotropy decay reflect the fast, segmental mobility of the fluorophore, characterized by a short correlation time $\tau_{c,1}$, and the global tumbling of the fluorophore with the macromolecule, with a larger correlation time $\tau_{c,2}$. From the amplitudes β_1 and β_2 of the two decay components, an order parameter S can be calculated according to

$$S = \sqrt{\frac{\beta_2}{\beta_1 + \beta_2}} \tag{8.15}$$

The order parameter S is related to the half opening angle θ of a cone within which the dye can move:

$$\theta = \arccos\left(\frac{1}{2}\left(\sqrt{1 + 8S} - 1\right)\right) \tag{8.16}$$

Values for θ of >60° indicate a nearly unrestricted motion of the dyes, with a narrow distribution of κ^2 around 2/3, justifying the 2/3 approximation [36]. Smaller values for θ point towards a restricted motion of the fluorophore. In this case, the actual distribution of κ^2 values has to be taken into account [37–40]. Incorrect assumption of $\kappa^2 = 2/3$ introduces distance errors of <26 %, or <10 % when one of the dyes rotates freely [38].

8.2.4 Analysis of Fluorescence Equilibrium Titrations

Any physical property X of a ligand A that changes linearly upon binding to an interaction partner B can be used to determine the K_d value of the respective AB complex. The (fluorescence) signal X of the ligand A is measured in the absence and in the presence of increasing concentrations of its interaction partner B (Fig. 8.3a). From this binding curve in terms of X as a function of the concentration of the interaction partner, K_d is determined. The spectroscopic signal X at each concentration of B can be expressed as

$$X = X_0 + \Delta X \tag{8.17}$$

where X_0 is the signal of the free ligand and ΔX is the change upon addition of the interaction partner at the respective concentration. The signal change ΔX can be expressed in terms of the binding degree ν, the ratio of the concentration of the complex [AB] in equilibrium and the total concentration of A, A_{tot} [Eq. (8.18)]

$$\nu = \frac{[AB]}{A_{tot}} \tag{8.18}$$

and the amplitude of the signal reached upon saturation, ΔX_{max}, as

$$\Delta X = \Delta X_{max} \cdot \nu = \Delta X_{max} \cdot \frac{[AB]}{A_{tot}} \tag{8.19}$$

The complex concentration [AB] in equilibrium at each concentration of B can be described by the solution of a quadratic equation as

$$[AB] = \frac{A_{tot} + B_{tot} + K_d}{2} - \sqrt{\left(\frac{A_{tot} + B_{tot} + K_d}{2}\right)^2 - A_{tot} \cdot B_{tot}} \tag{8.20}$$

B_{tot} denotes the total concentration of the interaction partner B at each point of the titration. Altogether, the signal X as a function of A_{tot}, B_{tot}, and K_d is obtained as

$$X = X_0 + \Delta X_{max} \cdot \frac{\frac{A_{tot}+B_{tot}+K_d}{2} - \sqrt{\left(\frac{A_{tot}+B_{tot}+K_d}{2}\right)^2 - A_{tot} \cdot B_{tot}}}{A_{tot}} \tag{8.21}$$

To determine the K_d value, the binding curve is described by Eq. (8.21), with K_d as a fitting parameter. Binding curves can be obtained by titrating a fluorescent ligand with its binding partner and measuring either the fluorescence intensity, the fluorescence anisotropy, or FRET as a fluorescent probe for binding.

Fluorescent moieties frequently affect binding, leading to an increase or decrease in affinity. In these cases, K_d values from fluorescence titrations can still be useful for relative comparisons, e.g., to determine ligand affinities of different variants of a protein, where the effect of the fluorophore will be comparable. The K_d value for a complex of the interaction partner with the unmodified ligand can be determined in displacement titrations. Upon titration of a fluorescent ligand/protein complex with unmodified ligand (Fig. 8.3a), the K_d of the unlabeled ligand/protein complex is determined by describing the experimentally obtained titration curve with the solution of a cubic equation [41, 42], keeping the previously determined K_d value for the fluorescent ligand complex constant during the fit.

8.2.5 Analysis of Kinetic Data

Any spectroscopic property that changes upon complex formation can be used to follow binding events as a function of time, and kinetic transients can be analyzed by single (or multiple) exponential functions to extract (observed) rate constants. The simplest example is the determination of a dissociation rate constant by adding an excess of unlabeled ligand to a preformed complex of the fluorescent ligand with its interaction partner and following the displacement of the fluorescent ligand by measuring fluorescence intensity or anisotropy (or intermolecular FRET) as a function of time. The dissociation curve can usually be described by a single exponential function, yielding the rate constant for the slowest dissociation step. To determine rate constants for binding, and to deduce the binding mechanism, a fluorescent ligand is mixed with increasing concentrations of its interaction partner. An excess of the binding partner in all experiments generates pseudo-first-order conditions and allows for the description of each transient by a single exponential to extract the observed rate constant k_{obs}. From a plot of k_{obs} as a function of the concentration of the interaction partner, individual rate constants can be obtained. For single-step binding, the observed rate constant k_{obs} depends linearly on the concentration of the binding partner B:

$$k_{obs} = k_{on}[B] + k_{off} \tag{8.22}$$

The rate constant for binding, k_{on}, is determined from the slope, the dissociation rate constant, k_{off}, from the y-axis intercept. The k_{off} obtained from extrapolation is affected with large errors, and an independent determination in displacement experiments is recommended. If binding occurs in two steps, the plot of k_{obs} as a function of the concentration of the interaction partner follows a hyperbola:

$$k_{obs} = k_{-2} + k_2 \frac{[B]}{[B] + K_{d1}} \tag{8.23}$$

where the y-axis intercept corresponds to k_{-2}, the amplitude of the hyperbola reflects k_2, and the curvature is related to the K_{d1} of the first rapid equilibrium.

In more complex binding mechanisms, and depending on the signal changes of the spectroscopic probe in each step, binding and dissociation of ligands may not be monophasic. From the concentration dependence of the observed rate constants and the associated amplitudes, possible binding models can be deduced. Similarly, competition kinetics with mixtures of labeled and unlabeled ligands reveal binding mechanisms and allow for the determination of the underlying rate constants. For details on more elaborate analysis procedures, the reader is referred to kinetics textbooks [43, 44].

Kinetic data can also be obtained from single-molecule experiments. For events that can be triggered, the reaction is initiated at the beginning of the data acquisition. At different time points, the populations present are quantified, either from single-molecule experiments on freely diffusing molecules over a short time window or by imaging surface-immobilized single molecules at different time points. The concentration of each species as a function of time can then be described by the appropriate model. This approach is in principle equivalent to ensemble experiments. For conformational changes that cannot be triggered, kinetic constants can be determined from FRET time traces of immobilized single molecules. Dwell times in a respective conformational state are calculated from these time traces, and kinetic constants for the conformational transition are obtained from the envelope of dwell time histograms (Fig. 8.5, [34]). This approach is based on the hypothesis that a single molecule, observed over extended periods of time, will reflect the behavior of an ensemble of molecules (ergodicity).

8.2.6 Fluorescent Labeling

Fluorescent approaches necessitate the site-specific introduction of suitable fluorophores into biomolecules (reviewed in [45]). While fluorescent nucleotides and a large variety of end- or internally labeled synthetic nucleic acids are commercially available, site-specific fluorescent labeling of larger nucleic acid molecules and of proteins is more challenging.

Fluorophores can be introduced into oligonucleotides during chemical synthesis. By hybridizing this oligonucleotide to several small, non-labeled fragments, the structure of interest can be formed. Larger nucleic acid molecules can be generated by ligation of the synthetic labeled fragment to other synthetic, in vitro transcribed (RNA) or PCR-generated (DNA) fragments. RNA molecules beyond the scope of chemical synthesis can be end-labeled by using fluorescently labeled nucleotides as starters for in vitro transcription or labeled internally by hybridizing synthetic, labeled oligonucleotides to single-stranded loop regions. These approaches also allow for the site-specific introduction of two fluorophores for FRET.

Proteins can be labeled selectively at the amino- and carboxyl termini [46, 47] or at surface exposed cysteines that react with maleimide derivates of dyes. Alternatively, unique functional groups can be introduced into proteins by enzymatic modifications ([48–50], reviewed in [51]), by affinity labeling [52, 53], or by incorporation of unnatural amino acids via suppression of stop codons by modified tRNA synthetases ([54, 55], reviewed in [56, 57]). By expressed protein ligation (reviewed in [58]), a recombinant protein can be coupled to a fluorescently labeled synthetic peptide or a second recombinant protein, or two differently labeled recombinant proteins can be linked. Combinations of these methods allow for the site-specific introduction of two (or more) fluorophores for smFRET experiments and allow the realization of complex labeling strategies for large molecules.

8.3 Probing the Catalytic Cycle of DEAD-Box Helicases

The following section illustrates the application of fluorescence techniques to dissect individual steps in the catalytic cycle of DEAD-box proteins and to investigate the overall mechanism of coupling ATP-dependent conformational changes to RNA unwinding (reviewed in [28]).

8.3.1 Nucleotide Binding

Fluorescently labeled methylanthraniloyl (mant)-nucleotides show a 2–3-fold increase in fluorescence upon binding to their protein partners, with a concomitant blueshift of the emission maximum. Mant-fluorescence (excitation at 360 nm, fluorescence maximum at 440 nm) and mant-nucleotides have been used to follow nucleotide binding to a variety of DEAD-box proteins (Fig. 8.3a, [10, 11, 26, 41, 59–61]). In some cases, the energy transfer from intrinsic tryptophans of the protein to the mant-group has been used to selectively detect the protein-bound nucleotide [10, 61]. Trinitrophenyl (TNP)-modified nucleotides serve as an alternative [62]. TNP fluorescence (excitation at 408 nm, emission at 514 nm) increases 4–5-fold upon binding to proteins. A variety of TNP- and mant adenine nucleotides are commercially available. The fluorescent moiety is coupled to the hydroxyl groups of the ribose and can alternate between the 2′- and 3′-hydroxyl groups by transesterification [63]. This slow interconversion can give rise to an additional phase in kinetic experiments, but can be avoided by using 3′-deoxy nucleotides.

Binding curves obtained from titration of mant-ADP with the respective DEAD-box protein can be analyzed for the K_d value of the complex with Eq. (8.21). For an accurate determination of the K_d, the curvature of the binding curve needs to be well defined, necessitating concentrations of the fluorescent ligand in the range of the K_d value. Although mant-ATP is commercially available, ATP hydrolysis during the time course of the titration may interfere with K_d determination, and non-hydrolyzable ATP analogues such as ADPNP, ATPγS, and ADPCP are often used instead. The pre- and post-hydrolysis state analogues ADP·BeF_x and ADP·MgF_x have been instrumental in the analysis of the mechanism of coupling ATP hydrolysis to duplex unwinding in DEAD-box proteins [64]. In principle, their fluorescent counterparts can be generated by incubating mant-ADP with the appropriate salts [65]. Since the reporter group can contribute to or interfere with binding, K_d values for unmodified nucleotides should be determined directly in displacement titrations of a preformed mant-nucleotide/DEAD-box protein with unmodified nucleotides (Fig. 8.3a, Sect. 8.2.4). Nucleotide binding to the DEAD-box protein YxiN from *B. subtilis* has been dissected in fluorescence equilibrium titrations of mant-ADPNP [11, 59] and in displacement titrations of YxiN/mant-ADP complexes with the non-modified nucleotides ADPNP, ADP·BeF_x and ADP·MgF_x, and ADP [26]. Mant-nucleotides have been instrumental in obtaining a kinetic and thermodynamic framework describing the interactions of the DEAD-box helicases DbpA (*E. coli*) and Mss116 (*S. cerevisiae*) with nucleotides and RNA [41, 60, 66]. DbpA and Mss116 populate the same intermediates in identical order in their nucleotide-dependent catalytic cycle. ATP hydrolysis and phosphate release are the rate-limiting steps, and the ADP·P_i state is thus predominantly populated during the catalytic cycle under in vitro conditions. At nucleotide concentrations prevalent in vivo, however, DbpA mostly populates the ADP state that exhibits a low RNA affinity, whereas Mss116 is predominantly in the ADP·P_i state with high

RNA affinity. It has been hypothesized that these differences may be linked to the in vivo functions of the two DEAD-box proteins [60].

In many DEAD-box proteins, the helicase core is regulated by flanking domains or by interaction partners. The translation factor eIF4A is part of a functional network in translation initiation (reviewed in [67, 68]). Fluorescence equilibrium titrations and kinetic studies using mant-nucleotides have revealed that the translation initiation factor eIF4G, by stabilizing eIF4A in a half-open conformation [10, 69], accelerates nucleotide exchange and reduces nucleotide affinity [10].

8.3.2 RNA Binding

RNA binding to DEAD-box helicases can be followed by fluorescence intensity or anisotropy changes of fluorescently labeled RNA substrates. A change in fluorescence intensity is indicative of direct interactions of the protein with the fluorophore, which should be avoided. Preferentially, RNA affinities are determined using fluorescence anisotropy as a read-out for binding. RNA binding has been followed using Alexa488- [70] or fluorescein-labeled RNA [19, 66, 71–73], or TAMRA-labeled protein [74]. Fluorescence equilibrium titrations of RNAs with DEAD-box proteins can be hampered by protein aggregation at larger concentrations and light scattering. Scattered light is polarized and thus affects the shape of the anisotropy titration curves and the extracted K_d values. For Mss116, this problem was circumvented by determining the K_d value of the Mss116/RNA complex from diffusion times using fluorescence correlation spectroscopy (FCS) [60] (Sect. 8.2.3.2).

To maximize the anisotropy change upon binding, the smaller binding partner should be fluorescently labeled. When binding of large RNAs is studied, the DEAD-box helicase is fluorescently labeled and titrated with increasing concentration of the RNA substrate. The RNA affinity of the *E. coli* DEAD-box protein SrmB has been determined in titrations of SrmB, labeled with Alexa488 at the N-terminus, with 23*S* ribosomal RNA [70]. In contrast, affinities for small model RNA substrates have been determined by anisotropy titrations of end-labeled RNA with unlabeled DEAD-box proteins [66, 72–74].

Several DEAD-box proteins carry C-terminal domains that confer RNA affinity and/or specificity [16, 18, 75–77]. Fluorescent anisotropy titrations can also be used to characterize the RNA affinity of these domains and to identify the key residues for RNA binding. Hera, a DEAD-box protein from *Thermus thermophilus* [77, 78], contains a C-terminal extension composed of a dimerization domain and an RNA binding domain (RBD) [79, 80] that mediates binding of Hera to hairpin 92 of 23*S* rRNA and to RNase P RNA [81]. From fluorescence anisotropy titrations of an end-labeled 32mer ribosomal RNA fragment with a series of RBD variants, helix α_1 and β-strand β_2, as well the C-terminal tail, were identified as crucial elements for RNA binding to the RBD [73].

8.3.3 RNA Unwinding

RNA unwinding has conventionally been monitored discontinuously by incubating a radioactively labeled double-stranded RNA with the DEAD-box protein, stopping the reaction at different time points, and separating the double-stranded substrate and single-stranded product by native gel electrophoresis, followed by quantification. Duplex separation can also be followed continuously with labeled RNA substrates monitoring fluorescence intensity, fluorescence anisotropy, or FRET (Fig. 8.4).

8.3.3.1 Discontinuous RNA Unwinding Assay: Product Separation by Native Gel Electrophoresis

A discontinuous unwinding assay with radiolabeled substrates and quantification of reaction products after separation in native gels is a standard procedure in many laboratories. The high detection sensitivity of phosphoimagers allows for detection of RNA down to femtomoles. With increased sensitivity of gel documentation systems using fluorescence detection, this assay can alternatively be performed using fluorescently labeled RNA or can exploit FRET when a donor-/acceptor-labeled duplex is used. Depending on the sensitivity of the imaging system, less than a picomole of RNA can be detected.

We have used a fluorescently labeled 32/9mer RNA, formed by annealing a 3′-fluorescein-labeled 9mer with a unlabeled 32mer, as unwinding substrate for the DEAD-box proteins YxiN and Hera, and visualized single- and double-stranded RNA in gels by fluorescence [11, 26, 27, 59, 73, 81, 82] (Fig. 8.4a). Wild-type YxiN unwinds this RNA in an ATP-dependent reaction [59]. A YxiN variant carrying a mutation in Walker A motif is ATPase-deficient and does not catalyze duplex unwinding [27]. Interestingly, this variant can adopt the closed conformation upon ATP and RNA binding [27], indicating that the conformational change is not sufficient for unwinding activity (Sect. 8.3.4). When ATP was replaced by the non-hydrolyzable ATP analogue ADPNP, the pre-hydrolysis state analogue ADP·BeF$_x$ [83], and the post-hydrolysis state ADP·MgF$_x$ [83–85], only ADP·BeF$_x$ supported duplex unwinding [26]. All three nucleotides allow for closure of the cleft in the helicase core, however [11, 26, 27], further indicating that formation of the closed conformer is only one step towards duplex unwinding (Sect. 8.3.4).

8.3.3.2 RNA Unwinding Monitored by 2-AP Fluorescence Intensity

Incorporation of the fluorescent adenine analogue 2-aminopurine (2-AP) into RNA slightly decreases duplex stability without altering the RNA structure ([86, 87] reviewed in [88]). 2-AP forms a base pair with thymine in Watson–Crick geometry and a wobble base pair with cytosine [89–91]. 2-AP fluorescence (excitation

303 nm, emission maximum 370 nm) is quenched when 2-AP is part of a duplex, rendering it a useful reporter for duplex unwinding (Fig. 8.4b). When the 32/9mer substrate is formed from the 9mer and a 32mer containing 2-AP, the 2-AP fluorescence decreases 12-fold [92]. The increase in 2-AP fluorescence upon single-strand release during eIF4A-catalyzed unwinding has been monitored in real time [92]. eIF4A on its own is a poor ATPase and helicase, but RNA unwinding is accelerated in the presence of eIF4B, and ATP hydrolysis is accelerated by eIF4G. When both factors are present, RNA unwinding and ATP hydrolysis by eIF4A are accelerated 13-fold, indicating increased coupling of ATP hydrolysis to duplex separation [92].

8.3.3.3 RNA Unwinding Monitored by Fluorescence Anisotropy

Fluorescence anisotropy provides an alternative signal to follow single-strand release and duplex unwinding. The fluorescence anisotropy of a 32/9mer that contains a 3′-fluorescein-labeled 9mer decreases upon duplex unwinding and 9mer release (Fig. 8.4c). RNA unwinding by the DEAD-box helicases YxiN [27, 59], DbpA [66], Hera [73, 81], and Mss116 [60] has been followed using anisotropy as a probe. For the *T. thermophilus* DEAD-box protein Hera, following unwinding of RNA by fluorescein anisotropy or monitoring product formation conventionally by separation with gel electrophoresis yields the same rate constants, confirming that the anisotropy decrease reflects unwinding [73]. The fluorescein-labeled 32/9mer RNA substrate is unwound by YxiN in the presence of ATP [11]. Incubation of labeled RNA with a mixture of the two RecA domains of YxiN does not lead to an anisotropy decrease, indicating that a covalent linkage between the two RecA-like domains of the helicase core is crucial for RNA unwinding [59].

8.3.3.4 RNA Unwinding Monitored by FRET

Unwinding can also be monitored by FRET, using a duplex RNA substrate that is labeled with a donor fluorophore on one strand and an acceptor fluorophore on the opposite strand. In the duplex, the FRET efficiency and thus the acceptor fluorescence will be high, whereas unwinding and strand separation lead to a decrease in FRET that can be monitored as an acceptor fluorescence decrease and a concomitant increase in donor fluorescence (Fig. 8.4d).

RNA unwinding by *T. thermophilus* Hera has been followed by FRET, using a 32/9mer RNA formed by annealing a 3′-Cy3-labeled 9mer and a 5′-labeled Cy5 32mer. Rate constants of unwinding determined from a decrease in acceptor fluorescence were identical to rate constants obtained from quantification of products after separation by gel electrophoresis, confirming that the acceptor fluorescence decrease reflects unwinding [73].

8.3.4 Conformational Changes of the Helicase Core

Crystal structures of DEAD-box proteins in the absence of ligands have shown that the two RecA domains can adopt a variety of different relative positions [12–15], pointing towards a central role of conformational changes in the catalytic cycle. The conformational changes in the helicase core of the DEAD-box helicase YxiN throughout the nucleotide cycle have been delineated using smFRET (reviewed in [28]). Two of the four intrinsic cysteines were identified as solvent-accessible and were replaced by serines or alanines. In a next step, cysteines for fluorophores attachment were introduced into each RecA domain, on opposite sides of the inter-domain cleft [11, 26, 27]. Donor and acceptor fluorophores were introduced by statistical labeling of a double-cysteine variant with Alexa488- and Alexa546-maleimides. The ratios of the dyes were optimized such as to maximize the fraction of molecules carrying both fluorophores and minimize the fraction of molecules carrying only the donor fluorophores. A flow chart on how to prepare proteins for smFRET experiments by donor/acceptor labeling via cysteines is found in [28].

Unimodal FRET histograms from single-molecule experiments on freely diffusing YxiN, labeled with donor and acceptor in different positions, demonstrated that YxiN adopts a unique conformation in solution. From the conversion of the FRET efficiencies into donor/acceptor distances (Sect. 8.2.3.3), the relative arrangement of the two RecA domains and hence the global conformation of the helicase core were deduced [11] (Sect. 8.3.6). The conformation of the YxiN helicase core in the absence of ATP and RNA resembles the structure of the DeaD protein from *Methanococcus jannaschii* [13]. Wide-open conformations, as represented by the crystal structure of eIF4A [12], are not significantly populated by YxiN in solution. By using the ATP analogues ADPNP, ADP·BeF$_x$, ADP·MgF$_x$, as well as ADP to populate individual nucleotide states, the conformational changes in the helicase core of YxiN throughout its catalytic cycle were followed [11, 26]. The low-FRET state observed in the absence of ligands converts into a high-FRET state in the presence of both RNA and ADPNP. Distance restraints from smFRET experiments with YxiN labeled in different positions [11] are consistent with a global conformation similar to the closed conformation in the RNA- and ADPNP-bound state observed in crystal structures [11, 15, 17–23]. The pre-hydrolysis state, mimicked by the ADP·BeF$_x$-bound state, and the post-hydrolysis state, mimicked by the ADP·MgF$_x$-bound state, are also high-FRET states, demonstrating that the helicase core remains in the closed conformation during ATP hydrolysis [26]. In the ADP-bound state, however, the FRET efficiency returns to the low value indicative of the open conformation. Hence, phosphate release must be linked to reopening of the inter-domain cleft in the helicase core.

In a YxiN variant carrying a mutation at the interface between the two RecA domains, the closure of the helicase core is hampered [27]. This variant is unwinding-deficient, linking closure of the helicase core to duplex separation. However, ADPNP leads to closure of the inter-domain left in YxiN, but does not support RNA unwinding [11, 27]. Similarly, an ATPase-deficient variant carrying a mutation in the Walker A motif adopts the closed conformation of the helicase core when ATP and RNA bind, but cannot unwind the RNA duplex [27]. Both observations indicate that the closure of the inter-domain cleft in the helicase core is not sufficient for RNA unwinding to occur. Overall, these smFRET data provide a model for the conformational cycle that forms the mechanistic basis for duplex separation by DEAD-box helicases (Fig. 8.1, reviewed in [3, 28]).

Interestingly, although considered to be a prototype for DEAD-box proteins, the closed conformation of eIF4A has eluded experimental detection for a long time. In smFRET experiments, we have recently detected a high-FRET state of eIF4A reflecting the closed conformation in the presence of eIF4B, eIF4G, ATP, and RNA for the first time [92].

8.3.5 Conformational Changes of the RNA Substrate

Conformational changes in the helicase core of DEAD-box proteins are coupled to structural changes in the RNA substrate. DEAD-box protein-mediated structural changes in the RNA substrate have been followed in smFRET TIRF experiments with surface-immobilized donor-/acceptor-labeled group II intron RNA and Mss116 [93]. Fluorophores were introduced into the RNA by hybridizing short donor- and acceptor-labeled oligonucleotides to single-stranded loop regions. The group II intron exists predominantly in an extended intermediate state in the absence of Mss116. In the presence of Mss116, transitions between three RNA conformations are observed, an extended intermediate, a compact intermediate, and the native state. The compact intermediate is stabilized by Mss116 in the absence of ATP, or by other RNA binding proteins. In contrast, formation of the native state requires Mss116 and ATP [93]. It has been suggested that Mss116 promotes group II intron folding by ATP-dependent unwinding of kinetic traps.

8.3.6 Mapping Global Conformations

Distance restraints from FRET experiments can be used to model global conformations and to define relative orientations of domains within multi-domain proteins. This approach has already been illustrated for the global architecture of the YxiN helicase core in the open and closed conformations ([11], Sect. 8.3.4). smFRET has also been used to probe the open conformation of eIF4A in solution [10], revealing that two domains of eIF4A are not as far apart as captured in the crystal structure [12]. In contrast, the distances obtained by smFRET for the eIF4A/eIF4G complex are in agreement with the half-open conformation observed the crystal structure [69].

Despite the extensive structural characterization of the helicase core of DEAD-box proteins, structural information on full-length DEAD-box proteins, including

N- and C-terminal extensions, is still scarce. Structural models have been constructed for the complete *T. thermophilus* Hera dimer [78], including the helicase core, the dimerization domain, and the C-terminal RNA binding domain. The structures of Mss116 and Cyt-19, comprising a helicase core, a C-terminal helical domain, and a flexible C-tail, have been determined by small-angle X-ray scattering [16]. *B. subtilis* YxiN contains a C-terminal RNA binding domain that adopts an RRM fold. The structures of the C-terminal RecA domain [94], of the isolated RBD [95], and of the isolated RBD in complex with RNA [96] have been determined, but the architecture of the full-length protein has remained elusive. Using YxiN carrying donor and acceptor fluorophores, one on the helicase core and one on the RBD in different positions, the relative position of the RBD with respect to the helicase core has been determined using distance restraints from smFRET experiments [82]. The resulting model for full-length YxiN is consistent with biochemical data and is in agreement with the RNA binding geometry by the RBD in the crystal structure of the RBD/RNA complex [96]. Mapping of global conformations by FRET is not limited by molecular size, and FRET is therefore a powerful tool in delineating the architecture of DEAD-box helicases and their complexes with RNAs and interaction partners.

8.4 Conclusions and Perspectives

An array of fluorescence-based methods has been used to monitor individual events in the catalytic cycle of DEAD-box helicases and to delineate their unwinding mechanism. Despite the few representatives characterized comprehensively so far, similarities and differences have already become apparent. Future studies on other DEAD-box proteins will help to further dissect common mechanistic principles that are shared among DEAD-box proteins and to identify specific features that are linked to specialized functions of individual representatives.

With the advent of single-molecule fluorescence methods, conformational changes of DEAD-box proteins and of their RNA substrates have become amenable to direct investigation. Single-molecule FRET studies in solution have identified the conformational states of DEAD-box proteins throughout the catalytic cycle. The molecular events in the catalytic cycle of the helicase core are now well understood, providing a basis for investigation of the modulation of the helicase core activity by flanking domains and interaction partners. Most studies have investigated DEAD-box helicase activity on short model RNA duplexes. Many laboratories now focus their efforts on the identification of in vivo RNA substrates and interaction partners of DEAD-box proteins to define their functional context in the cell, using cross-linking approaches in combination with mass-spectrometric and deep sequencing analyses. Once the cellular context of a particular DEAD-box protein has been identified, fluorescence methods can contribute to define the architecture of larger functional complexes with its interaction partners and to delineate structural rearrangements of DEAD-box proteins and their RNA substrates in the context of these functional networks. With advances in in vivo fluorescence spectroscopy and imaging, the monitoring of cellular localization, of interactions, and of conformational dynamics of DEAD-box helicases and their RNA substrates within their cellular context should now be within reach.

References

1. Herschlag D (1995) RNA chaperones and the RNA folding problem. J Biol Chem 270 (36):20871–20874
2. Henn A, Bradley MJ, De La Cruz EM (2012) ATP utilization and RNA conformational rearrangement by DEAD-box proteins. Annu Rev Biophys 41:247–267. doi:10.1146/annurev-biophys-050511-102243
3. Hilbert M, Karow AR, Klostermeier D (2009) The mechanism of ATP-dependent RNA unwinding by DEAD box proteins. Biol Chem 390(12):1237–1250. doi:10.1515/BC.2009.135
4. Jankowsky E, Fairman ME (2007) RNA helicases–one fold for many functions. Curr Opin Struct Biol 17(3):316–324. doi:10.1016/j.sbi.2007.05.007
5. Linder P, Jankowsky E (2011) From unwinding to clamping - the DEAD box RNA helicase family. Nat Rev Mol Cell Biol 12(8):505–516. doi:10.1038/nrm3154
6. Gorbalenya AE, Koonin EV (1993) Helicases - Amino-Acid-Sequence Comparisons and Structure-Function-Relationships. Curr Opin Struct Biol 3(3):419–429
7. Fairman-Williams ME, Guenther UP, Jankowsky E (2010) SF1 and SF2 helicases: family matters. Curr Opin Struct Biol 20(3):313–324. doi:10.1016/j.sbi.2010.03.011
8. Gorbalenya AE, Koonin EV, Donchenko AP, Blinov VM (1989) Two related superfamilies of putative helicases involved in replication, recombination, repair and expression of DNA and RNA genomes. Nucleic Acids Res 17(12):4713–4730
9. Linder P, Lasko PF, Ashburner M, Leroy P, Nielsen PJ, Nishi K, Schnier J, Slonimski PP (1989) Birth of the D-E-A-D box. Nature 337(6203):121–122. doi:10.1038/337121a0
10. Hilbert M, Kebbel F, Gubaev A, Klostermeier D (2011) eIF4G stimulates the activity of the DEAD box protein eIF4A by a conformational guidance mechanism. Nucleic Acids Res 39 (6):2260–2270. doi:10.1093/nar/gkq1127
11. Theissen B, Karow AR, Kohler J, Gubaev A, Klostermeier D (2008) Cooperative binding of ATP and RNA induces a closed conformation in a DEAD box RNA helicase. Proc Natl Acad Sci USA 105(2):548–553. doi:10.1073/pnas.0705488105
12. Caruthers JM, Johnson ER, McKay DB (2000) Crystal structure of yeast initiation factor 4A, a DEAD-box RNA helicase. Proc Natl Acad Sci USA 97(24):13080–13085. doi:10.1073/pnas.97.24.13080
13. Story RM, Li H, Abelson JN (2001) Crystal structure of a DEAD box protein from the hyperthermophile *Methanococcus jannaschii*. Proc Natl Acad Sci USA 98(4):1465–1470. doi:10.1073/pnas.98.4.1465
14. Cheng Z, Coller J, Parker R, Song H (2005) Crystal structure and functional analysis of DEAD-box protein Dhh1p. RNA 11(8):1258–1270. doi:10.1261/rna.2920905
15. Andersen CB, Ballut L, Johansen JS, Chamieh H, Nielsen KH, Oliveira CL, Pedersen JS, Seraphin B, Le Hir H, Andersen GR (2006) Structure of the exon junction core complex with a trapped DEAD-box ATPase bound to RNA. Science 313(5795):1968–1972. doi:10.1126/science.1131981
16. Mallam AL, Jarmoskaite I, Tijerina P, Del Campo M, Seifert S, Guo L, Russell R, Lambowitz AM (2011) Solution structures of DEAD-box RNA chaperones reveal conformational changes and nucleic acid tethering by a basic tail. Proc Natl Acad Sci USA 108(30):12254–12259. doi:10.1073/pnas.1109566108

17. Bono F, Ebert J, Lorentzen E, Conti E (2006) The crystal structure of the exon junction complex reveals how it maintains a stable grip on mRNA. Cell 126(4):713–725. doi:10.1016/j.cell.2006.08.006
18. Del Campo M, Lambowitz AM (2009) Structure of the Yeast DEAD box protein Mss116p reveals two wedges that crimp RNA. Mol Cell 35(5):598–609. doi:10.1016/j.molcel.2009.07.032
19. Montpetit B, Thomsen ND, Helmke KJ, Seeliger MA, Berger JM, Weis K (2011) A conserved mechanism of DEAD-box ATPase activation by nucleoporins and InsP6 in mRNA export. Nature 472(7342):238–242. doi:10.1038/nature09862
20. Sengoku T, Nureki O, Nakamura A, Kobayashi S, Yokoyama S (2006) Structural basis for RNA unwinding by the DEAD-box protein *Drosophila* Vasa. Cell 125(2):287–300. doi:10.1016/j.cell.2006.01.054
21. Collins R, Karlberg T, Lehtio L, Schutz P, van den Berg S, Dahlgren LG, Hammarstrom M, Weigelt J, Schuler H (2009) The DEXD/H-box RNA helicase DDX19 is regulated by an {alpha}-helical switch. J Biol Chem 284(16):10296–10300. doi:10.1074/jbc.C900018200
22. Napetschnig J, Kassube SA, Debler EW, Wong RW, Blobel G, Hoelz A (2009) Structural and functional analysis of the interaction between the nucleoporin Nup214 and the DEAD-box helicase Ddx19. Proc Natl Acad Sci USA 106(9):3089–3094. doi:10.1073/pnas.0813267106
23. von Moeller H, Basquin C, Conti E (2009) The mRNA export protein DBP5 binds RNA and the cytoplasmic nucleoporin NUP214 in a mutually exclusive manner. Nat Struct Mol Biol 16 (3):247–254. doi:10.1038/nsmb.1561
24. Cordin O, Banroques J, Tanner NK, Linder P (2006) The DEAD-box protein family of RNA helicases. Gene 367:17–37. doi:10.1016/j.gene.2005.10.019
25. Russell R, Jarmoskaite I, Lambowitz AM (2013) Toward a molecular understanding of RNA remodeling by DEAD-box proteins. RNA Biol 10(1):44–55. doi:10.4161/rna.22210
26. Aregger R, Klostermeier D (2009) The DEAD box helicase YxiN maintains a closed conformation during ATP hydrolysis. Biochemistry 48(45):10679–10681. doi:10.1021/bi901278p
27. Karow AR, Klostermeier D (2009) A conformational change in the helicase core is necessary but not sufficient for RNA unwinding by the DEAD box helicase YxiN. Nucleic Acids Res 37 (13):4464–4471. doi:10.1093/nar/gkp397
28. Andreou AZ, Klostermeier D (2012) Conformational changes of DEAD-box helicases monitored by single molecule fluorescence resonance energy transfer. Methods Enzymol 511:75–109. doi:10.1016/B978-0-12-396546-2.00004-8
29. Putnam AA, Jankowsky E (2013) DEAD-box helicases as integrators of RNA, nucleotide and protein binding. Biochim Biophys Acta 1829(8):884–893. doi:10.1016/j.bbagrm.2013.02.002
30. O'Connor DV, Phillips D (1984) Time-correlated single photon counting. Academic, London
31. Parker CA, Rees WT (1960) Correction of Fluorescence Spectra and Measurement of Fluorescence Quantum Efficiency. Analyst 85(1013):587–600
32. Williams ATR, Winfield SA, Miller JN (1983) Relative Fluorescence Quantum Yields Using a Computer-Controlled Luminescence Spectrometer. Analyst 108(1290):1067–1071. doi:10.1039/An9830801067
33. Förster T (1948) Intermolecular energy migration and fluorescence test. Annalen der Physik 2:55–75
34. Gubaev A, Klostermeier D (2013) Fluorescence resonance energy transfer as a tool to investigate RNA structure and folding. In: Klostermeier D, Hammann C (eds) RNA structure and folding. de Gruyter, Berlin, pp 181–213
35. Lamichhane R, Solem A, Black W, Rueda D (2010) Single-molecule FRET of protein-nucleic acid and protein-protein complexes: surface passivation and immobilization. Methods 52 (2):192–200. doi:10.1016/j.ymeth.2010.06.010
36. Parkhurst LJ, Parkhurst KM, Powell R, Wu J, Williams S (2001) Time-resolved fluorescence resonance energy transfer studies of DNA bending in double-stranded oligonucleotides and in DNA-protein complexes. Biopolymers 61(3):180–200. doi:10.1002/bip.10138

37. Albaugh S, Lan JQ, Steiner RF (1989) The effect of a distribution of separations upon intramolecular distances in biopolymers, as determined by radiationless energy transfer. Biophys Chem 33(1):71–76
38. Haas E, Katchalskikatzir E, Steinberg IZ (1978) Effect of Orientation of Donor and Acceptor on Probability of Energy-Transfer Involving Electronic-Transitions of Mixed Polarization. Biochemistry 17(23):5064–5070
39. Klostermeier D, Millar DP (2001) Time-resolved fluorescence resonance energy transfer: a versatile tool for the analysis of nucleic acids. Biopolymers 61(3):159–179. doi:10.1002/bip.10146
40. Klostermeier D, Millar DP (2001) RNA conformation and folding studied with fluorescence resonance energy transfer. Methods 23(3):240–254. doi:10.1006/meth.2000.1135
41. Talavera MA, De La Cruz EM (2005) Equilibrium and kinetic analysis of nucleotide binding to the DEAD-box RNA helicase DbpA. Biochemistry 44(3):959–970. doi:10.1021/bi048253i
42. Thrall SH, Reinstein J, Wohrl BM, Goody RS (1996) Evaluation of human immunodeficiency virus type 1 reverse transcriptase primer tRNA binding by fluorescence spectroscopy: specificity and comparison to primer/template binding. Biochemistry 35(14):4609–4618. doi:10.1021/bi9526387
43. Bernasconi CF (1976) Relaxation kinetics. Academic Press, New York
44. Gutfreund H (1995) Kinetics for the life sciences: receptors, transmitters and catalysts. Cambridge University Press, Cambridge
45. Toseland CP (2013) Fluorescent labeling and modification of proteins. J Chem Biol 6(3):85–95. doi:10.1007/s12154-013-0094-5
46. Volkmann G, Liu XQ (2009) Protein C-terminal labeling and biotinylation using synthetic peptide and split-intein. PLoS One 4(12):e8381. doi:10.1371/journal.pone.0008381
47. Schuler B, Pannell LK (2002) Specific labeling of polypeptides at amino-terminal cysteine residues using Cy5-benzyl thioester. Bioconjug Chem 13(5):1039–1043
48. Lin CW, Ting AY (2006) Transglutaminase-catalyzed site-specific conjugation of small-molecule probes to proteins in vitro and on the surface of living cells. J Am Chem Soc 128 (14):4542–4543. doi:10.1021/ja0604111
49. Duckworth BP, Zhang Z, Hosokawa A, Distefano MD (2007) Selective labeling of proteins by using protein farnesyltransferase. Chembiochem 8(1):98–105. doi:10.1002/cbic.200600340
50. Carrico IS, Carlson BL, Bertozzi CR (2007) Introducing genetically encoded aldehydes into proteins. Nat Chem Biol 3(6):321–322. doi:10.1038/nchembio878
51. Jager M, Nir E, Weiss S (2006) Site-specific labeling of proteins for single-molecule FRET by combining chemical and enzymatic modification. Protein Sci 15(3):640–646. doi:10.1110/ps.051851506
52. Kapanidis AN, Ebright YW, Ebright RH (2001) Site-specific incorporation of fluorescent probes into protein: hexahistidine-tag-mediated fluorescent labeling with (Ni(2+): nitrilotriacetic Acid (n)-fluorochrome conjugates. J Am Chem Soc 123(48):12123–12125
53. Adams SR, Campbell RE, Gross LA, Martin BR, Walkup GK, Yao Y, Llopis J, Tsien RY (2002) New biarsenical ligands and tetracysteine motifs for protein labeling in vitro and in vivo: synthesis and biological applications. J Am Chem Soc 124(21):6063–6076
54. Noren CJ, Anthony-Cahill SJ, Griffith MC, Schultz PG (1989) A general method for site-specific incorporation of unnatural amino acids into proteins. Science 244(4901):182–188
55. Brustad EM, Lemke EA, Schultz PG, Deniz AA (2008) A general and efficient method for the site-specific dual-labeling of proteins for single molecule fluorescence resonance energy transfer. J Am Chem Soc 130(52):17664–17665. doi:10.1021/ja807430h
56. Lemke EA (2011) Site-Specific Labeling of Proteins for Single-Molecule FRET Measurements Using Genetically Encoded Ketone Functionalities. Methods Mol Biol 751:3–15. doi:10.1007/978-1-61779-151-2_1
57. Ellman J, Mendel D, Anthonycahill S, Noren CJ, Schultz PG (1991) Biosynthetic Method for Introducing Unnatural Amino-Acids Site-Specifically into Proteins. Method Enzymol 202:301–336

58. Blaschke UK, Silberstein J, Muir TW (2000) Protein engineering by expressed protein ligation. Methods Enzymol 328:478–496
59. Karow AR, Theissen B, Klostermeier D (2007) Authentic interdomain communication in an RNA helicase reconstituted by expressed protein ligation of two helicase domains. FEBS J 274 (2):463–473. doi:10.1111/j.1742-4658.2006.05593.x
60. Cao W, Coman MM, Ding S, Henn A, Middleton ER, Bradley MJ, Rhoades E, Hackney DD, Pyle AM, De La Cruz EM (2011) Mechanism of Mss116 ATPase reveals functional diversity of DEAD-Box proteins. J Mol Biol 409(3):399–414. doi:10.1016/j.jmb.2011.04.004
61. Henn A, Cao W, Hackney DD, De La Cruz EM (2008) The ATPase cycle mechanism of the DEAD-box rRNA helicase, DbpA. J Mol Biol 377(1):193–205. doi:10.1016/j.jmb.2007.12.046
62. Bi X, Ren J, Goss DJ (2000) Wheat germ translation initiation factor eIF4B affects eIF4A and eIFiso4F helicase activity by increasing the ATP binding affinity of eIF4A. Biochemistry 39 (19):5758–5765. doi:10.1021/bi992322p
63. Moore KJ, Lohman TM (1994) Kinetic mechanism of adenine nucleotide binding to and hydrolysis by the *Escherichia coli* Rep monomer. 1. Use of fluorescent nucleotide analogues. Biochemistry 33(48):14550–14564
64. Liu F, Putnam A, Jankowsky E (2008) ATP hydrolysis is required for DEAD-box protein recycling but not for duplex unwinding. Proc Natl Acad Sci USA 105(51):20209–20214. doi:10.1073/pnas.0811115106
65. Del Toro DY, Klostermeier D (2011) Nucleotide-driven conformational changes in the reverse gyrase helicase-like domain couple the nucleotide cycle to DNA processing. Phys Chem Chem Phys 13(21):10009–10019. doi:10.1039/c0cp02859b
66. Henn A, Cao W, Licciardello N, Heitkamp SE, Hackney DD, De La Cruz EM (2010) Pathway of ATP utilization and duplex rRNA unwinding by the DEAD-box helicase, DbpA. Proc Natl Acad Sci USA 107(9):4046–4050. doi:10.1073/pnas.0913081107
67. Marintchev A (2013) Roles of helicases in translation initiation: a mechanistic view. Biochim Biophys Acta 1829(8):799–809. doi:10.1016/j.bbagrm.2013.01.005
68. Andreou AZ, Klostermeier D (2013) The DEAD-box helicase eIF4A: paradigm or the odd one out? RNA Biol 10(1):19–32. doi:10.4161/rna.21966
69. Schutz P, Bumann M, Oberholzer AE, Bieniossek C, Trachsel H, Altmann M, Baumann U (2008) Crystal structure of the yeast eIF4A-eIF4G complex: an RNA-helicase controlled by protein-protein interactions. Proc Natl Acad Sci USA 105(28):9564–9569. doi:10.1073/pnas.0800418105
70. Trubetskoy D, Proux F, Allemand F, Dreyfus M, Iost I (2009) SrmB, a DEAD-box helicase involved in *Escherichia coli* ribosome assembly, is specifically targeted to 23S rRNA in vivo. Nucleic Acids Res 37(19):6540–6549. doi:10.1093/nar/gkp685
71. Mallam AL, Del Campo M, Gilman B, Sidote DJ, Lambowitz AM (2012) Structural basis for RNA-duplex recognition and unwinding by the DEAD-box helicase Mss116p. Nature 490 (7418):121–125. doi:10.1038/nature11402
72. Marintchev A, Edmonds KA, Marintcheva B, Hendrickson E, Oberer M, Suzuki C, Herdy B, Sonenberg N, Wagner G (2009) Topology and regulation of the human eIF4A/4G/4H helicase complex in translation initiation. Cell 136(3):447–460. doi:10.1016/j.cell.2009.01.014
73. Steimer L, Wurm JP, Linden MH, Rudolph MG, Wohnert J, Klostermeier D (2013) Recognition of two distinct elements in the RNA substrate by the RNA-binding domain of the *T. thermophilus* DEAD box helicase Hera. Nucleic Acids Res 41(12):6259–6272. doi:10.1093/nar/gkt323
74. Rajagopal V, Park EH, Hinnebusch AG, Lorsch JR (2012) Specific domains in yeast translation initiation factor eIF4G strongly bias RNA unwinding activity of the eIF4F complex toward duplexes with 5′-overhangs. J Biol Chem 287(24):20301–20312. doi:10.1074/jbc.M112.347278
75. Tsu CA, Kossen K, Uhlenbeck OC (2001) The *Escherichia coli* DEAD protein DbpA recognizes a small RNA hairpin in 23S rRNA. RNA 7(5):702–709

76. Kossen K, Karginov FV, Uhlenbeck OC (2002) The carboxy-terminal domain of the DExDH protein YxiN is sufficient to confer specificity for 23S rRNA. J Mol Biol 324(4):625–636
77. Morlang S, Weglohner W, Franceschi F (1999) Hera from *Thermus thermophilus*: the first thermostable DEAD-box helicase with an RNase P protein motif. J Mol Biol 294(3):795–805. doi:10.1006/jmbi.1999.3282
78. Klostermeier D (2013) Rearranging RNA structures at 75 degrees C? towards the molecular mechanism and physiological function of the *Thermus thermophilus* DEAD-box helicase hera. Biopolymers. doi:10.1002/bip.22316
79. Rudolph MG, Klostermeier D (2009) The *Thermus thermophilus* DEAD box helicase Hera contains a modified RNA recognition motif domain loosely connected to the helicase core. RNA 15(11):1993–2001
80. Klostermeier D, Rudolph MG (2009) A novel dimerization motif in the C-terminal domain of the *Thermus thermophilus* DEAD box helicase Hera confers substantial flexibility. Nucleic Acids Res 37(2):421–430. doi:10.1093/nar/gkn947
81. Linden MH, Hartmann RK, Klostermeier D (2008) The putative RNase P motif in the DEAD box helicase Hera is dispensable for efficient interaction with RNA and helicase activity. Nucleic Acids Res 36(18):5800–5811
82. Karow AR, Klostermeier D (2010) A structural model for the DEAD box helicase YxiN in solution: localization of the RNA binding domain. J Mol Biol 402(4):629–637. doi:10.1016/j.jmb.2010.07.049
83. Fisher AJ, Smith CA, Thoden JB, Smith R, Sutoh K, Holden HM, Rayment I (1995) X-ray structures of the myosin motor domain of *Dictyostelium discoideum* complexed with MgADP. BeFx and MgADP.AlF4. Biochemistry 34(28):8960–8972
84. Bigay J, Deterre P, Pfister C, Chabre M (1987) Fluoride complexes of aluminium or beryllium act on G-proteins as reversibly bound analogues of the gamma phosphate of GTP. EMBO J 6 (10):2907–2913
85. Chabre M (1990) Aluminofluoride and beryllofluoride complexes: a new phosphate analogs in enzymology. Trends Biochem Sci 15(1):6–10
86. Law SM, Eritja R, Goodman MF, Breslauer KJ (1996) Spectroscopic and calorimetric characterizations of DNA duplexes containing 2-aminopurine. Biochemistry 35(38):12329–12337. doi:10.1021/bi9614545
87. Eritja R, Kaplan BE, Mhaskar D, Sowers LC, Petruska J, Goodman MF (1986) Synthesis and properties of defined DNA oligomers containing base mispairs involving 2-aminopurine. Nucleic Acids Res 14(14):5869–5884
88. Putnam A, Jankowsky E (2012) Analysis of duplex unwinding by RNA helicases using stopped-flow fluorescence spectroscopy. Methods Enzymol 511:1–27
89. Ward DC, Reich E, Stryer L (1969) Fluorescence studies of nucleotides and polynucleotides. I. Formycin, 2-aminopurine riboside, 2,6-diaminopurine riboside, and their derivatives. J Biol Chem 244(5):1228–1237
90. Nordlund TM, Andersson S, Nilsson L, Rigler R, Graslund A, McLaughlin LW (1989) Structure and dynamics of a fluorescent DNA oligomer containing the EcoRI recognition sequence: fluorescence, molecular dynamics, and NMR studies. Biochemistry 28(23):9095–9103
91. Sowers LC, Fazakerley GV, Eritja R, Kaplan BE, Goodman MF (1986) Base pairing and mutagenesis: observation of a protonated base pair between 2-aminopurine and cytosine in an oligonucleotide by proton NMR. Proc Natl Acad Sci USA 83(15):5434–5438
92. Andreou AZ, Klostermeier D (2013) eIF4B and eIF4G Jointly Stimulate eIF4A ATPase and Unwinding Activities by Modulation of the eIF4A Conformational Cycle. J Mol Biol. doi:10.1016/j.jmb.2013.09.027
93. Karunatilaka KS, Solem A, Pyle AM, Rueda D (2010) Single-molecule analysis of Mss116-mediated group II intron folding. Nature 467(7318):935–939. doi:10.1038/nature09422

94. Caruthers JM, Hu Y, McKay DB (2006) Structure of the second domain of the *Bacillus subtilis* DEAD-box RNA helicase YxiN. Acta Crystallogr Sect F Struct Biol Cryst Commun 62 (Pt 12):1191–1195
95. Wang S, Hu Y, Overgaard MT, Karginov FV, Uhlenbeck OC, McKay DB (2006) The domain of the *Bacillus subtilis* DEAD-box helicase YxiN that is responsible for specific binding of 23S rRNA has an RNA recognition motif fold. RNA 12(6):959–967. doi:10.1261/rna.5906
96. Hardin JW, Hu YX, McKay DB (2010) Structure of the RNA binding domain of a DEAD-box helicase bound to its ribosomal RNA target reveals a novel mode of recognition by an RNA recognition motif. J Mol Biol 402(2):412–427. doi:10.1016/j.jmb.2010.07.040

Chapter 9
Use of Fluorescent Techniques to Study the In Vitro Movement of Myosins

Christopher Toepfer and James R. Sellers

Abstract Myosins are a large superfamily of actin-dependent molecule motors that carry out many functions in cells. Some myosins are cargo carriers that move processively along actin which means that a single molecule of myosin can take many ATP-dependent steps on actin per initial encounter. Other myosins are designed to work in large ensembles such as myosin thick filaments. In vitro motility assays are a powerful method for studying the function of myosins. These assays in general use small amounts of protein, are simple to implement, and can be done on microscopes commonly found in many laboratories. There are two basic versions of the assay which involve different geometries. In the sliding actin in vitro motility assay, myosin molecules are bound to a coverslip surface in a simply constructed microscopic flow chamber. Fluorescently labeled actin filaments are added to the flow chamber in the presence of ATP, and the movement of these actin filaments powered by the surface-bound myosins is observed. This assay has been used widely for a variety of myosins including both processive and nonprocessive ones. From this assay, one can easily measure the rate at which myosin is translocating actin. The single-molecule motility assay uses an inverted geometry compared to the sliding actin in vitro motility assay. It is most useful for processive myosins. Here, actin filaments are affixed to the coverslip surface. Fluorescently labeled single molecules of myosins (usually ones with processive kinetics) are introduced, and the movement of single molecules along the actin filaments is observed. This assay typically uses total internal reflection fluorescent (TIRF) microscopy to reduce the background signal arising from myosins in solution. From this assay, one can measure the velocity of movement, the frequency of movement, and the run length. If sufficient photons can be collected, one can use Gaussian fitting of the point spread function to determine the position of the labeled myosin to within a few nanometers which allows for measurement of the step size

C. Toepfer • J.R. Sellers (✉)
Laboratory of Molecular Physiology, National Heart, Lung and Blood Institute, National Institutes of Health, Bethesda, MD 20892, USA
e-mail: sellersj@nhlbi.nih.gov

C.P. Toseland and N. Fili (eds.), *Fluorescent Methods for Molecular Motors*,
Experientia Supplementum 105, DOI 10.1007/978-3-0348-0856-9_9,

and the stepping kinetics. Together, these two assays are powerful tools to elucidate myosin function.

Keywords In vitro motility • Myosin • Actin • Fluorescence • TIRF

Abbreviations

TIRF	Total internal refection fluorescence
CCD	Charged-coupled device
HMM	Heavy meromyosin
S1	Subfragment-1
FIONA	Fluorescence Imaging at One Nanometer with Accuracy
GFP	Green fluorescent protein

9.1 Introduction

Myosins are a superfamily of molecular motors that use the energy from ATP to interact with actin filaments in a variety of manners in muscle or nonmuscle cells [1]. The classic example of myosin function is its role in skeletal muscle fibers where myosin forms thick filaments which are interdigitated with actin filaments in the almost crystalline lattice of the sarcomere. These myosins are members of class 2, but now more than 35 classes of myosins present throughout phylogeny have been defined which are evolutionarily adapted to perform a plethora of intracellular tasks [2]. There are two simplistic methods of classifying members of the myosin superfamily. One historic classification system terms all class 2 myosins to be "conventional" and all other myosin "unconventional." Another functional classification system of the myosin superfamily divides the members into two categories based on their duty ratios. The "duty ratio" is defined as the fraction of time myosin is bound tightly to actin during a cycle of ATP hydrolysis. Low-duty ratio myosins work as large ensembles and only interact with actin strongly for a small percentage of their kinetic cycles, examples of which include skeletal muscle myosins. High-duty ratio myosins such as myosin 5a spend the majority of their kinetic cycle in a strongly actin-bound state. This type of kinetics is suitable for myosins that transport cargo and need to move along actin for several microns without detaching [3]. Prior to the early 1980s, in vitro assays applied to myosins measured the ability of actin to bind myosin and to accelerate its MgATPase activity in vitro. Transient kinetic studies revealed the rate and equilibrium constants of the individual steps involved in their enzymatic cycle. However, an in vitro assay to measure myosin's ability to translocate actin was missing until Sheetz and Spudich [4] developed the first functional in vitro motility assay. This assay used long actin cables exposed by dissection of cylindrical, multinucleated intermodal cells of the green alga, *Nitella axillaris*, as a "railroad" track to support the movement of micron-sized beads coated with myosin. This was a major innovation to myosin research, which yielded some very interesting insights. However, the assay had several limitations. *Nitella* had to be cultured in the lab, the dissection was not trivial, the preparations were typically only stable for an hour or so, and myosin had to be attached to a bead. The beads sometimes picked up the very fast endogenous myosin motors from the alga, and the movement was occurring on bundles of plant actins, with unknown specificity for mammalian myosins. Thus, while this assay was very useful, a truly in vitro assay that only used purified proteins was still needed.

Such an assay became available in 1986 when Kron and Spudich [5] introduced the sliding actin in vitro motility assay which required only two highly purified proteins, myosin and actin, to reconstitute the essence of a muscle contraction, the ability for myosin to translocate actin. This assay was made possible by the observations that single F-actin filaments were readily visible in solution by fluorescence microscopy, if labeled with rhodamine phalloidin [6]. Kron and Spudich described the movement of fluorescent rhodamine phalloidin labeled actin by myosin filaments from skeletal muscle or the amoeba, *Dictyostelium discoideum*, which had been bound to a glass coverslip surface [5]. This assay was easy to implement, used microscopy equipment already present in many labs, and was fast, reproducible, and quantitative. Furthermore, it could be performed with very small amounts of purified myosin and actin. The methodologies for this assay were further refined by Toyoshima et al. [7] who used a nitrocellulose-coated coverslip surface and demonstrated that not only myosin filaments but even the tailless proteolytic fragments of myosin, heavy meromyosin (HMM), and subfragment-1 (S1) moved actin. The motility assay provided a new way to classify and assess the characteristics of different classes of myosin by quantifying how fast they could move actin. The effects of varying ionic strength, magnesium, calcium, pH, and temperature on the speed of movement could be easily studied [8]. Below, we will describe this assay, highlight the evolution of this assay into versions that can measure mechanical properties of single molecules of myosin, and cover some of the major findings that have derived from the use of these elegant assays.

9.2 Discoveries Using the In Vitro Motility Assay

The development of the in vitro motility assay led to many important fundamental observations about the mechanism of actin and myosin interaction. The discovery that S1 moved actin filaments [7] proved fatal to several models for muscle contraction that required an organized sarcomere or the collapse of the coiled coil of myosin to power sarcomere shortening [9, 10]. Subsequent studies revealed important features of the assay. Single-headed myosins moved actin filaments demonstrating that a two-headed myosin was not necessary [7, 11]. The rate of movement of actin was largely independent of the concentration of myosin on the surface or on the length of the actin filament [7, 12]. Once the threshold of myosin density for movement was achieved, the movement occurred at the peak velocity. This meant that assays could be performed easily and quickly without having to control for actin filament length or myosin head density over a specific threshold.

The assay showed that there was amazing functional diversity among myosins and that the rate of actin filament sliding, even within the myosin class II family, varied from 0.02 μm/s for nonmuscle myosin 2B to 4 μm/s for fast skeletal muscle myosin [13]. Unconventional myosins, and not just class II myosins, were shown to be competent motors using this assay [14–17].

The best kinetic correlation for the rate at which the actin is moved in the motility assay is the rate at which ADP is released from actomyosin [18]. This means that the motility assay correlates well with the measurement of unloaded shortening velocity in muscle contraction, as they are both constrained by the rate of detachment of myosin from actin. The unloaded shortening velocity is a standard physiological parameter measured when a muscle mounted in a force transducer is allowed to shorten with no load applied. It is considered to be a measure of the maximal rate of myosin function in the muscle.

Structural data had proposed that myosin moved actin by swinging its neck region (also known as light-chain binding region or lever arm) [19]. This hypothesis was tested by showing a proportionality between the rate of movement and the neck length of mutant myosins created by inserting or deleting light-chain binding sites [20, 21].

9.3 Sliding Actin In Vitro Motility Assay

While the sliding actin in vitro motility assay has many variations in protocol, broadly, the assay observes fluorescently labeled actin filaments gliding over a cover slip that has been coated with myosin in the presence of ATP [22] (Fig. 9.1). This movement is captured with a sensitive video camera, often an ICCD (charged-coupled device) camera or other low-light imaging system. The movement of the labeled actin is tracked using software to estimate velocity.

9.3.1 General Design of the Motility Assay

9.3.1.1 Myosin Preparation

Ideally in vitro motility assays are performed with freshly prepared myosin. However, myosin that has been flash frozen and stored in liquid nitrogen is often used. Damaged myosin that no longer hydrolyzes ATP, but binds actin strongly (dead heads) can often interfere with motility and reduce the quality of actin movement. There are two techniques to minimize this problem. The myosin-coated coverslip

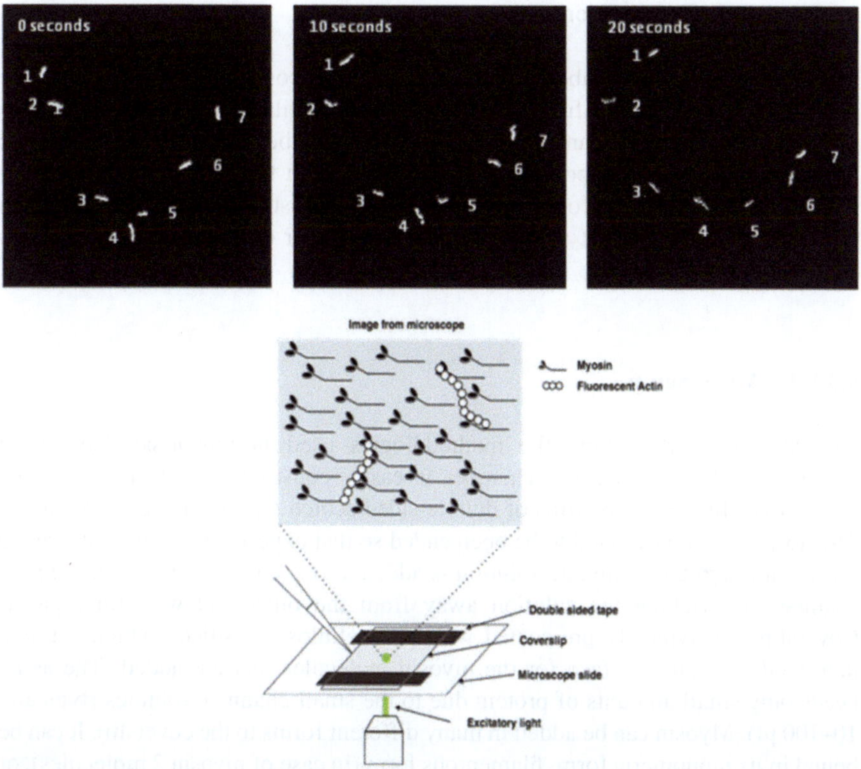

Fig. 9.1 Sliding actin in vitro motility assay. *Upper panel*, shown are the position of 7 actin filaments at 0, 10, and 20 s while moving over a surface coated with myosin 5. *Lower panel*, schematic of the assay flow chamber

can be washed with unlabeled actin (termed "black actin") in the presence of ATP that binds and blocks the damaged myosins before introducing the fluorescently labeled actin. Another technique, known as a "dead head spin," mixes stoichiometric amount of F-actin with myosin in the presence of ATP at high ionic strength, followed by sedimentation in an ultracentrifuge. Under these conditions, actively cycling myosins are dissociated from actin, and the dead heads remain tightly bound to actin and are removed from the supernatant. Finally, during the measurement, a viscosity-enhancing reagent, methylcellulose (at 0.7 % concentration), can be added to the motility buffer to increase the likelihood of continuous movement over surfaces sparsely coated with myosins or at higher ionic strength, where the affinity of myosin for actin is reduced. Methylcellulose restricts the Brownian diffusion of actin to its long axis and inhibits "escape" of the actin filament, if it is transiently free of any strong myosin interactions.

9.3.1.2 Actin Preparation

The actin is most often labeled with stoichiometric concentrations of rhodamine phalloidin. Phalloidin is a fungal toxin that stabilizes filamentous actin and can be conjugated to various organic dyes to fluorescently label the actin and reduce the critical concentration for polymerization, so that actin filaments remain polymerized even at nanomolar concentrations. Actin is most commonly purified from rabbit skeletal muscle, but actin from other species or different isoforms of actin can also be used.

9.3.1.3 Assay Setup

Several articles describing the methodologies used in this assay have been published [22–24]. An assay chamber is easily constructed with a microscope slide, coverslip, and two strips of double-sided scotch tape to create a flow track. The flow cell is constructed to be open ended so that experimental solutions can be flushed through the chamber. Solution is added at one end and drawn through the chamber by wicking the solution away from the other end with filter paper. Coverslips are typically pre-coated with nitrocellulose or silicon which act as a nonspecific binding surface for the myosin molecules that are added. The assay needs only small amounts of protein due to the small chamber volumes (typically 10–100 μl). Myosin can be added in many different forms to the coverslip. It can be bound in its monomeric form, filamentous form (in case of myosin 2 molecules), or can be in the form of proteolytic or truncation fragments such as HMM or S1. After myosin is bound, the flow cell is washed with blocking buffer containing a moderate concentration of bovine serum albumin to prevent nonspecific binding of actin filaments. Fluorescently labeled F-actin is added to the flow chamber. The actin is left for a short period to bind to myosin in the absence of ATP. Excess actin is washed from the flow chamber by the addition of a few volumes of buffer to the flow cell. Finally, a solution with ATP and an oxygen-scavenging cocktail containing glucose, glucose oxidase, and catalase is added to initiate motility. The oxygen-scavenging cocktail helps to reduce photobleaching of the fluorophore that occurs when free radicals are generated by the illumination of the sample. Video data can then be analyzed using tracking software to track individual filament movement and then extract sliding velocities and percentage of motile filaments.

9.3.2 Variations of the In Vitro Motility Assay

9.3.2.1 Myosin Filaments

Variations on the classical in vitro motility assay have been developed creating tailored techniques to investigate different forms and classes of myosin. Studies have shown that myosin 2 molecules bound to the coverslip surface in monomeric form move actin at the same speed as does this myosin when bound as filaments [7]. Several studies observed the sliding of actin over intact native thick filaments imaged by dark field microscopy [25, 26], demonstrating that actin moves most efficiently when moving toward the center of the thick filament.

9.3.2.2 Measurements Under Load

The movement of actin filaments in the standard sliding actin in vitro motility assay is under unloaded conditions. Several methods have been developed to measure the effect of load on myosin in this assay. Actin-cross-linking proteins, such as α-actinin, bound to the coverslip surface along with myosin create a frictional load by transiently interacting with actin filaments. This results in a slowing of the actin filament sliding speed [27, 28]. This technique allows the measurement of isotonic kinetics, force-velocity relationships, and power curves in the motility assay. Motility assays have been performed with mixtures of fast and slow myosins to produce relationships that describe the ability of the slower myosin to reduce the movement velocity of the actin over the surface of the mixture [29, 30].

9.3.2.3 Calcium Regulation

Calcium-dependent actin filament regulation was studied by the addition of the skeletal muscle regulatory proteins troponin and tropomyosin [31, 32] or the smooth muscle regulatory proteins caldesmon and calmodulin [33, 34]. Smooth muscle and nonmuscle myosin 2 are purified in an autoinhibited off state, which require phosphorylation by myosin light-chain kinase for activity. Molluscan muscle myosins require calcium for activity [25, 35, 36].

9.3.2.4 Nonrecombinant/Myosin from Cell Extracts

The motility assay can be tailored to select a specific myosin from an impure mixture by first binding an antibody that binds an epitope of a particular myosin to the coverslip. The surface is then coated with bovine serum albumin to block nonspecific binding of other proteins. The antibody selects for the desired myosin and unwanted myosins or other proteins are washed out [37, 38]. The selecting antibody should be one that does not interfere with the function of the desired myosin. Antibodies directed toward the tails (C-terminal regions) of myosin often are appropriate for this use. Such antibodies can also be used to attach the myosin to the surface rather than relying on nonspecific attachment of the myosin to the surface [39].

9.3.2.5 Polarity/Directionality

The F-actin filament is polar with a "+" end and a "−" end. Actin can be fluorescently labeled to demonstrate its polarity by using small actin "seeds" labeled with one fluorophore to nucleate the assembly of longer actin filaments which will then be labeled with another fluorophore [16, 40]. These types of actin filaments can be used to determine the directionality of the movement of F-actin by myosin. Thus far, all myosins move actin filaments with the + end leading except for myosin 6 [16].

9.4 Single-Molecule Motility Assays

The development of TIRF (total internal reflection fluorescent) microscopy allowed for the design of single-molecule motility assays where the movement of processive myosins along a surface-attached actin filament can be observed [41]. In TIRF microscopy, a laser beam is introduced to the coverslip at a critical angle, which results in complete internal reflection of the laser light at the coverslip-buffer boundary. This sets up an evanescent wave that penetrates the assay chamber to a depth of only 150–200 nm leaving all fluorophores above this in a dark, unexcited state. The result is a dramatic improvement in the signal-to-noise ratio to allow visualization of a single fluorophore.

9.4.1 Assay Setup

9.4.1.1 Myosin Labeling

There are several such ways of labeling myosins for this purpose. A fluorescently labeled light chain can be exchanged into myosin. The light chain is usually labeled with organic dyes [41] although light chains fused with fluorescent fusion proteins such as GFP can also be used [42]. Myosins with GFP fused to the heavy chain at either the amino (N-) or the carboxy (C-) terminus can also be used [43], or the myosin can be tagged with a quantum dot [44]. There are advantages and disadvantages to each of these approaches. Fusion of GFP molecules to the N- or C-termini of myosins does not seem to have a major effect on their activity, and these fluorescent fusion proteins are available in a variety of colors by expression of recombinant DNA in the baculovirus/Sf9 system [45]. However, the brightness and photostability of the fluorescent fusion proteins varies with the type of fusion protein selected, and these proteins, in general, are not as bright or as photostable as organic fluorophore dyes which can be introduced by chemical modification of surface lysines or cysteines residues. The use of quantum dots also has benefits and problems. Quantum dots come in different colors, are very bright, and maintain fluorescence over an extended period of time. However, they suffer from the blinking phenomena, which means that they periodically go dark and then emit light again. Their size is variable, but they are usually at least the size of a GFP molecule. The most common way to attach quantum dots to myosin is by introducing a biotin tag onto either the N-terminus or the C-terminus of a myosin and using streptavidin-functionalized quantum dots. If the quantum dot is attached to the N-terminus of a single head of myosin, the movement of that head is visualized, whereas if the quantum dot is attached to the C-terminus, the movement of the center of mass of the molecule is followed. Here, one must be very careful to only have one myosin attached per quantum dot, if single-molecule mechanics are to be explored. The way to accomplish this is to add only 1–4 myosin molecules per quantum dot in the coupling reaction. One can create an artificial "cargo"-bearing multiple myosins per quantum dot by increasing this labeling ratio.

Performing the Assay

In the single-molecule in vitro motility assay, actin filaments are bound to the coverslip surface via NEM-inactivated skeletal muscle HMM or via the use of lightly biotinylated actin filaments which can be tethered to a biotin-streptavidin-coated coverslip. Fluorescently labeled myosin is added and is visualized in a TIRF microscope. The images of the diffraction limited moving myosin spots are typically captured by a high-sensitivity camera (Fig. 9.2). Several parameters can be gleaned from this assay including speed, run length, and run frequency. Run length is usually determined by compiling a histogram of many events and fitting the curve with an exponential decay. This version of the single-molecule motility assay is now in routine use in many laboratories to study the processive movement of several classes of myosin, including myosin 5, myosin 6, myosin 7, and myosin 10 [17, 21, 41, 46, 47]. There is good agreement between rates that single processive molecules move along actin filaments and the rates at which these myosins slide actin filaments in the gliding assay.

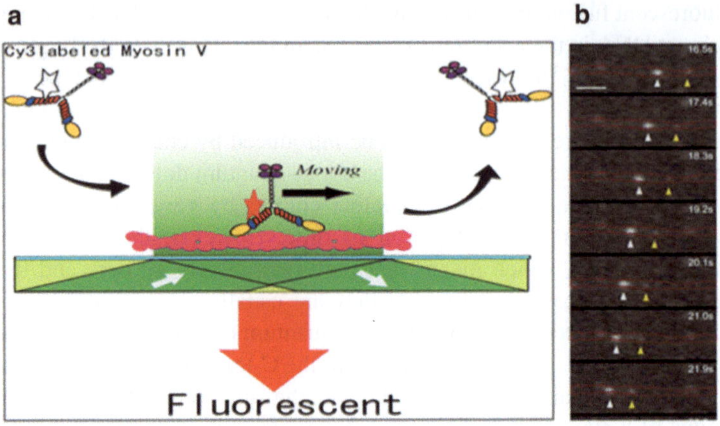

Fig. 9.2 Single-molecule in vitro motility assay. (**a**) Diagram of the geometry of the assay using myosin 5a. The *star* on myosin 5a represents a fluorophore bound to a calmodulin light chain. When it is in the evanescent field, it is depicted in *red* showing that the fluorophore is excited. The *dark green* shading shows the direction of the laser excitation. The actin filament bound to the surface is colored *red*. (**b**) The position of a moving fluorophore along actin as a function of time (*top* to *bottom*). In this case, the fluorophore is GFP-Rab27a which is bound to the motor, myosin 5a, via melanophilin. The image is taken from Wu et al. [63]. The position of the stationary actin filament is marked by *red lines*, and the positions of two moving spots are marked with *arrowheads*. The scale bar is 3.2 μm

9.5 Assay Variations

9.5.1 Attachment Lifetimes

There is at least one case, where this assay geometry was used to measure the lifetime of attachment of a nonprocessive myosin. Nagy et al. [48] used the single-molecule motility assay with nonmuscle myosin 2B HMM which was shown to be nonprocessive as a single molecule in optical trapping experiments, but to have a long attachment lifetime due to its very slow kinetics. GFP-tagged nonmuscle myosin 2B HMM molecules could be observed to bind to and then detach from surface-bound actin filaments, and the kinetics of these events matched those observed in the optical trapping experiments. Full-length nonmuscle myosin 2B molecules form short bipolar filaments in vitro at low ionic strengths and in cells. When GFP-labeled filaments of nonmuscle myosin 2B are formed, they move along the surface-attached actin filament for several microns at a slow rate consistent with the rate of actin filament translocation in the sliding actin assay.

9.5.2 Types of Actin Filaments

Actin filaments exist in pleiotropic forms in cells including single actin filaments, branched arbor-like networks, meshworks, loose bundles, and very tightly packed bundles [49]. These various states are formed and maintained by a slew of actin-binding proteins. Myosins are capable of interacting with different forms of actin, and different myosin isoforms are specifically localized to different areas of the cell or are found in actin-containing areas of specialized cells such as in the sarcomere of muscle fibers or the stereocilia of inner ear hair cells. The single-molecule motility assay can be performed on bundles of actin [50, 51] or with actin-containing regulatory proteins such as tropomyosin [52]. Actin filament meshworks can be created in vitro, and it is seen that myosin 5a can readily make abrupt turns and switch from one filament to another or can "step over" an intersecting filament [53]. Myosin 6, in contrast, prefers to stay on the initial filament [51]. Several studies have "deroofed" tissue culture cells by washing with a detergent which exposed the underlying actin cytoskeleton and then have added fluorescently tagged myosins to these cells to examine their motility in this more cellular-like in vitro setting [54, 55]. Interestingly, different classes of myosins prefer to move on different actin structures in these assays.

9.5.3 Calcium Regulation

Regulation can also be studied in the single-molecule motility assay. Purified full-length myosin 5a exists in an autoinhibited state where the C-terminal cargo-binding domain folds back and contacts the motor domain to adopt a state with very low actin-activated MgATPase and inability to move on actin [56–58]. This is clearly seen by examining the frequency of movement of full-length myosin 5a compared to that of the constitutively active myosin 5a HMM [57]: the latter moves much more frequently than the former. The ATPase activity of full-length myosin 5a in the presence of actin is activated by calcium ions, but paradoxically, this impairs the ability of the myosin to move processively on actin [59]. The reason for this apparent contradiction is that some calmodulin molecules dissociate from myosin 5a upon addition of calcium. This leaves the alpha-helical region of the IQ motif exposed and unstable, so the myosin is unable to convert its enzymatic output into a mechanical output [59, 60].

9.5.4 Cargo Regulation

Myosin 5a is involved in the actin-dependent movement of melanosomes which are membrane-bound pigment granules found in melanocytes. The linkage of myosin

5a to melanocytes is via two other proteins (see [3] for review). The cargo-binding domain of myosin 5a interacts with the C-terminus of a protein called melanophilin which in turn interacts through its N-terminus with Rab27a, a resident melanosome membrane-binding protein. Two studies have examined aspects of this complex. Wu et al. [61] mixed GFP-tagged Rab27a with melanophilin and myosin 5a and observed that fluorescent spots moved processively along actin filaments. No movement was seen if melanophilin or myosin 5a was omitted, and thus, it was concluded that the tripartite complex could be reconstituted in vitro at low concentrations. Melanophilin also has an actin-binding site which is important for enhancing the frequency of processive runs. A more recent study used fluorescently labeled myosin 5a and melanophilin to examine this in more detail [62]. It was shown that melanophilin significantly enhanced the run frequency and slowed the rate of movement of myosin 5a by acting as a tether.

Several classes of myosin including myosins 6, 7, and 10 do not appear to form stable dimers as purified from expression systems, despite the presence of a predicted coiled-coil motif in their tail sequence [46, 63–66]. Some of these myosins appear to be able to dimerize and move processively in single-molecule TIRF motility assays when "clustered" by first binding to actin filaments at high density in the absence of ATP or by interaction with certain binding partners that induce dimerization [46, 67, 68].

9.6 Super-Resolution Tracking of Moving Myosins

The single-molecule in vitro motility assay can be used to measure the step size of myosins through the use of super-resolution microscopy. The resolution of the conventional epifluorescence microscope is limited by the wavelength of light and is typically 200–300 nm. However, if sufficient photons can be collected from a single fluorophore, its point spread function can be fitted to a two-dimensional Gaussian whose peak can be localized with 2–3 nm resolution (Fig. 9.3). This technique was first used for myosins by Yildiz et al. [69] who coined the term FIONA (Fluorescence Imaging at One Nanometer Accuracy) for the technique. It allows the user to measure the size of the steps taken by myosin and to get information on the stepping kinetics. By lowering the ATP concentration to decrease the rate of stepping, which allows for the collection of more photons at each time point, Yildiz et al. [69] showed that a Cy3-labeled myosin 5a took 36 nm steps moving along actin, consistent with values obtained by optical trapping and EM imaging; see [70] for review. Examining the stepping distance and kinetics of molecules in which only one of the two heads was labeled provided evidence that the molecule stepped in a hand-over-hand mechanism where the trailing head dissociated, moved forward to a new actin-binding site 72 nm away. Interestingly, this stepping pattern precisely matches the crossover repeat of the helical actin filament and allows the myosin 5a to walk straight along an actin filament. The ability to observe single steps of myosin 5a was subsequently shown to be

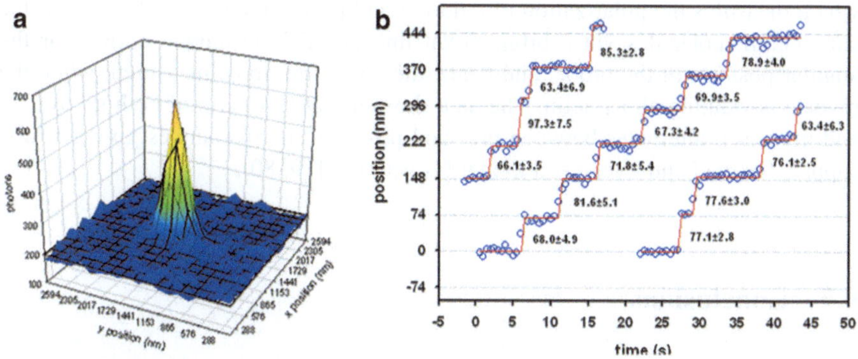

Fig. 9.3 **Nanometer** tracking of a myosin's movement by FIONA. (**a**) Cartoon demonstrating the fitting of a point spread function of a single fluorophore to a two-dimensional Gaussian. (**b**) Three traces of GFP-myosin 5a moving along actin with the position determined at each time point by the FIONA technique. Only a single head of each myosin is labeling with GFP. The *numbers* in between steps show the distance between successive steps along with the standard deviation. The data is taken from Snyder et al. [43]

applicable to GFP-labeled myosins [43] and to myosins with bound quantum dots [71]. The latter were used in an elegant study in which each head of a myosin 5a molecule was tagged with a quantum dot of a different color [44]. The two colors could be shown to alternate positions (separated by 36 nm) as the molecule moved processively along actin. The assay was used to confirm that the neck region acted as a rigid lever arm by examining the step size of the same set of myosin 5a lever mutants as was used for the in vitro motility studies described earlier [72]. Another study used simultaneous imaging of a labeled myosin 5a molecule and a fluorescent ATP analog, deac-aminoATP, whose fluorescence was enhanced 15-fold upon binding to myosin, to demonstrate that one ATP molecule was hydrolyzed for each step, demonstrating tight coupling between the enzymology and mechanics of this myosin [73].

The FIONA assay has also been applied to artificially dimerized myosin 6 [74], myosin 7a [46], and myosin 10 [75] molecules. The results show that myosin 6 moves along actin with a much more variable step size and that myosin 7a takes somewhat shorter steps than myosin 5a. FIONA measurements of myosin 10 are somewhat controversial with a range of values reported [75–77]. The longer step size of 36 nm is surprisingly similar in magnitude to that of myosin 5a as the lever arm of myosin 10 binds only three calmodulins. However, this myosin has a stable single alpha-helical (or SAH) domain which serves to extend the lever arm as determined by electron microscopy [65]. A chimeric molecule in which the motor domain plus first three IQ motifs of myosin 5a was fused to a SAH domain with similar length and dimerized by a portion of the coiled-coil domain took 36 nm steps consistent with the SAH domain being part of a rigid lever arm [78].

Another variation of the single-molecule motility assay allows one to gain insight into the detailed mechanisms of myosin movement. polTIRF is a technique which measures the polarization of a fluorescent probe rigidly bound to one of the calmodulin molecules via a bifunctional linkage. This technique allows for the angular position of the probe, and hence the lever arm, to be measured, while the myosin is walking along actin. The data showed that lever arm of both myosin 5a and myosin 6 alternates between pre- and post-powerstroke angles giving strong support for the hand-over-hand model for walking [79, 80].

9.7 Conclusions

As can be seen from the above, in vitro motility assays have dramatically increased our understanding of molecular mechanism of myosin-actin interactions and have given insight into the regulation of this interaction and to the functional diversity found in the myosin superfamily. The assay has been adapted to study myosins from 13 of the known classes. The fundamental principles involved in the assay, the ability to observe fluorescently labeled actin filaments and to bind myosin to a surface, led to the development of optical trapping techniques that allows one to measure lifetimes of interaction, power stroke size, and force production of single-myosin molecules [81]. Variants of the assay are being continually developed. These include using zero-mode waveguides to further increase the sensitivity of the assay at higher fluorophore concentrations [82] and the use of DNA origami to create scaffolds for binding more than one motor [83].

Acknowledgments We thank Yasuharu Takagi for the images shown in Fig. 9.1.

References

1. Sellers JR (2000) Myosins: a diverse superfamily. Biochim Biophys Acta 1496:3–22
2. Odronitz F, Kollmar M (2007) Drawing the tree of eukaryotic life based on the analysis of 2269 manually annotated myosins from 328 species. Genome Biol 8:R196
3. Hammer JA III, Sellers JR (2012) Walking to work: roles for class V myosins as cargo transporters. Nat Rev Mol Cell Biol 13:13–26
4. Sheetz MP, Spudich JA (1983) Movement of myosin-coated fluorescent beads on actin cables in vitro. Nature 303:31–35
5. Kron SJ, Spudich JA (1986) Fluorescent actin filaments move on myosin fixed to a glass surface. Proc Natl Acad Sci USA 83:6272–6276
6. Yanagida T, Nakase M, Nishiyama K, Oosawa F (1984) Direct observation of motion of single F-actin filaments in the presence of myosin. Nature 307:58–60
7. Toyoshima YY, Kron SJ, McNally EM, Niebling KR, Toyoshima C, Spudich J (1987) Myosin subfragment-1 is sufficient to move actin filaments in vitro. Nature 328:536–539
8. Homsher E, Wang F, Sellers JR (1992) Factors affecting movement of F-actin filaments propelled by skeletal muscle heavy meromyosin. Am J Physiol Cell Physiol 262:C714–C723
9. Harrington WF (1971) A mechanochemical mechanism for muscle contraction. Proc Natl Acad Sci USA 68:685–689

10. Oplatka A (1994) The role of water in the mechanism of muscular contraction. FEBS Lett 355:1–3
11. Harada Y, Noguchi A, Kishino A, Yanagida T (1987) Sliding movement of single actin filaments on one-headed myosin filaments. Nature 326:8005–808
12. Harada Y, Yanagida T (1988) Direct observation of molecular motility by light microscopy. Cell Motil Cytoskeleton 10:71–76
13. Sellers JR (1999) Myosins, 2nd edn. Oxford University Press, Oxford
14. Collins K, Sellers JR, Matsudaira P (1990) Calmodulin dissociation regulates brush border myosin I 110-kD-calmodulin) mechanochemical activity in vitro. J Cell Biol 110:1137–1147
15. Nascimento AAC, Cheney RE, Tauhata SBF, Larson RE, Mooseker MS (1996) Enzymatic characterization and functional domain mapping of brain myosin-V. J Biol Chem 271:17561–17569
16. Wells A, Lin AW, Chen L-Q, Safer D, Cain SM, Hasson T, Carragher BO, Milligan RA, Sweeney HL (1999) Myosin VI is an actin-based motor that moves backwards. Nature 401:505–508
17. Nagy S, Ricca BL, Norstrom MF, Courson DS, Brawley CM, Smithback PA, Rock RS (2008) A myosin motor that selects bundled actin for motility. Proc Natl Acad Sci USA 105:9616–9620
18. Yengo CM, Takagi Y, Sellers JR (2012) Temperature dependent measurements reveal similarities between muscle and non-muscle myosin motility. J Muscle Res Cell Motil 33:385–394
19. Rayment I, Holden HM, Whittaker M, Yohn CB, Lorenz M, Holmes KC, Milligan RA (1993) Structure of the actin-myosin complex and its implications for muscle contraction. Science 261:58–65
20. Warshaw DM, Guilford WH, Freyzon Y, Krementsova E, Palmiter KA, Tyska MJ, Baker JE, Trybus KM (2000) The light chain binding domain of expressed smooth muscle heavy meromyosin acts as a mechanical lever. J Biol Chem 275:37167–37172
21. Sakamoto T, Wang F, Schmitz S, Xu YH, Xu Q, Molloy JE, Veigel C, Sellers JR (2003) Neck length and processivity of myosin V. J Biol Chem 278:29201–29207
22. Sellers JR, Cuda G, Wang F, Homsher E (1993) Myosin-Specific Adaptations of Motility Assays. In: Scholey JM (ed) Motility Assays for Motor Proteins. CA, Academic Press, Inc., San Diego, pp 23–49
23. Sellers JR (2006) In vitro motility assays with actin. In: Celis JE (ed) Cell biology. A laboratory handbook, vol 2. Academic, Amsterdam, pp 387–392
24. Kron SJ, Toyoshima YY, Uyeda TQP, Spudich JA (1991) Assays for actin sliding movement over myosin-coated surfaces. Methods Enzymol 196:399–416
25. Sellers JR, Kachar B (1990) Polarity and velocity of sliding filaments. Control of direction by actin and of speed by myosin. Science 249:406–408
26. Previs MJ, Beck PS, Gulick J, Robbins J, Warshaw DM (2012) Molecular mechanics of cardiac myosin-binding protein C in native thick filaments. Science 337:1215–1218
27. Janson LW, Sellers JR, Taylor DL (1992) Actin-binding proteins regulate the work performed by myosin II motors on single actin filaments. Cell Motil Cytoskeleton 22:274–280
28. Greenberg MJ, Moore JR (2010) The molecular basis of frictional loads in the in vitro motility assay with applications to the study of the loaded mechanochemistry of molecular motors. Cytoskeleton (Hoboken) 67:273–285
29. Harris DE, Work SS, Wright RK, Alpert NR, Warshaw DM (1994) Smooth, cardiac and skeletal muscle myosin force and motion generation assessed by cross-bridge mechanical interactions *in vitro*. J Muscle Res Cell Motil 15:11–19
30. Cuda G, Pate E, Cooke R, Sellers JR (1997) In vitro actin filament sliding velocities produced by mixtures of different types of myosin. Biophys J 72:1767–1779
31. Lin D, Bobkova A, Homsher E, Tobacman LS (1996) Altered cardiac troponin T in vitro function in the presence of a mutation implicated in familial hypertrophic cardiomyopathy. J Clin Invest 97:2842–2848

32. Fraser IDC, Marston SB (1995) *In vitro* motility analysis of actin-tropomyosin regulation by troponin and calcium. The thin filament is switched as a single cooperative unit. J Biol Chem 270:7836–7841
33. Shirinsky VP, Biryukov KG, Hettasch JM, Sellers JR (1992) Inhibition of the relative movement of actin and myosin by caldesmon and calponin. J Biol Chem 267:15886–15892
34. Fraser IDC, Marston SB (1995) *In vitro* motility analysis of smooth muscle caldesmon control of actin-tropomyosin filament movement. J Biol Chem 270:19688–19693
35. Umemoto S, Sellers JR (1990) Characterization of in vitro motility assays using smooth muscle and cytoplasmic myosins. J Biol Chem 265:14864–14869
36. Kurzawa-Goertz SE, Perreault-Micale CL, Trybus KM, Szent-Gyorgyi AG, Geeves MA (1998) Loop I can modulate ADP affinity, ATPase activity, and motility of different scallop myosins. Transient kinetic analysis of S1 isoforms. Biochemistry 37:7517–7525
37. Cuda G, Fananapazir L, Zhu W-S, Sellers JR, Epstein ND (1993) Skeletal muscle expression and abnormal function of b-myosin in hypertrophic cardiomyopathy. J Clin Invest 91:2861–2865
38. Reck-Peterson SL, Tyska MJ, Novick PJ, Mooseker MS (2001) The yeast class V myosins, Myo2p and Myo4p, are nonprocessive actin- based motors. J Cell Biol 153:1121–1126
39. Sweeney HL, Rosenfeld SS, Brown F, Faust L, Smith J, Xing J, Stein LA, Sellers JR (1998) Kinetic tuning of myosin via a flexible loop adjacent to the nucleotide binding pocket. J Biol Chem 273:6262–6270
40. Liao JC, Elting MW, Delp SL, Spudich JA, Bryant Z (2009) Engineered myosin VI motors reveal minimal structural determinants of directionality and processivity. J Mol Biol 392:862–867
41. Sakamoto T, Amitani I, Yokota E, Ando T (2000) Direct observation of processive movement by individual myosin V molecules. Biochem Biophys Res Commun 272:586–590
42. Kengyel A, Wolf WA, Chisholm RL, Sellers JR (2010) Nonmuscle myosin IIA with a GFP fused to the N-terminus of the regulatory light chain is regulated normally. J Muscle Res Cell Motil 31:163–170
43. Snyder GE, Sakamoto T, Hammer JA III, Sellers JR, Selvin PR (2004) Nanometer localization of single green fluorescent proteins: evidence that myosin V walks hand-over-hand via telemark configuration. Biophys J 87:1776–1783
44. Warshaw DM, Kennedy GG, Work SS, Krementsova EB, Beck S, Trybus KM (2005) Differential labeling of myosin V heads with quantum dots allows direct visualisation of hand-over-hand processivity. Biophys J 88: L30–L32
45. Lippincott-Schwartz J, Patterson GH (2003) Development and use of fluorescent protein markers in living cells. Science 300:87–91
46. Yang Y, Kovacs M, Sakamoto T, Zhang F, Kiehart DP, Sellers JR (2006) Dimerized Drosophila myosin VIIa: a processive motor. Proc Natl Acad Sci USA 103:5746–5751
47. Rock RS, Rice SE, Wells AL, Purcell TJ, Spudich JA, Sweeney HL (2001) Myosin VI is a processive motor with a large step size. Proc Natl Acad Sci USA 98:13655–13659
48. Nagy A, Takagi Y, Billington N, Sun SA, Hong DK, Homsher E, Wang A, Sellers JR (2012) Kinetic Characterization of Nonmuscle Myosin IIB at the Single Molecule Level. J Biol Chem 288:709–722
49. Pollard TD, Cooper JA (2009) Actin, a central player in cell shape and movement. Science 326:1208–1212
50. Nagy S, Rock RS (2010) Structured post-IQ domain governs selectivity of myosin X for fascin-actin bundles. J Biol Chem 285:26608–26617
51. Ali MY, Previs SB, Trybus KM, Sweeney HL, Warshaw DM (2013) Myosin VI has a one track mind versus myosin Va when moving on actin bundles or at an intersection. Traffic 14:70–81
52. Hodges AR, Krementsova EB, Bookwalter CS, Fagnant PM, Sladewski TE, Trybus KM (2012) Tropomyosin is essential for processive movement of a class v Myosin from budding yeast. Curr Biol 22:1410–1416

53. Ali MY, Krementsova EB, Kennedy GG, Mahaffy R, Pollard TD, Trybus KM, Warshaw DM (2007) From the Cover: Myosin Va maneuvers through actin intersections and diffuses along microtubules. Proc Natl Acad Sci USA 104:4332–4336
54. Brawley CM, Rock RS (2009) Unconventional myosin traffic in cells reveals a selective actin cytoskeleton. Proc Natl Acad Sci USA 106:9685–9690
55. Sivaramakrishnan S, Spudich JA (2009) Coupled myosin VI motors facilitate unidirectional movement on an F-actin network. J Cell Biol 187:53–60
56. Krementsov DN, Krementsova EB, Trybus KM (2004) Myosin V: regulation by calcium, calmodulin, and the tail domain. J Cell Biol 164:877–886
57. Thirumurugan K, Sakamoto T, Hammer JA III, Sellers JR, Knight PJ (2006) The cargo-binding domain regulates structure and activity of myosin 5. Nature 442:212–215
58. Li XD, Jung HS, Mabuchi K, Craig R, Ikebe M (2006) The globular tail domain of myosin Va functions as an inhibitor of the myosin Va motor. J Biol Chem 281:21789–21798
59. Lu H, Krementsova EB, Trybus KM (2006) Regulation of myosin V processivity by calcium at the single molecule level. J Biol Chem 281:31987–31994
60. Koide H, Kinoshita T, Tanaka Y, Tanaka S, Nagura N, Meyer zu HG, Miyagi A, Ando T (2006) Identification of the single specific IQ motif of myosin V from which calmodulin dissociates in the presence of Ca2+. Biochemistry 45:11598–11604
61. Wu X, Sakamoto T, Zhang F, Sellers JR, Hammer JA III (2006) In vitro reconstitution of a transport complex containing Rab27a, melanophilin and myosin Va. FEBS Lett 580:5863–5868
62. Sckolnick M, Krementsova EB, Warshaw DM, Trybus KM (2013) More Than Just a Cargo Adapter: Melanophilin Prolongs and Slows Processive Runs of Myosin Va. J Biol Chem 288:29313–291322
63. Lister I, Schmitz S, Walker M, Trinick J, Buss F, Veigel C, Kendrick-Jones J (2004) A monomeric myosin VI with a large working stroke. EMBO J 23:1729–1738
64. Umeki N, Jung HS, Watanabe S, Sakai T, Li XD, Ikebe R, Craig R, Ikebe M (2009) The tail binds to the head-neck domain, inhibiting ATPase activity of myosin VIIA. Proc Natl Acad Sci USA 106:8483–8488
65. Knight PJ, Thirumurugan K, Yu Y, Wang F, Kalverda AP, Stafford WF III, Sellers JR, Peckham M (2005) The predicted coiled-coil domain of myosin 10 forms a novel elongated domain that lengthens the head. J Biol Chem 280:34702–34708
66. Umeki N, Jung HS, Sakai T, Sato O, Ikebe R, Ikebe M (2011) Phospholipid-dependent regulation of the motor activity of myosin X. Nat Struct Mol Biol 18:783–788
67. Park H, Ramamurthy B, Travaglia M, Safer D, Li-Qiong C, Franzini-Armstrong C, Selvin PR, Sweeney HL (2006) Full-length myosin VI dimerizes and moves processively along actin filaments upon monomer clustering. Mol Cell 21:331–336
68. Phichith D, Travaglia M, Yang Z, Liu X, Zong AB, Safer D, Sweeney HL (2009) Cargo binding induces dimerization of myosin VI. Proc Natl Acad Sci USA 106:17320–17324
69. Yildiz A, Forkey JN, McKinney SA, Ha T, Goldman YE, Selvin PR (2003) Myosin V walks hand-over-hand: single fluorophore imaging with 1.5-nm localization. Science 300:2061–2065
70. Sellers JR, Veigel C (2006) Walking with myosin V. Curr Opin Cell Biol 18:68–73
71. Ali MY, Lu H, Bookwalter CS, Warshaw DM, Trybus KM (2008) Myosin V and Kinesin act as tethers to enhance each others' processivity. Proc Natl Acad Sci USA 105:4691–4696
72. Sakamoto T, Yildiz A, Selvin PR, Sellers JR (2005) Step-size is determined by neck length in myosin V. Biochemistry 44:16203–16210
73. Sakamoto T, Webb MR, Forgacs E, White HD, Sellers JR (2008) Direct observation of the mechanochemical coupling in myosin Va during processive movement. Nature 455:128–132
74. Yildiz A, Park H, Safer D, Yang Z, Chen LQ, Selvin PR, Sweeney HL (2004) Myosin VI steps via a hand-over-hand mechanism with its lever arm undergoing fluctuations when attached to actin. J Biol Chem 279:37223–37226
75. Ricca BL, Rock RS (2010) The stepping pattern of myosin X is adapted for processive motility on bundled actin. Biophys J 99:1818–1826

76. Sun Y, Sato O, Ruhnow F, Arsenault ME, Ikebe M, Goldman YE (2010) Single-molecule stepping and structural dynamics of myosin X. Nat Struct Mol Biol 17:485–491
77. Bao J, Huck D, Gunther LK, Sellers JR, Sakamoto T (2013) Actin Structure-Dependent Stepping of Myosin 5a and 10 during Processive Movement. PLoS One 8:e74936
78. Baboolal TG, Sakamoto T, Forgacs E, White HD, Jackson SM, Takagi Y, Farrow RE, Molloy JE, Knight PJ, Sellers JR, Peckham M (2009) The SAH domain extends the functional length of the myosin lever. Proc Natl Acad Sci USA 106:22193–22198
79. Syed S, Snyder GE, Franzini-Armstrong C, Selvin PR, Goldman YE (2006) Adaptability of myosin V studied by simultaneous detection of position and orientation. EMBO J 25:1795–1803
80. Sun Y, Schroeder HW III, Beausang JF, Homma K, Ikebe M, Goldman YE (2007) Myosin VI walks "wiggly" on actin with large and variable tilting. Mol Cell 28:954–964
81. Finer JT, Simmons RM, Spudich JA (1994) Single myosin molecule mechanics: Piconewton forces and nanometre steps. Nature 368:113–119
82. Elting MW, Leslie SR, Churchman LS, Korlach J, McFaul CM, Leith JS, Levene MJ, Cohen AE, Spudich JA (2013) Single-molecule fluorescence imaging of processive myosin with enhanced background suppression using linear zero-mode waveguides (ZMWs) and convex lens induced confinement (CLIC). Opt Express 21:1189–1202
83. Lu H, Efremov AK, Bookwalter CS, Krementsova EB, Driver JW, Trybus KM, Diehl MR (2012) Collective dynamics of elastically coupled myosin V motors. J Biol Chem 287:27753–27761

Chapter 10
Fluorescence Tracking of Motor Proteins In Vitro

Mark DeWitt, Thomas Schenkel, and Ahmet Yildiz

Abstract Motor proteins convert the chemical energy of adenosine triphosphate (ATP) hydrolysis into directed movement along filamentous tracks, such as DNA, microtubule, and actin. The motile properties of motors are essential to their wide variety of cellular functions, including cargo transport, mitosis, cell motility, nuclear positioning, and ciliogenesis. Detailed understanding of the biophysical mechanisms of motor motility is therefore essential to understanding the physical basis of these processes. In which direction is the motor going? How fast and how far can a single motor walk down its track? How is ATP hydrolysis coupled to directed motion? How do multiple subunits of a motor coordinate with each other during motility? These questions can be addressed directly by tracking motors at a single-molecule level. This chapter will focus on high-resolution fluorescence tracking techniques of the processive cytoskeletal motors: myosins, kinesins, and cytoplasmic dynein. We outline the theoretical and practical considerations for studying these motors in vitro using fluorescence tracking at nanometer precision.

Keywords Molecular motors • Fluorescence tracking • Single molecule • Microscopy techniques

Abbreviations

ATP	Adenosine triphosphate
BSA	Bovine serum albumin
CCD	Charge-coupled device

M. DeWitt • A. Yildiz (✉)
Biophysics Graduate Group and Physics Department, University of California, Berkeley, CA 94720, USA
e-mail: yildiz@berkeley.edu

T. Schenkel
Accelerator and Fusion Research Division, Lawrence Berkeley National Laboratory, Berkeley, CA, USA

C.P. Toseland and N. Fili (eds.), *Fluorescent Methods for Molecular Motors*, Experientia Supplementum 105, DOI 10.1007/978-3-0348-0856-9_10,

Cy3	Cyanine-3
Cy5	Cyanine-5
DFT	Dark-field tracking
EM	Electron multiplication
FP	Fluorescent protein
FRET	Fluorescence resonance energy transfer
GFP	Green fluorescent protein
PCA/PCD	Protocatechuic acid/protocatechuate-3,4-dioxygenase system
PSF	Point spread function
QD	Quantum dot nanocrystal
SHRImP	Single-molecule high-resolution imaging with photobleaching
SNR	Signal-to-noise ratio
TIR	Total internal reflection
TMR	Tetramethylrhodamine

10.1 Imaging of Single Molecules

Direct observation of the motility of subcellular structures, including molecular motors, long remained elusive due to lack of appropriate microscopy techniques. The development of transmission light microscopy methods (e.g., bright field, differential interference contrast, phase contrast methods) and digital video microscopy allowed tracking the movement of large (>~300 nm) organelles and microtubule filaments [1]. However, most organelles and all individual proteins were invisible, because they do not have sufficient contrast to be detected using these methods. Early attempts to track single molecules involved attachment of large beads or long polymers to make their motility visible under a microscope [2]. However, these methods are usually limited to in vitro applications because of the large probe size and provide information about the overall motility of a motor (e.g., speed and run length and directionality). They are not suitable for in vivo labeling or high-resolution tracking, because of the Brownian motion of the large probe under thermal fluctuations. Optical trapping methods have frequently been used to study the movement of molecular motors under an applied load [3], in which the Brownian motion of the bead is greatly reduced and the bead position can be tracked with sub-nanometer resolution.

Single-molecule fluorescent detection [4] has opened new opportunities for particle tracking studies, because of their small size, availability of site-specific labeling strategies for protein conjugation, and high signal-to-noise ratio (SNR). Due to the large Stokes shift between their absorption and fluorescence emission spectra, fluorophores can be excited by intense laser illumination and fluorescent photons are readily isolated using optical filters, allowing detection of extremely weak signals. However, the detection of single fluorophores in solution posed a bigger challenge because the signal of a single fluorophore is usually weaker than the Raman scattering of 1 μm^3 of water, which overlaps with the spectrum of the fluorescence emission. In order to minimize the background, several attempts centered on reducing the excitation volume, effectively reducing the Raman signal. Confocal microscopy focuses light into a diffraction-limited spot and utilizes a small pinhole in front of a single-channel detector to block out-of-focus light, allowing 3D imaging of whole cells at single-molecule sensitivity [5]. However, in order to create an image using confocal microscopy, the focused spot needs to be scanned over the region of interest. As a result, the time resolution of scanning confocal microscopy is lower than wide-field excitation, which limits its suitability for tracking fast-moving particles. The development of spinning disk confocal greatly improved the time resolution of the confocal method, at the cost of reduced signal detection and high cross talk between adjacent pixels due to spurious emission from out-of-focus planes, which can increase background in thick samples [6].

Axelrod and coworkers used total internal reflection (TIR) of the excitation light to greatly improve the sensitivity of fluorescence detection without sacrificing time resolution [7]. TIR creates an evanescent field at the glass/water interface of a sample chamber (Fig. 10.1a). The evanescent wave does not propagate through the sample and illuminates to a depth of ~100 nm in water, greatly reducing the Raman cross section. The method is only applicable when the imaged object is bound to the coverslip surface (for in vitro imaging) or at or near the cell surface (in vivo). Use of TIR alone greatly reduces the background, but SNR still suffers from the glow of objective lenses and scattered light bypassing fluorescence filters, when the TIR beam is sent directly through the objective. In order to further reduce the background, the TIR beam can be directed to the top surface of the sample chamber using a prism. As a result, the excitation path is fully separated from the detection path, which results in further improvement in SNR. The use of TIR microscopy allowed detection of single fluorescently labeled motors in solution [8, 9]. Due to the combined ease of use, superior SNR, and high time resolution, TIR is the most widely used method for tracking motor proteins in vitro.

10.2 Particle Tracking Beyond the Diffraction Limit: Theory and Practical Considerations

While the development of fluorescent TIRF video microscopy allowed routine tracking of individual fluorescent molecules, the ability of conventional light microscopy to resolve fine details in the image plane is limited by the diffraction of detected light. Particles smaller than the diffraction-limited spot, the so-called Point Spread Function (PSF, Fig. 10.1b), cannot be observed [10]. The width of the PSF is equal to $\lambda/(2NA)$, where NA is the numerical aperture of the microscope

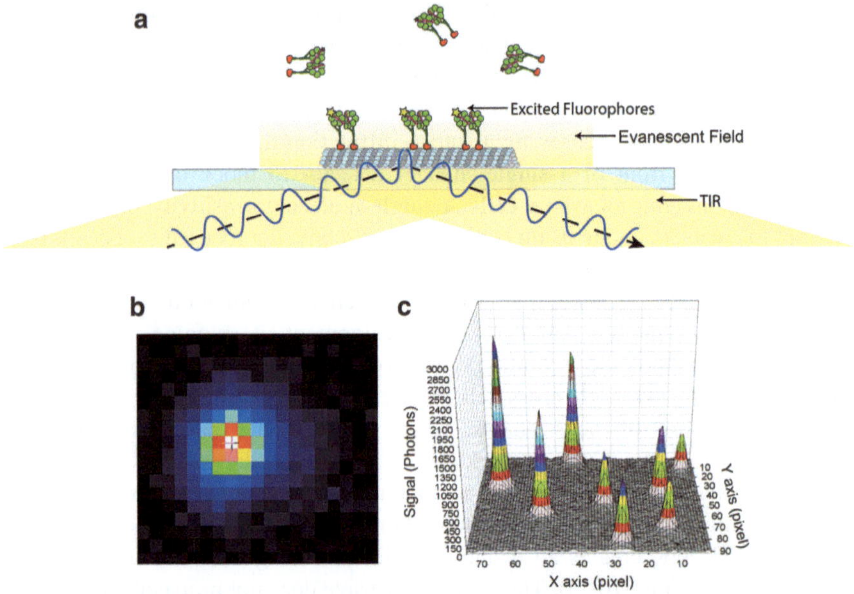

Fig. 10.1 Fluorescence tracking of molecular motors. (**a**) Schematic of typical in vitro single-molecule motility assay. Cytoskeletal tracks are immobilized on a glass coverslip. Fluorescently labeled motors are flown into the sample chamber. Excitation light is total internally reflected at the glass water boundary of a coverslip. The evanescent field created by the TIR excitation decays exponentially with 100 nm decay constant. Therefore, fluorescently labeled motors bound to cytoskeletal filaments are selectively excited, even in the presence of diffusing motors in the chamber. (**b**) Images of individual fluorophores are diffraction limited and generate PSFs. (**c**) Three-dimensional representation of a camera image shows that individual PSFs can be approximated by a two-dimensional Gaussian. While the width of the Gaussian is large (~300 nm), its center can be tracked with nanometer precision if a sufficient number of photons per image is collected

objective (1.4–1.65 NA objectives are available). For visible light, this makes resolution of distances smaller than 200–300 nm impossible [11].

High-resolution fluorescence tracking is a means of getting around the diffraction limit and determining the "true" location of a fluorophore with nanometer accuracy. However, it only works under certain conditions. First, a fluorophore must act as an isotropic emitter during the acquisition of each image [12] and the microscope must be free of aberrations to ensure that the center of the PSF represents the position of a dye [13]. Second, the fluorophores must be physically separated so as to produce distinct spots, and each spot must be sufficiently bright. This makes imaging of fluorescent molecules in crowded environments (e.g., in cells) very challenging, unless photoactivation [14] or superresolution imaging [15] techniques are employed.

Another challenge to tracking single fluorophores is the precise localization of particle position. Under appropriate conditions, the center of the PSF represents the particle position. To find this center, the spot intensity profile is fitted to a peak function, and the center of this fit is taken as the true location of the emitter

(Fig. 10.1c). Most particle tracking algorithms fit the intensity profile of a spot to a two-dimensional Gaussian function, which is a good approximation of the central peak of the PSF [16]. However, other algorithms have been employed, such as centroid tracking or Gaussian mask fitting, although these can introduce biases into tracking data and/or reduce tracking accuracy [17]. Recently developed maximum-likelihood fitting with empirical functions that better approximate the PSF can improve the fitting by up to one third, without the biases introduced by other methods [18]. In our experience, least-squares fitting with a 2D Gaussian function yields the best results for bright probes [19], while a specially designed Gaussian mask algorithm is more suitable for dim probes [20].

The accuracy of particle tracking is limited by several noise factors. Statistical noise in photon counting (shot noise) is a fundamental limitation. In theory, in a photon-shot noise-limiting case, the precision of Gaussian center localization, or the standard error of the mean, is equal to $\sigma/\sqrt{N}$, where σ is the standard deviation of a Gaussian (approximately one half of the PSF width, ~100 nm) and N is the total number of detected photon per image [10]. Using this equation, ~10^4 photons need to be detected to achieve 1 nm precision.

In most cases, the actual noise of the experiment is higher than the photon-shot noise. Background fluorescence caused by the excitation light is discussed above. The choice of camera for TIRF imaging is also critical to reducing detection-related noise, such as camera pixilation and readout noise. The ideal pixel size of the magnified image should be between 80 and 150 nm [16], depending on the SNR. The readout noise of the camera can be reduced by amplifying the signal before readout. The most common amplification technique is electron multiplication (EM), wherein the photon signal (photoelectric current) is amplified prior to digital readout. EM-CCD (EM charge-coupled device) cameras currently predominate, because they almost completely eliminate camera readout noise without creating artifacts in the pixelated image. However, EM is a random process which multiplies the photon-shot noise by ~$\sqrt{2}$ along with true signal. As a result, EM is ideal for low (<6) SNR imaging but reduces the precision in particle tracking at high SNR. Newly developed scientific CMOS cameras are promising, because they have a very low readout noise and hence do not require amplification. However, the most advanced front-illuminated (where the amplification electronics lie in front of the photo-detection surface, reflecting some of the incident light) CMOS chips have 70 % photon detection efficiency, which is low compared to 92 % of back-illuminated (where the electronics are behind the photo-detection surface) EM-CCDs in the market. Furthermore, pixel-to-pixel variation in detection efficiency and digitization creates artifacts in CMOS images, which are particularly problematic for two-color tracking experiments (see below). The anticipated development of back-illuminated CMOS cameras, as well as the better control of pixel-pixel variation, may overcome some of these difficulties.

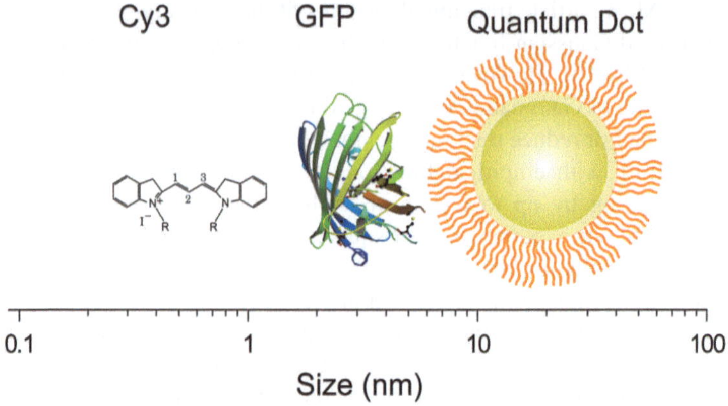

Fig. 10.2 Probes for fluorescent tracking of molecular motors. Organic dyes such as Cy3 are exceptionally small (1 nm) but are dim compared to QD probes. Fluorescent proteins such as GFP are of intermediate (2–4 nm) size and are ideally suited for in vivo studies but are too dim for high-resolution tracking. QD probes are exceptionally bright but are very large (>15 nm). Synthesis and successful implementation of "small" (<10 nm) QDs will open new possibilities for future tracking studies

10.3 Tracking Fluorescently Labeled Molecular Motors In Vitro

Motor proteins are well suited to single-molecule tracking, because they walk along polymer tracks, such as microtubules and actin, which can be readily fixed to a glass surface. High-resolution tracking of motors requires that two conditions be met: Motors must be spatially isolated from one another on a filament and labeled with a bright probe to estimate the center of an isolated spot (Figs. 10.1 and 10.2). Single-molecule assays reveal the motile properties of processive motors, such as speed, run length, and directionality [21]. For non-processive motors, filament-gliding assays, in which the motors are immobilized on a coverslip and fluorescent tracks glide on motor-coated surfaces [22], are more suitable (see Chap. 9).

10.3.1 Choosing the Right Fluorophore for the Application

The choice of fluorescent probe is critical for tracking time and precision and dye photostability in solution must be optimized to achieve high precision (Fig. 10.2). For tracking in vivo, where labeling may be difficult, genetically encoded fluorescent proteins (FPs) such as green fluorescent protein (GFP) can be used, but they are not bright or photostable enough to accurately resolve individual steps taken by motors, unless the target can be labeled with >10 FPs [23] (Chap. 13, single-molecule imaging in cells). Most in vitro studies use organic dyes, because they have higher photostability compared to FPs. These dyes are small in size, are available in many colors and can be covalently linked to a variety of biological targets. The three primary pathways of dye photobleaching (irreversible loss of fluorescence under intense illumination) are photooxidation by singlet oxygen species, triplet excited-state formation, and charge-separated states by electron transfer reactions [24]. Several enzymatic approaches have been used to scavenge free oxygen from solution. Glucose oxidase and catalase have been commonly used, but this method produces peroxides that slowly acidify the solution [25]. The mixture of PCA/PCD (protocatechuic acid/protocatechuate-3,4-dioxygenase) system yields superior results without affecting pH and has been used more widely in recent years [26]. Anti-blinking reagents such as Trolox and β-mercaptoethanol (BME) have been used to prevent triplet-state arrest of dyes [27]. The best performing organic dyes, such as tetramethylrhodamine (TMR) and cyanine-3 (Cy3), have been successfully used for tracking the stepping of cytoskeletal motors [19, 28, 29]. Single fluorophores were tracked with 1 nm precision for more than a minute. Red-shifted dyes such as Cy5 suffer from poor photostability, making the selection of a second color difficult [27]. Alexa and Atto dyes may offer improved photostability at these wavelengths [30]. Alternatively, newly developed triplet-state quencher conjugates significantly improve the photostability of Cy5 [31], which may allow minutes-long tracking of Cy5, comparable to Cy3.

Quantum dot (QD) nanocrystals offer unprecedented photostability and brightness, allowing tracking at nanometer accuracy for several minutes without photobleaching [32] (Fig. 10.2). QDs have been used to track kinesin, myosin, and dynein motors, providing detailed information about the step size and kinetics at very high temporal resolution (up to 500 images per second). Unfortunately, commercially available QDs are large (~15–20 nm diameter), which can hinder the dynamics of many motor proteins, especially when two QDs are used. Several groups have developed smaller QDs for in vivo labeling [33, 34], but reduction in the thickness of the polymer coat around the semiconductor core of a QD reduces their solubility and photostability over extended periods of time. In addition, QDs "blink" on and off frequently. Blinking can be reduced by addition of thiols such as BME [35]. QDs are multivalent, making cross-linking of different motor proteins a possibility. This can be limited by using a large molar excess of dots or attaching QDs to motors that are already immobilized on tracks. QD labeling is often poor, with ~20 % or less of motors are labeled when using ~10–100 molar excess of QDs. In future, small multicolor dots with controllable valency for both one- and two-color studies must be developed specifically for use in in vitro motility assays.

10.3.2 Preparing Motor Proteins for In Vitro Tracking Experiments

To prepare fluorescently labeled motor proteins, standard heterologous expression and purification techniques are used. Different strategies can be used to label cytoskeletal motors. In kinesin, native solvent-accessible cysteines are removed with site-directed mutagenesis, and the fluorophore is attached directly using maleimide or similar chemistry [36]. A unique cysteine can be labeled with biotin, allowing attachment of a streptavidin-coated QD. In myosin, most investigators labeled the calmodulin accessory proteins that coat the heavy chain lever arm first and then exchanged them onto the motor complex [37]. In cytoplasmic dynein, specially engineered genetic tags (such as SNAPTag from Promega Inc. and HaloTag from New England Biolabs) were fused to the N- or C-terminus and protein was labeled with ligands bearing fluorescent head groups (such as SNAP-TMR or HaloTag-rhodamine) [29]. To attach QDs or other large probes, streptavidin-coated QDs [38] or QDs coated with a ligand that attaches directly to a genetic tag have been used [20].

10.3.3 Sample Preparation

Cleaning of glass coverslips is necessary to reduce background fluorescence from dirt on glass surface. Organic substrates can be effectively cleaned using Piranha etch (a mixture of sulfuric acid and hydrogen peroxide). However, Piranha solutions are extremely corrosive, and routine use is not recommended. Routine cleaning can be done using a combination of 1 M KOH and water washes followed by oxidative cleaning, using a plasma cleaner or a similar device.

Once motor proteins are labeled with a suitable probe, the sample is prepared by immobilizing tracks on glass. Large bundles of filaments such as sea urchin axonemes can be nonspecifically adsorbed to a glass surface. Individual cytoskeletal filaments, such as actin and microtubules, can be immobilized via a biotin–streptavidin linkage. In these assays, unlabeled tubulin (or actin monomers) is mixed with 5–10 % biotinylated tubulin in polymerization reaction. The glass surface is treated with 1 mg/ml biotinylated bovine serum albumin (BSA) and then with 0.2 mg/ml streptavidin. Biotinylated filaments attach tightly to the streptavidin-coated surface. After filament immobilization, the glass surface is passivated with 1–2 mg/ml BSA or casein to prevent nonspecific adsorption of motors to the glass surface. If BSA or casein passivation is not sufficient, the glass surface can be passivated with a mixture of polyethylene glycol (PEG) and PEG-biotin before track immobilization. After the track is stably bound to a surface, a dilute (~1 nM) solution of labeled motor proteins are added to the flow chamber and washed several times with assay buffer, leaving only motors immobilized along the tracks. Stepping is induced by addition of ATP. In addition to ATP, oxygen scavengers, triplet-state quenchers, and anti-blinking reagents (see above) are commonly added to final imaging buffer to improve fluorophore performance.

10.3.4 Imaging and Image Analysis

The sample is typically excited under TIR illumination and the fluorescent signal of labeled motors is detected with an EM-CCD camera. Selection of proper image acquisition parameters is important to obtain the best quality data for a given application. The frame transfer mode of a CCD is used to prevent inter-frame dead time and optimize light detection. For most EM-CCD cameras, the EM gain and preamplifier gain should be set according to manufacturer specifications. Magnification of the microscope detection path is adjusted to obtain a 85–150 nm pixel size.

Depending on the stepping rate of the motor, the frame rate and laser power must be optimized to resolve individual steps. A general rule of thumb is that on average 10 frames should elapse for each step the motor takes. For example, yeast cytoplasmic dynein takes 10 steps per second at saturating ATP [29], and hence ~100 frames per second acquisition is necessary to detect ~90 % of the steps. Once a target frame rate is selected, the illuminating laser intensity should be adjusted to offer the best trade-off between the resulting fluorescent spot intensity and photobleaching. To achieve 1.5 nm localization, the laser intensity should be adjusted to collect 10,000 photons per frame from a single molecule [19]. For tracking of small-molecule dyes such as TMR and Cy3, illumination with 10–50 mW of laser intensity is sufficient to track a motor for ~1 min at 2 frames per second [19]. For tracking of bright QD probes, tracking with acquisition rates as high as 500 frames per second has been reported [20]. Rather than increasing the frame rate and laser intensity, the amount of ATP can be reduced to slow the motor stepping rate for in vitro applications. Data acquisition at saturating ATP is often difficult, because motors take 10–250 steps per second. Tracking these motors at full speed requires more than 1,000 Hz acquisition rate to observe individual steps, which can be achieved with dark-field tracking methods (Sect. 10.6).

Videos of moving fluorescent spots are processed into traces using single-particle tracking algorithms discussed above. These algorithms find or are given a starting location of a spot to begin tracking, localize the spot center, and continue to track the spot until it photobleaches or the video ends. The output of the tracking algorithm is a trace of motor *x*–*y* position in time (Fig. 10.3).

10.3.5 Analysis of Traces

Analysis of the traces of single motors gives insight into their mechanism and dynamics. Motors "step" in a stochastic fashion and a plot of motor position versus

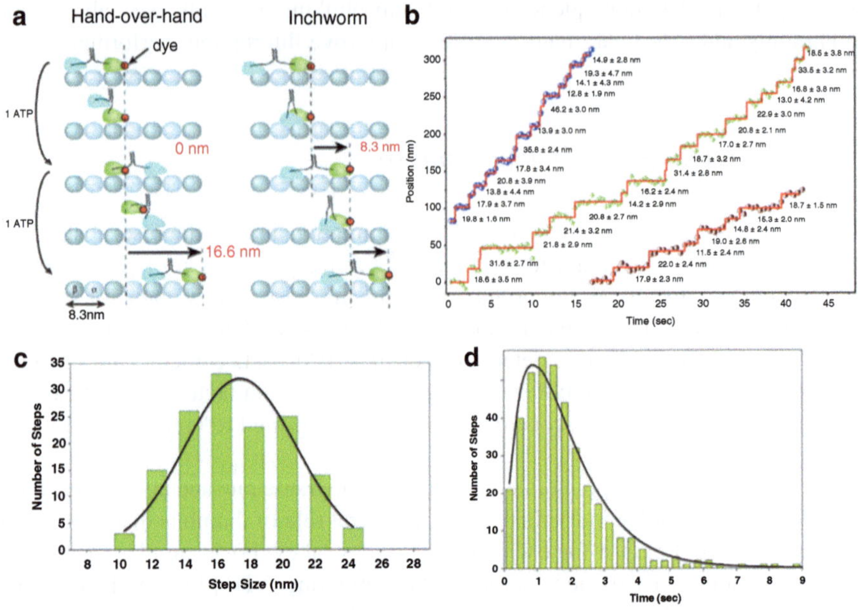

Fig. 10.3 Using single-color fluorescence tracking (FIONA) to determine the stepping mechanism of kinesin. (**a**) Models for processive stepping of kinesin. In the hand-over-hand mechanism, the two identical heads step past each other in an alternating manner, each taking a 16 nm step, then waiting for the other head to step. In the inchworm mechanism, the two heads step 8 nm simultaneously. One head always leads, and the other follows. (**b**) Using FIONA, the step size and dwell time of a kinesin head is determined by fitting the traces to a "staircase" function. (**c**) The step size was ~16 nm on average. (**d**) The dwell time distribution was consistent with a two-step process for a given head's step. These data clearly show that kinesin steps through a hand-over-hand and not an inchworm mechanism. Figure is reprinted from [45]

time has a "staircase" appearance. The two parameters of interest from step fitting are the step size of the labeled site, and the "dwell time" between individual steps (Fig. 10.3). If the motor takes a constant step size, such as kinesin-1, we recommend using a model-free pairwise method to determine the average step size [39]. In order to discern individual steps, fluorescent traces must be fit with a staircase function (Fig. 10.3b). Selection of a step-fitting method is important for accurate fitting. Several fitting algorithms are available to automatically detect steps and minimize user bias. T-tests can be employed to estimate when the steps occur [40] or more detailed statistical analyses such as hidden Markov models to compare alternate models for stepping [41]. Least-squares fitting of staircases with a user-specified number of steps [42] fits steps well but are biased toward the user's estimation of the number of steps. The appropriate number of steps can be estimated using the Schwarz information criterion [43], thus eliminating the main source of bias. Importantly, even with the best step-fitting methods, it is essential to verify all steps by eye. Issues with step finding are a major source of systematic error in tracking experiments. If the average number of data points between adjacent steps is <10 or the standard deviation of spot localization is higher than 25 % of the step size, a significant portion of steps will be missed, skewing the data toward high step sizes and long dwell times. Step fitters assume constant uncorrelated Gaussian noise, while in practice traces may have correlative or time-dependent noise.

The resulting step size and dwell time data from many individual traces are binned into histograms and further analyzed to infer key structure-function relationships. Step size histograms reveal the average step size, its standard deviation, and the presence of additional stepping modes, as well as the percentage of backward steps. By comparing the step size and dwell time histograms of fluorophores attached at different sites on the motor, the mechanism of a motor can be inferred (Sect. 10.4). The distribution of dwell times can give insight into the number of rate-limiting steps in the mechanochemical cycle. If an observed step of a motor is limited by a single rate-limiting step, such as ATP binding at limiting ATP concentrations, the dwell time histogram fits well to a single exponential decay. If there is more than one identical rate-limiting step, such as multiple ATP hydrolysis events, the histogram fits well to a gamma distribution [19]. The order of the gamma distribution reveals the number of rate-limiting steps. The number of rate-limiting steps can sometimes be deduced directly from the traces by calculating the randomness parameter, which is defined as the ratio of the squared mean of dwell times over the variance of the dwell times [44].

10.4 Applications of Single-Color Tracking of Molecular Motors In Vitro

Fluorescence tracking has been used in a variety of contexts to gain insight into the stepping mechanism of molecular motors. High-resolution fluorescence tracking was first developed to study myosin V [19] and kinesin-1 [45]. A primary goal of these studies is to use step size and step timing at different label locations to determine how the heads of a molecular motor communicate to achieve processive motion. For example, Yildiz et al. [45] used single-color tracking measurements to determine whether kinesin walks via a "hand-over-hand" or "inchworm" mechanism (Fig. 10.3a). From previous studies, it was already known that kinesin takes 8 nm steps along a microtubule filament [39]. If kinesin steps in a "hand-over-hand" manner, the heads alternate taking 16 nm steps to generate 8 nm center-of-mass steps. Alternately, if kinesin steps in an "inchworm" manner [46], the heads do not alternate, and each takes an 8 nm step at every cycle. Kinesin was labeled with a single Cy3 on the motor domain, and the step size and dwell time were measured using Gaussian tracking, to ~±2 nm resolution at 500 ms frame time, at limiting ATP (340 nM, Fig. 10.3b). By comparing the step size of motors labeled at different locations, it was found that the step size of one head was exactly twice that of the tail (16 nm vs. 8 nm, Fig. 10.3c, d). The dwell time distribution of motors labeled at the head fit well to a Gamma distribution with a two-step process. In contrast, the dwell time distribution of motors labeled at the tail fit well to a single exponential decay. These data showed that the heads of a kinesin dimer take twice the center-of-mass step at every other step of a motor. Therefore, these studies conclusively showed that the heads of kinesin alternately take steps in a "hand-over-hand" manner and not in an "inchworm" manner [47, 48] (Fig. 10.3), similar to myosin V [19, 49] and VI [28, 50].

Similar approaches have been used to answer a variety of questions about the stepping mechanism. Tracking of kinesin mutants with extended polypeptide linkers showed that loosening the mechanical connection between the heads of a kinesin dimer reduced the apparent amount of head–head coordination [51]. The step size of these loosely connected kinesin heads was highly variable, more closely resembling the results from a similar tracking study of yeast cytoplasmic dynein [29]. Subsequent tracking of yeast dynein at high time resolution revealed that the heads likely do not alternate either, indicating that cytoplasmic dynein likely steps via a unique uncoordinated mechanism [20]. It is possible that these loosely connected kinesin dimers achieve robust processivity with a similar lack of coordination. Comparable studies on mammalian dynein did not show high variability in step size [52]. Tracking has also been used to determine the complex mixed hand-over-hand-inchworm mechanism of myosin VI stepping [53].

10.5 Two-Color Fluorescence Tracking: Theory and Practice

If two different colors of fluorescent label are used to track a single motor, entirely new dimensions of analysis become accessible. Simultaneous tracking of two fluorophores allows measurement of dwell time and step size at two distinct locations, allowing more definitive characterization of the mechanism of a dimeric motor. Using two-color mapping techniques, the relative location of the two fluorophores in space can be determined with 3–5 nm accuracy [54]. This allows simultaneous tracking and distance measurement, analogous to single-molecule fluorescence resonance energy transfer (FRET) [55] but for larger distances (>10 nm for mapping vs. <10 nm for FRET). Two-color experiments yield exceptionally rich information, but they are more challenging than one-color experiments.

10.5.1 Selection of Probes

Additional care must be taken in the selection of probes for a two-color assay than for a one-color assay. Both probes must be spectrally distinct so as not to cause bleedthrough between fluorescent channels. Because two-color measurement is impossible when one of the probes photobleaches, both probes must be exceptionally photostable. Very few small-molecule probe pairs meet these requirements [30, 54]. For the highest resolution imaging, investigators have primarily used QD probes [20, 38, 53, 56]; however, the large size of commercial QDs renders measurement of short distances (<15 nm) difficult.

10.5.2 Two-Color Labeling and Protein Preparation

Two-color tracking requires attachment of two spectrally distinct probes at two distinct sites on a motor protein. Many processive motors are homodimeric, and one can randomly label these sites with a mixture of differently colored probes and limit analysis to two-color dual-labeled motors. Where two chemically distinct sites are needed, the multimeric nature of processive motors can be used to engineer artificial heterodimers, allowing specific labeling of monomers and subsequent dimerization. For example, to label heterodimeric dynein motors, Dewitt et al. generated monomeric forms of cytoplasmic dynein fused with a HaloTag at the motor domain [20]. One of the monomers is labeled with biotinylated HaloTag ligand, while the other was left unlabeled. After the heterodimerization of the monomers, two distinct sites were labeled with a mixture of streptavidin-coated and HaloTag-coated QDs. A number of enzyme-directed tags have been developed that ligate probes to small moieties within proteins [57–59], but these labeling strategies have not yet been applied to fluorescent tracking studies of motor proteins.

10.5.3 Two-Color Fluorescence Imaging

The microscope setup employed is identical to that used for one-color imaging above, with a few modifications. For different emission but similar excitation spectra (e.g., QDs), both probes can be excited with a single laser beam. For probes with different excitation spectra, two laser beams must be aligned together for excitation (e.g., a 532 laser for Cy3 imaging and a 633 laser for Cy5 imaging). In fluorescence detection, the goal is to image the two probes with perfect separation of their signal, either in space or in time. For separation of imaging in time, two laser beams can be alternately turned on and off to excite only one type of fluorophore at a time [30], and images can be collected with a single CCD camera. For separation of imaging in space, both fluorophores are excited at the same time and the signal of the fluorophores is separated onto two different detection channels [54] using a set of dichroic and emission filters (Fig. 10.4a). Splitting the fluorescence in space allows high-speed acquisition with QD probes over a full CCD surface, while splitting in time may make subpixel map construction more reliable. For multichannel detection, we recommend EM-CCD cameras over sCMOS

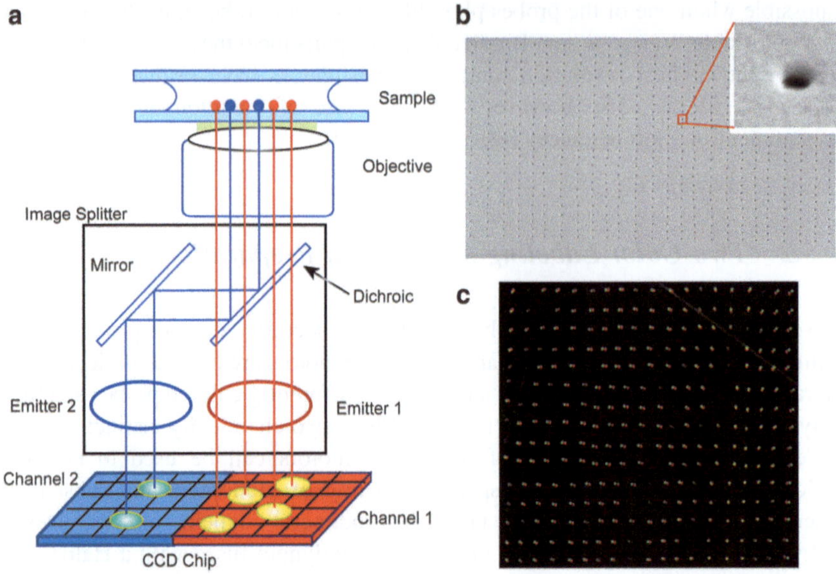

Fig. 10.4 Multicolor fluorescent tracking and image registration. (**a**) Fluorescence emission of different-colored probes is split to different detection channels on a CCD chip, using a set of dichroic mirrors and emission filters in the detection path. (**b**) To register the two channels with respect to each other, a gold thin film decorated with 180 nm diameter holes is imaged under bright field. (**c**) Superimposed images of nanohole pairs show that relative positions of the pairs differ depending on where they are located on the CCD. Image registration requires correction for various sources of image distortion and aberration. Precise mapping requires a secondary map of probes isospectral to the labels used in the motility assay. Finally, corrective maps are applied to the traces to measure relative position of the dye labels

cameras, because currently available sCMOS cameras have significant variation in pixel sensitivity, making map construction difficult. In either case, the output of the experiment will be two sets of images, one for each color.

10.5.4 Construction of Subpixel Maps

To simultaneously track two fluorophores at distinct locations on a walking motor and measure the distance between them, the relative positions of the fluorophores must be registered with nanometer accuracy. For image registration, the sample is decorated with fiducial markers that appear in both channels. To map the CCD surface, fiducials are localized in both fluorescent channels to generate a set of colocalizations [54]. Coordinates of colocalized spots are fitted using nonlinear regression with Taylor polynomials to make a subpixel map from these colocalizations.

Image registration corrects for several distortions and optical aberrations between the channels. There are three primary contributing factors to the mapping accuracy: First, the channels may differ in terms of actual position, rotation and magnification. These types of errors can be easily fixed by polynomial fits of the fiducial markers. Second, optical aberrations (spherical, astigmatism) slightly distort one channel image relative to the other. These aberrations, while substantial, can also be modeled using polynomial regression of relatively limited number of points. Remarkably, using the subpixel map generated by one set of fiducials fails to fully register another set of fiducial markers, emitting at a slightly different wavelength. Ideally the same set of fluorophores should be used as fiducials to correct for chromatic aberrations. Third, the CCD surface often bears slight pixel-to-pixel imperfections [60]. This error can be only be corrected by collecting multiple colocalizations per pixel and/or separately registering small parts of the CCD surface. Using active feedback of a nanopositioning stage, the Chu group achieved mapping accuracies of ~1 nm [60] in a small part of a CCD chip. Alternatively, mapping can be bypassed entirely using confocal imaging and multicolor QDs [61, 62]. The key principle is that differently colored QDs have similar excitation spectra, allowing simultaneous excitation and multicolor detection. This approach is appealing, as nanoscale mapping procedures are not required. However, the time resolution of scanning confocal microscopy limits its application in dynamic systems, such as tracking of molecular motors.

It is also important to calibrate the mapping accuracy by measuring short end-to-end distances (e.g., short DNA oligonucleotides at different lengths) using the same set of fluorophores [20, 54]. The mapping and localization errors lead to systematic overestimation of the measured distance, which can be corrected for two fluorophores that remain at a fixed distance. However, calibration of measured distances is problematic when the distance between the fluorophores fluctuates stochastically, as is the case for motor proteins and other dynamic systems. Therefore, mapping and localization error must be kept significantly lower than the required resolution of the experiment.

10.5.5 Methods for Generation of Subpixel Maps

Exact approaches to generate a map vary. We will describe the approach used in our laboratory in more detail. In a two-color study of dynein using QD probes, DeWitt et al. [20] constructed maps of the entire CCD surface using two rounds of image registration (Fig. 10.4). First, a subpixel map was generated using an array of diffraction-limited nanoholes (Fig. 10.4b) [63]. The 25 × 25 array of nanoholes was formed in a 150 nm thick gold layer deposited on a glass slide by sputter deposition. A 12.5 nm thick titanium layer served to enhance the adhesion of the gold film to the glass surface. Holes were formed by focused ion beam drilling using the 30 keV gallium ion beam. The hole diameter was 180 nm and the hole spacing was ~2 μm (Fig. 10.4b).

The sample was illuminated under bright field and moved at random intervals and directions across the surface. Each nanohole produces a diffraction-limited spot in both fluorescent channels (Fig 10.4c). These spots were localized to ~1 nm accuracy and paired to create a list of >1,000,000 colocalizations. The subpixel map created from these colocalizations can achieve image registration with ~3 nm accuracy. To correct for chromatic aberrations, it was necessary to generate a smaller set of colocalizations using custom-made fluorescent fiducial markers. For the QD585–QD655 pair used in the study, the fiducial was an amino-polymer QD655 coated with Rhodamine Red X, whose spectrum is nearly identical to the QD585. A second corrective map of the entire CCD chip was able to map pairs of QDs with 3–5 nm accuracy [20].

10.5.6 Data Analysis

Once maps are constructed, movies of labeled motors are acquired in an analogous manner to the one-color experiments described above, producing movies of dual-labeled motors. Colocalized fluorophores are tracked using the same algorithms. The stepping trace of one dye is overlaid upon the other using the mapping function created from the set of colocalizations of the fiducials, as described above. The overlaid two-color trace (Fig. 10.5) is then fit to staircase functions as described above.

Two-color traces yield substantially more information than one-color traces (Fig. 10.5). In addition to the simultaneous detection of step size and dwell time of the two labeling sites, these traces reveal the distance between the sites, which site is in the lead and the orientation of the sites relative to the cytoskeletal track. Two-color tracking methods were first used to observe the hand-over-hand movement of the two motor domains of myosin V [38, 54]. Yanagida and coworkers showed that myosin VI steps via a combination of hand-over-hand and inchworm stepping during processive motility [53].

The information that can be gleaned from a two-color experiment is especially useful for studying motors with more complex stepping mechanisms, such as cytoplasmic dynein [20, 30]. The results of a two-color study on cytoplasmic dynein by DeWitt et al. are illustrated in Fig. 10.5. The investigators measured the distance between the two heads at 30 ms time resolution as the motor walked along microtubules at ~20 nm/s. This led to several interesting findings, all of which are hard to access using one-color measurements alone. First, the heads remained widely separated during stepping (~30 nm). Second, the leading and trailing heads both stepped with a nearly identical frequency. Third, one head could step multiple times before the other. The results showed that the dynein heads do not step in a coordinated manner, which challenged established views of how motors achieve processive movement. The ability of motors to move processively in the absence of interhead coordination has also been observed in myosin or kinesin motors [64–66].

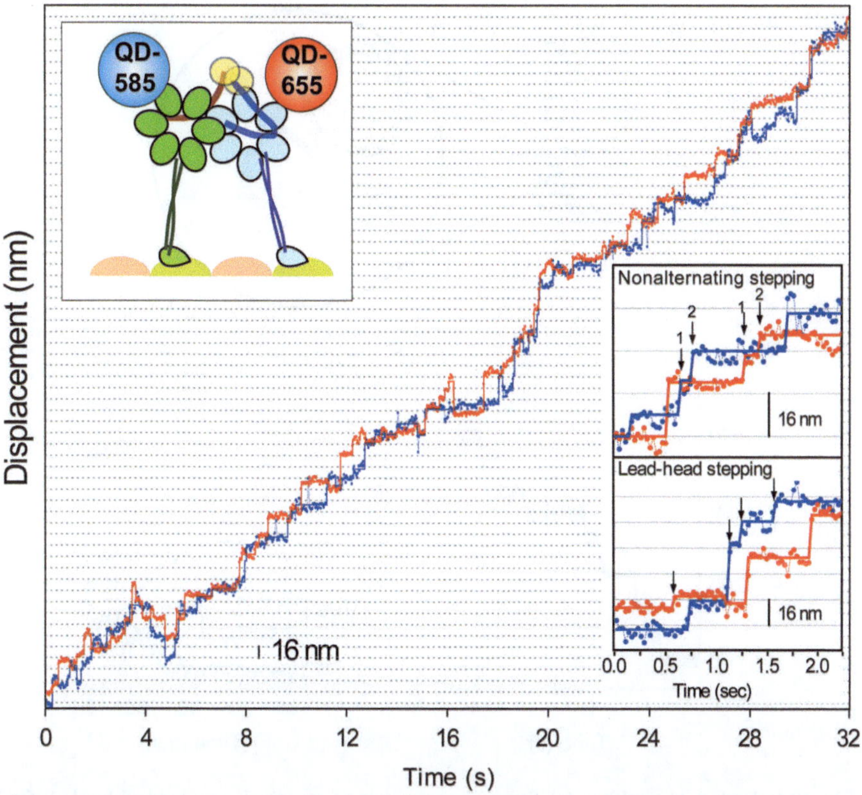

Fig. 10.5 Using two-color tracking to determine the stepping mechanism of cytoplasmic dynein. Cytoplasmic dynein is labeled with QD585–QD655 pair (*top insert*). Simultaneous tracking of the two heads not only allows their stepping motion but also provides information about their orientation relative to the microtubule. Subpixel registered trace of a dual-labeled dynein shows that the heads do not strictly step in alternating fashion. Instead, the same head can take multiple steps before its partner moves, and the leading head is almost as likely to step as the trailing head (*bottom insert*). These data indicate that cytoplasmic dynein steps in an uncoordinated manner, in contrast to the "hand-over-hand" mechanism of kinesin and myosin (adapted from [20])

10.6 Other Tracking Methods for Studying Processive Cytoskeletal Motors

Dark-field microscopy, which detects scattered light from objects with high index of refraction, is a promising tool to track motor proteins (Fig. 10.6). Dark-field tracking (DFT) works by imaging gold nanoparticles under dark-field illumination [67]. Since the image arises from light scattering and not fluorescence, probe photostability is not a concern, and very high incident light flux can be used. This allows the probe to be localized with arbitrarily high spatiotemporal resolution (~1 nm spatial and ~100 μs time resolution). DFT can be used to study extremely

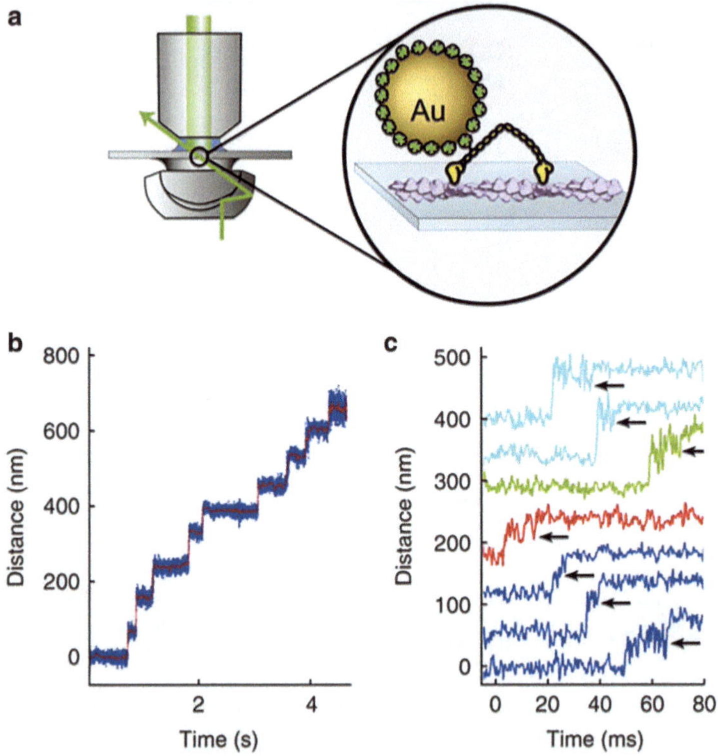

Fig. 10.6 Dark-field tracking. (**a**) Stepwise movement of myosin V was tracked by dark-field imaging of ~40 nm gold nanoparticles, which allows extremely fast and accurate position detection with zero applied force. (**b**) Individual steps taken by myosin V were clearly discernible. (**c**) Rapid conformational changes, such as the myosin head fluctuations (arrow heads), were detected between adjacent steps (reproduced from [69])

rapid events at zero applied force, such as the release and rebinding of a motor head during a step [68] or the step size of rapidly moving motors [69]. To get sufficient scattering, the gold nanoparticle must be relatively large (>40 nm diameter), and thus DFT is not suitable for applications that require a small probe.

Polarization-based methods provide information about the orientation of the probe attached to a molecular motor (Fig. 10.7). The key principle is that fluorophores have an excitation dipole moment and absorption efficiency of linearly polarized excitation light is given by $\cos^2\varphi$, where φ is the angle between the excitation dipole moment and the photon polarization. Therefore, it is possible to obtain the orientation of the dye from its intensity when illuminated with different polarizations of excitation light [70]. However, polarization measurements are limited by the high rotational mobility of probes attached through a single linkage. If the rotational correlation time is shorter than the image acquisition rate, the probe acts as an isotropic emitter. This can be circumvented by attaching the dye at two locations, using bifunctional rhodamine [71] or a multivalent QD nanorod [72].

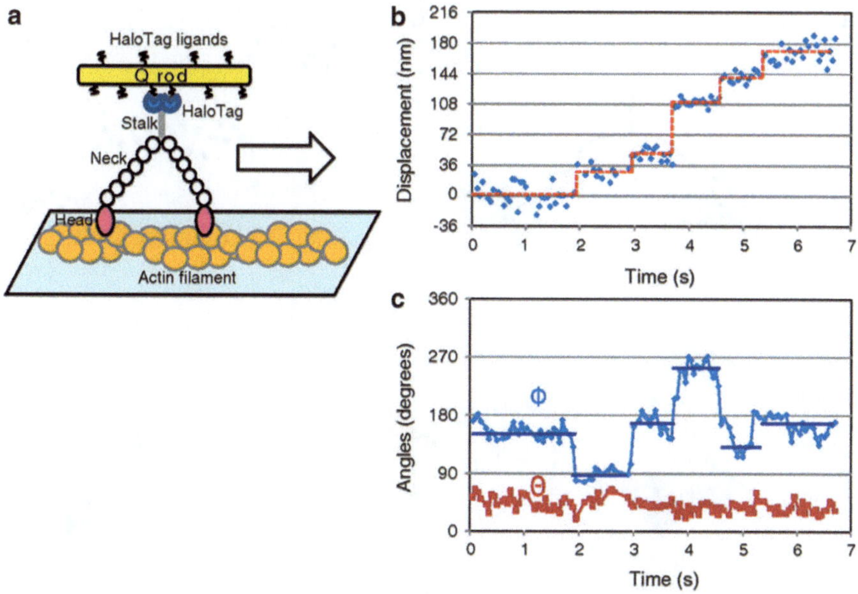

Fig. 10.7 Tracking both the position and orientation of a motor with a nanorod. (**a**) QD nanorods are attached at two locations on myosin V to track rotational and translational steps of the motor. (**b**, **c**) Simultaneous position (**b**) and orientation (**c**) detection allows measurement of angular changes along with position changes (reproduced from [72])

In a single-molecule fluorescence polarization microscope, the sample is excited with linearly polarized light under TIR illumination and the fluorescence emission is recorded with a single channel. An alternative strategy is to excite the sample with circularly polarized light and split the fluorescence emission into two orthogonal polarizations [73]. In order to circumvent twofold degeneracy of in-plane angles (φ vs. $-\varphi$), the sample can be excited with four different polarization angles (0°, 45°, 90°, 135°) [74].

Goldman and coworkers excited the sample with two orthogonal incident beams in a prism-type TIR to determine the three-dimensional orientation of a fluorophore. The emission pathway was split into two orthogonal channels. By labeling the myosin V lever arm with bifunctional rhodamine, they observed that the myosin V light chain alternates between two well-defined orientations as the motor walks, providing evidence that myosin V walks hand over hand [49].

Single-molecule high-resolution imaging with photobleaching (SHRImP) is an alternative to two-color mapping that can be used to measure distances between two sites of motor protein without the construction of subpixel maps [75]. SHRImP has been applied to myosin VI labeled with two GFPs, one on each motor head (Fig. 10.8). The fluorescent spot of two GFPs was localized using Gaussian fitting. When one of the GFPs photobleaches, the resulting spot from the remaining GFP is localized. The distance between the centers of the two spots (categorized based on spot intensity) can be used to calculate the original GFP–GFP distance with ~5 nm

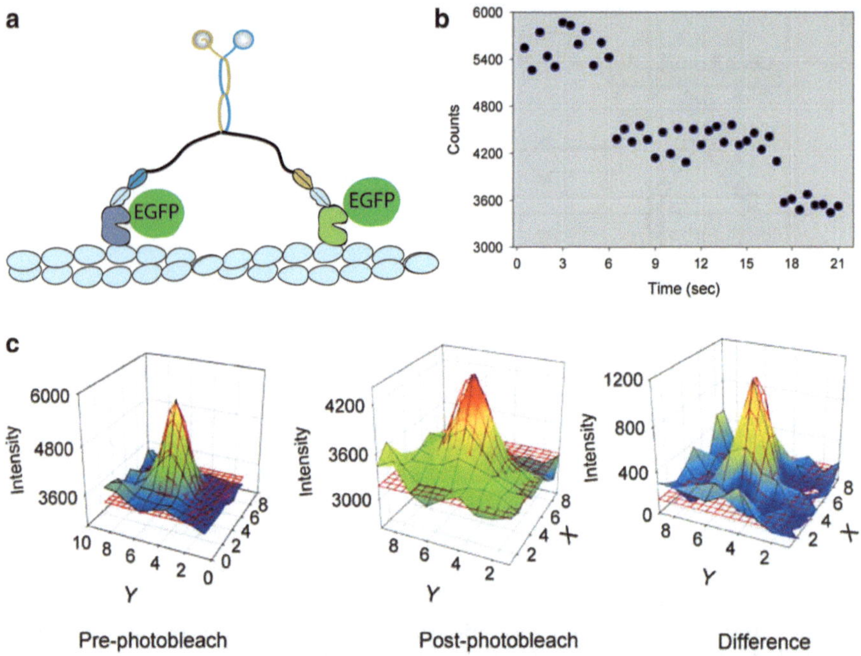

Fig. 10.8 SHRImP to measure head to head distance. (**a**) A myosin VI motor is labeled with an eGFP per head. (**b**) Intensity of a diffraction-limited spot disappears in two steps, evidence for the presence of two probes on a motor. (**c**) (*Left*) PSFs of eGFPs overlap, when they are both fluorescent. (*Middle*) As one of the eGFPs bleach, the center of the PSF shifts. This reveals the PSF of the remaining fluorescent eGFP. (*Right*) The PSF of the bleached eGFP can be calculated by subtracting the fluorescent image from each other. Localization of both eGFPs measures the interhead distance (adapted from [76])

accuracy [76]. Further studies showed that the photobleaching-based distance measurements are not limited to two sites [15] and can potentially be extended to obtain detailed structural information about the motor proteins as they walk along their respective track.

10.7 Conclusion

Fluorescent tracking is a powerful method for determining the mechanism of motor proteins. Compared to optical trapping methods, high-resolution fluorescence tracking methods require no applied force to observe steps and they can track hundreds of motor molecules at a time. Advances in probe technology, commercialization of TIRF microscopes, and the latest generation of detectors have made single-molecule tracking in vitro accessible to nonspecialists. Using these probes, data can be acquired at high time resolution, at multiple sites of attachment. Using two-color tracking methods and the construction of nanoscale CCD maps, the relative movements of different parts of a protein can be followed at every step, yielding data of striking richness for mechanistic analysis. Future improvements in fluorescent probes and detection ability will be required to study more complex systems such as identification and characterization of motor regulatory proteins and protein complexes.

References

1. Horio T, Hotani H (1986) Visualization of the dynamic instability of individual microtubules by dark-field microscopy. Nature 321(6070):605–607. doi:10.1038/321605a0
2. Gelles J, Schnapp BJ, Sheetz MP (1988) Tracking kinesin-driven movements with nanometre-scale precision. Nature 331(6155):450–453. doi:10.1038/331450a0
3. Finer JT, Simmons RM, Spudich JA (1994) Single myosin molecule mechanics: piconewton forces and nanometre steps. Nature 368(6467):113–119
4. Moerner WE, Kador L (1989) Optical detection and spectroscopy of single molecules in a solid. Phys Rev Lett 62(21):2535–2538
5. Inoue S (2006) Foundations of confocal scanned imaging in light microscopy, Chap 1. In: Pawley JB (ed) Handbook of biological confocal microscopy, 3rd edn. Springer, New York, NY
6. Shimozawa T, Yamagata K, Kondo T, Hayashi S, Shitamukai A, Konno D, Matsuzaki F, Takayama J, Onami S, Nakayama H, Kosugi Y, Watanabe TM, Fujita K, Mimori-Kiyosue Y (2013) Improving spinning disk confocal microscopy by preventing pinhole cross-talk for intravital imaging. Proc Natl Acad Sci USA 110(9):3399–3404. doi:10.1073/pnas.1216696110
7. Burghardt TP, Axelrod D (1981) Total internal reflection/fluorescence photobleaching recovery study of serum albumin adsorption dynamics. Biophys J 33(3):455–467, doi: S0006-3495(81)84906-5[pii]
8. Funatsu T, Harada Y, Tokunaga M, Saito K, Yanagida T (1995) Imaging of single fluorescent molecules and individual ATP turnovers by single myosin molecules in aqueous solution. Nature 374:555–559
9. Vale RD, Funatsu T, Pierce DW, Romberg L, Harada Y, Yanagida T (1996) Direct observation of single kinesin molecules moving along microtubules. Nature 380(6573):451–453
10. Bobroff N (1986) Position measurement with a resolution and noise-limited instrument", Rev. Sci. Instrum. 57(6), 1152–1157(1986). Rev Sci Instrum 57(6):1152–1157
11. Yildiz A, Selvin PR (2005) Fluorescence imaging with one nanometer accuracy: application to molecular motors. Acc Chem Res 38(7):574–582. doi:10.1021/ar040136s
12. Enderlein J, Toprak E, Selvin PR (2006) Polarization effect on position accuracy of fluorophore localization. Opt Express 14(18):8111–8120
13. Bartko AP, Xu K, Dickson RM (2002) Three-dimensional single molecule rotational diffusion in glassy state polymer films. Phys Rev Lett 89(2):026101
14. Betzig E, Patterson GH, Sougrat R, Lindwasser OW, Olenych S, Bonifacino JS, Davidson MW, Lippincott-Schwartz J, Hess HF (2006) Imaging intracellular fluorescent proteins at nanometer resolution. Science 313(5793):1642–1645, doi: 1127344 [pii]
15. Rust MJ, Bates M, Zhuang X (2006) Sub-diffraction-limit imaging by stochastic optical reconstruction microscopy (STORM). Nat Methods 3(10):793–795, doi: nmeth929[pii]
16. Thompson RE, Larson DR, Webb WW (2002) Precise nanometer localization analysis for individual fluorescent probes. Biophys J 82(5):2775–2783
17. Cheezum MK, Walker WF, Guilford WH (2001) Quantitative comparison of algorithms for tracking single fluorescent particles. Biophys J 81(4):2378–2388

18. Mortensen KI, Churchman LS, Spudich JA, Flyvbjerg H (2010) Optimized localization analysis for single-molecule tracking and super-resolution microscopy. Nat Methods 7 (5):377–381. doi:10.1038/nmeth.1447
19. Yildiz A, Forkey JN, McKinney SA, Ha T, Goldman YE, Selvin PR (2003) Myosin V walks hand-over-hand: single fluorophore imaging with 1.5-nm localization. Science 300 (5628):2061–2065
20. DeWitt MA, Chang AY, Combs PA, Yildiz A (2012) Cytoplasmic dynein moves through uncoordinated stepping of the AAA+ring domains. Science 335(6065):221–225, doi:science.1215804 [pii]
21. Vale RD, Reese TS, Sheetz MP (1985) Identification of a novel force-generating protein, kinesin, involved in microtubule-based motility. Cell 42(1):39–50
22. Toyoshima YY, Kron SJ, Spudich JA (1990) The myosin step size: measurement of the unit displacement per ATP hydrolyzed in an in vitro assay. Proc Natl Acad Sci USA 87 (18):7130–7134
23. Kural C, Kim H, Syed S, Goshima G, Gelfand VI, Selvin PR (2005) Kinesin and dynein move a peroxisome in vivo: a tug-of-war or coordinated movement? Science 308(5727):1469–1472
24. Vogelsang J, Kasper R, Steinhauer C, Person B, Heilemann M, Sauer M, Tinnefeld P (2008) A reducing and oxidizing system minimizes photobleaching and blinking of fluorescent dyes. Angew Chem 47(29):5465–5469. doi:10.1002/anie.200801518
25. Tokunaga M, Kitamura K, Saito K, Iwane AH, Yanagida T (1997) Single Molecule Imaging of Fluorophores and Enzymatic Reactions Achieved By Objective-Type Total Internal Reflection Fluorescence Microscopy. Biochem Biophys Res Commun 235(1):47–53
26. Aitken CE, Marshall RA, Puglisi JD (2008) An oxygen scavenging system for improvement of dye stability in single-molecule fluorescence experiments. Biophys J 94(5):1826–1835. doi:10.1529/biophysj.107.117689
27. Rasnik I, McKinney SA, Ha T (2006) Nonblinking and long-lasting single-molecule fluorescence imaging. Nat Methods 3(11):891–893
28. Okten Z, Churchman LS, Rock RS, Spudich JA (2004) Myosin VI walks hand-over-hand along actin. Biol, Nat Struct Mol
29. Reck-Peterson SL, Yildiz A, Carter AP, Gennerich A, Zhang N, Vale RD (2006) Single-molecule analysis of dynein processivity and stepping behavior. Cell 126(2):335–348. doi:10.1016/j.cell.2006.05.046
30. Qiu W, Derr ND, Goodman BS, Villa E, Wu D, Shih W, Reck-Peterson SL (2012) Dynein achieves processive motion using both stochastic and coordinated stepping. Nat Struct Mol Biol 19(2):193–200, doi:nsmb.2205 [pii]
31. Altman RB, Zheng Q, Zhou Z, Terry DS, Warren JD, Blanchard SC (2012) Enhanced photostability of cyanine fluorophores across the visible spectrum. Nat Methods 9 (5):428–429. doi:10.1038/nmeth.1988
32. Lacoste TD, Michalet X, Pinaud F, Chemla DS, Alivisatos AP, Weiss S (2000) Ultrahigh-resolution multicolor colocalization of single fluorescent probes. Proc Natl Acad Sci USA 97 (17):9461–9466
33. Howarth M, Liu W, Puthenveetil S, Zheng Y, Marshall LF, Schmidt MM, Wittrup KD, Bawendi MG, Ting AY (2008) Monovalent, reduced-size quantum dots for imaging receptors on living cells. Nat Methods 5(5):397–399. doi:10.1038/nmeth.1206
34. Smith AM, Nie S (2008) Minimizing the hydrodynamic size of quantum dots with multifunctional multidentate polymer ligands. J Am Chem Soc 130(34):11278–11279. doi:10.1021/ja804306c
35. Hohng S, Ha T (2004) Near-complete suppression of quantum dot blinking in ambient conditions. J Am Chem Soc 126(5):1324–1325. doi:10.1021/ja039686w
36. Rice S, Lin AW, Safer D, Hart CL, Naber N, Carragher BO, Cain SM, Pechatnikova E, Wilson-Kubalek EM, Whittaker M, Pate E, Cooke R, Taylor EW, Milligan RA, Vale RD (1999) A structural change in the kinesin motor protein that drives motility. Nature 402 (6763):778–784

37. Irving M, Allen TS, Sabido-David C, Craik JS, Brandmeier B, Kendrickjones J, Corrie JET, Trentham DR, Goldman YE (1995) Tilting of the light-chain region of myosin during step length changes and active force generation in skeletal muscle. Nature 375(6533):688–691
38. Warshaw DM, Kennedy GG, Work SS, Krementsova EB, Beck S, Trybus KM (2005) Differential labeling of myosin V heads with quantum dots allows direct visualization of hand-over-hand processivity. Biophys J 88(5):L30–32
39. Svoboda K, Schmidt CF, Schnapp BJ, Block SM (1993) Direct observation of kinesin stepping by optical trapping interferometry. Nature 365(6448):721–727
40. Moffitt JR, Chemla YR, Aathavan K, Grimes S, Jardine PJ, Anderson DL, Bustamante C (2009) Intersubunit coordination in a homomeric ring ATPase. Nature 457(7228):446–450, doi: nature07637 [pii]
41. Mullner FE, Syed S, Selvin PR, Sigworth FJ (2010) Improved hidden Markov models for molecular motors, part 1: basic theory. Biophys J 99(11):3684–3695. doi:10.1016/j.bpj.2010.09.067
42. Kerssemakers JW, Munteanu EL, Laan L, Noetzel TL, Janson ME, Dogterom M (2006) Assembly dynamics of microtubules at molecular resolution. Nature 442(7103):709–712
43. Kalafut B, Visscher K (2008) An objective, model-independent method for detection of non-uniform steps in noisy signals. Comput Phys Commun 179(10):716–723
44. Schnitzer MJ, Block SM (1995) Statistical kinetics of processive enzymes. Cold Spring Harb Symp Quant Biol 60:793–802
45. Yildiz A, Tomishige M, Vale RD, Selvin PR (2004) Kinesin walks hand-over-hand. Science 303(5658):676–678. doi:10.1126/science.1093753
46. Hua W, Chung J, Gelles J (2002) Distinguishing inchworm and hand-over-hand processive kinesin movement by neck rotation measurements. Science 295(5556):844–848
47. Kaseda K, Higuchi H, Hirose K (2003) Alternate fast and slow stepping of a heterodimeric kinesin molecule. Nat Cell Biol 5(12):1079–1082
48. Asbury CL, Fehr AN, Block SM (2003) Kinesin moves by an asymmetric hand-over-hand mechanism. Science 302(5653):2130–2134
49. Forkey JN, Quinlan ME, Shaw MA, Corrie JE, Goldman YE (2003) Three-dimensional structural dynamics of myosin V by single-molecule fluorescence polarization. Nature 422 (6930):399–404
50. Yildiz A, Park H, Safer D, Yang Z, Chen LQ, Selvin PR, Sweeney HL (2004) Myosin VI Steps via a Hand-over-Hand Mechanism with Its Lever Arm Undergoing Fluctuations when Attached to Actin. J Biol Chem 279(36):37223–37226
51. Yildiz A, Tomishige M, Gennerich A, Vale RD (2008) Intramolecular strain coordinates kinesin stepping behavior along microtubules. Cell 134(6):1030–1041, doi:S0092-8674(08)00936-7 [pii]
52. Toba S, Watanabe TM, Yamaguchi-Okimoto L, Toyoshima YY, Higuchi H (2006) Overlapping hand-over-hand mechanism of single molecular motility of cytoplasmic dynein. Proc Natl Acad Sci USA 103(15):5741–5745. doi:10.1073/pnas.0508511103
53. Nishikawa S, Arimoto I, Ikezaki K, Sugawa M, Ueno H, Komori T, Iwane AH, Yanagida T (2010) Switch between large hand-over-hand and small inchworm-like steps in myosin VI. Cell 142(6):879–888, doi: S0092-8674(10)01002-0[pii]
54. Churchman LS, Okten Z, Rock RS, Dawson JF, Spudich JA (2005) Single molecule high-resolution colocalization of Cy3 and Cy5 attached to macromolecules measures intramolecular distances through time. Proc Natl Acad Sci USA 102(5):1419–1423
55. Ha T, Enderle T, Ogletree DF, Chemla DS, Selvin PR, Weiss S (1996) Probing the interaction between two single molecules: Fluorescence resonance energy transfer between a single donor and a single acceptor. Proc Natl Acad Sci 93(13):6264–6268
56. Ikezaki K, Komori T, Sugawa M, Arai Y, Nishikawa S, Iwane AH, Yanagida T (2012) Simultaneous observation of the lever arm and head explains myosin VI dual function. Small 8(19):3035–3040. doi:10.1002/smll.201200765

57. Howarth M, Takao K, Hayashi Y, Ting AY (2005) Targeting quantum dots to surface proteins in living cells with biotin ligase. Proc Natl Acad Sci USA 102(21):7583–7588. doi:10.1073/pnas.0503125102
58. Fernandez-Suarez M, Baruah H, Martinez-Hernandez L, Xie KT, Baskin JM, Bertozzi CR, Ting AY (2007) Redirecting lipoic acid ligase for cell surface protein labeling with small-molecule probes. Nat Biotechnol 25(12):1483–1487. doi:10.1038/nbt1355
59. Kaiser CM, Goldman DH, Chodera JD, Tinoco I Jr, Bustamante C (2011) The ribosome modulates nascent protein folding. Science 334(6063):1723–1727. doi:10.1126/science.1209740
60. Pertsinidis A, Zhang Y, Chu S (2010) Subnanometre single-molecule localization, registration and distance measurements. Nature 466(7306):647–651, doi: 10.1038/nature09163[pii]
61. Agrawal A, Deo R, Wang GD, Wang MD, Nie S (2008) Nanometer-scale mapping and single-molecule detection with color-coded nanoparticle probes. Proc Natl Acad Sci USA 105 (9):3298–3303. doi:10.1073/pnas.0712351105
62. Antelman J, Wilking-Chang C, Weiss S, Michalet X (2009) Nanometer distance measurements between multicolor quantum dots. Nano Lett 9(5):2199–2205. doi:10.1021/nl901163k
63. Moberly Chan WJ, Adams DP, Aziz MJ, Hobler G, Schenkel T (2007) Fundamentals of focused ion beam nanostructural processing: below, at, and above the surface. MRS Bull 32 (424)
64. Thoresen T, Gelles J (2008) Processive movement by a kinesin heterodimer with an inactivating mutation in one head. Biochemistry 47(36):9514–9521. doi:10.1021/bi800747e
65. Elting MW, Bryant Z, Liao JC, Spudich JA (2011) Detailed tuning of structure and intramolecular communication are dispensable for processive motion of myosin VI. Biophys J 100 (2):430–439. doi:10.1016/j.bpj.2010.11.045
66. Liao JC, Elting MW, Delp SL, Spudich JA, Bryant Z (2009) Engineered myosin VI motors reveal minimal structural determinants of directionality and processivity. J Mol Biol 392 (4):862–867, doi: 10.1016/S0022-2836(09)00899-7[pii]
67. Braslavsky I, Amit R, Jaffar Ali BM, Gileadi O, Oppenheim A, Stavans J (2001) Objective-type dark-field illumination for scattering from microbeads. Appl Opt 40(31):5650–5657
68. Dunn AR, Spudich JA (2007) Dynamics of the unbound head during myosin V processive translocation. Nat Struct Mol Biol 14(3):246–248. doi:10.1038/nsmb1206
69. Dunn AR, Chuan P, Bryant Z, Spudich JA (2010) Contribution of the myosin VI tail domain to processive stepping and intramolecular tension sensing. Proc Natl Acad Sci USA 107 (17):7746–7750, doi: 1002430107[pii]
70. Ha T, Glass J, Enderle T, Chemla DS, Weiss S (1998) Hindered rotational diffusion and rotational jumps of single molecules. Phys Rev Lett 80(10):2093–2096
71. Corrie JET, Brandmeier B, Ferguson R, Trentham D, Kendrick-Jones J, Hopkins S, van der Heide U, Goldman Y, Sabido-David C, Dale R, Criddle S, Irving M (1999) Dynamic measurement of myosin light-chain-domain tilt and twist in muscle contraction. Nature 400 (6743):425–430
72. Ohmachi M, Komori Y, Iwane AH, Fujii F, Jin T, Yanagida T (2012) Fluorescence microscopy for simultaneous observation of 3D orientation and movement and its application to quantum rod-tagged myosin V. Proc Natl Acad Sci USA 109(14):5294–5298. doi:10.1073/pnas.1118472109
73. Sase I, Miyata H, Corrie JET, Craik JS, Kinosita K Jr (1995) Real time imaging of single fluorophores on moving actin with an epifluorescence microscope. Biophys J 69:323–328
74. Sosa H, Peterman EJG, Moerner WE, Goldstein LSB (2001) ADP-induced rocking of the Kinesin motor domain revealed by single-molecule fluorescence polarization microscopy. Nat Struct Biol 8(6):540–544
75. Gordon MP, Ha T, Selvin PR (2004) Single-molecule high-resolution imaging with photobleaching. Proc Natl Acad Sci USA 101(17):6462–6465
76. Balci H, Ha T, Sweeney HL, Selvin PR (2005) Interhead distance measurements in myosin VI via SHRImP support a simplified hand-over-hand model. Biophys J 89(1):413–417

Chapter 11
Measuring Transport of Motor Cargos

Leslie Conway and Jennifer L. Ross

Abstract In this chapter, we describe experimental techniques used in vitro to illuminate how small teams of motors can work to translocate cargos. We will focus on experiments utilizing in vitro reconstitution, artificial or ex vivo purified cargos, and fluorescence imaging. A number of studies have been able to recapitulate the activities of cargo transport driven by small teams of motors elucidating how multiple motors can work together to transport cargos within the cell. Here, we describe some of the methods employed and highlight important experimental details needed to perform these experiments.

Keywords Cargo transport • Motor protein • Kinesin • Dynein • Myosin • Microscopy • Beads • Quantum dot • DNA scaffold • Multiple motors • Tug-of-war

Abbreviations

AFM	Atomic force microscopy
ATP	Adenosine triphosphate
DIC	Differential interference contrast
DNA	Deoxyribonucleic acid
FIONA	Fluorescence imaging with one nanometer accuracy
GFP	Green fluorescent protein
mRNA	Messenger ribonucleic acid
Qdots	Quantum dots
TIRF	Total internal reflection fluorescence

L. Conway
Molecular and Cellular Biology Graduate Program, University of Massachusetts Amherst, Amherst, MA, USA

J.L. Ross (✉)
Department of Physics, University of Massachusetts Amherst, Amherst, MA, USA
e-mail: rossj@physics.umass.edu

C.P. Toseland and N. Fili (eds.), *Fluorescent Methods for Molecular Motors*, Experientia Supplementum 105, DOI 10.1007/978-3-0348-0856-9_11,

11.1 Introduction

In cells, motor proteins are used to transport cellular cargo throughout the cell. Cargo transport is an essential process in cells, which can range in diameter from 10 to 100 μm, to ensure that membrane-encapsulated organelles or vesicles, protein complexes, or small macromolecules are placed in the correct cellular location in a timely manner. Long, extended cellular processes, such as axons, can extend up to 1 m in length and require intracellular transport for differentiation and maintenance of these structures. New material made in the cell body is needed along and at the end of the axon to replenish degraded protein, and degraded proteins need to be removed efficiently. Dysfunction of intracellular transport results in axon retraction that causes diseases, such as amyotrophic lateral sclerosis (ALS) [1]. Thus, understanding the fundamental aspects of this important process has implications for disease.

Small teams of motors often transport organelle and vesicular cargo. These teams can be of the same type of motor that walk in the same direction or of different motor species that can walk in the same or opposite directions on polarized cytoskeletal filaments, resulting in a tug-of-war. Systematic studies of how small teams of motors are able to work together are essential to understanding intracellular transport. One straightforward method to determine the mechanisms of a biological process is in vitro reconstitution of that process using purified and minimal components. Intracellular transport has been successfully reconstituted in vitro using a variety of techniques. Artificial cargos with teams of motors can be created using glass or polymer beads, quantum dots, or even scaffolds of protein or DNA. Endogenous cargos purified from tissue can be used to probe more complex systems of motors that better mimic the cellular scenario. Here, we will discuss the techniques employed to probe multi-motor in vitro reconstitution experiments on cargos and some helpful hints to help with performing these types of experiments.

11.2 Beads as Artificial Cargos

Beads were one of the first artificial cargos used to study cargo transport by molecular motors. Beads used in these studies are made of a variety of materials including latex, polystyrene, silica, glass, and gold and can be conjugated to fluorescent dyes for imaging. Alternatively, nonfluorescent beads can be used and imaged via differential interference contrast (DIC) microscopy. Beads also represent a flexible system to study a range of cargo sizes, as they are available in diameters in the range of 20 nm to 10 μm.

One of the pioneering studies of cellular transport was performed by incubating beads with squid axoplasm [2]. We now know that the cellular components that caused the motion were motor proteins. Since then, such bead assays have been used to characterize the motility properties of cargos carried by single or multiple motors [3–9]. More systematically, complex experiments have been performed to study the motility of beads transported by multiple species of motors [2, 10–14]. Since beads are often made from materials with a high index of refraction, they are easily trapped using an optical tweezer (optical trap). Such experiments characterized the forces produced by single motors and revealed with high resolution the stepping behavior of motors [15–22]. While the majority of studies with beads have been performed using in vitro systems, beads can also be microinjected into live cells or incorporated into live cells by endocytosis. In the cell, beads nonspecifically bind endogenous motors and are transported along the cell's cytoskeletal tracks [10, 12, 23].

To study motor transport of bead cargos, motors must be attached to beads through specific or nonspecific interactions. In the case of either attachment mechanism, the number of motors bound to each bead can be tuned by altering the concentration of motors incubated with the beads [3]. This allows the study of transport by single or multiple motors. In the following paragraphs, we will discuss schemes for attaching motors nonspecifically and specifically.

The nonspecific attachment of full-length kinesin, dynein, and myosin motors to beads can be accomplished by simply incubating beads with purified motors or cell extract [2, 12, 13, 15, 16, 22, 24–29] (Fig. 11.3). In the case of kinesin, these nonspecific attachments have been shown to be enhanced by using carboxylated beads [2, 4, 5, 10, 12, 14, 18, 23, 30, 31] or by coating beads with casein prior to incubation with motors [3, 6, 19, 21]. Carboxylation increases the binding of kinesin to beads by forming a negatively charged surface on the bead [23], which promotes the binding of the positively charged kinesin-1 tail [32]. The exact mechanism by which casein functions to promote kinesin binding is unclear. However, it is thought to help prevent denaturation of motors upon binding to the bead surface, as well as promote binding of motors in the proper orientation for better function, both of which increase the number of active motors bound to beads [3]. Dynein and myosin motors are typically adsorbed onto beads with no coating, and casein is often used after incubation with motors to prevent additional motors from binding [13, 29]. These motors have hydrophobic and charged tail groups that likely bind to the polymer or glass bead surface.

Specific attachment of kinesin and myosin motors to beads is typically achieved through designed molecular interactions (Fig. 11.3). Expressed and purified motors with specific target tags genetically incorporated into the protein sequence will bind to certain moieties to cause binding to beads. The most common binding technique uses antibodies to attach motors to beads. For instance, beads conjugated to an anti-His antibody mediate binding of 6xHis-tagged kinesin or myosin motors. The anti-His antibody is bound to beads either nonspecifically [33] or through streptavidin by using a biotinylated anti-His antibody [7, 29]. A similar method uses an antibody against GFP to bind GFP-tagged motors [11, 34]. In addition, antibodies against the C-terminal, or tail region, of the motor can be used [22, 35]. Biotinylated and His-tagged motors can also be directly linked to streptavidin-conjugated beads and NTA-conjugated beads, respectively, without the use of antibodies [17, 36].

Here, we describe many points to consider when performing experiments with beads as cargos.

11.2.1 Low Numbers of Events

Large beads diffuse relatively slowly and are therefore slow to bind cytoskeletal filaments in the chamber. This, coupled with the inability to see many filaments in the same location, results in low numbers of association events while imaging. In addition, beads are often heavy enough to settle on the chamber bottom and not associate with filaments. This further reduces the number of association and walking events that are observed. Increasing the number of beads and/or filaments can improve the likelihood of observing an event, but can lead to crowding or other difficulties, and is often challenging to control. A more controlled method that allows for the observation of more association events is the use of optical tweezers. Optical tweezers can be used not only for force measurements but also to capture beads and place them on cytoskeletal tracks, allowing for a higher rate of association while observing. Further, the percentage of associating beads can inform about the number of motors attached to the bead, as described previously [20].

11.2.2 Number and Location of Bound Motors

When incubating beads with motors, one can systematically alter the concentration of motors to achieve increased or decreased average numbers of bound motors. Unfortunately, the motors will associate randomly on the surface of the bead, making it difficult to determine the placement, and therefore the number of motors capable of simultaneously interacting with the microtubule. Even when starting with an initial dilute solution of motors, there is some probability, albeit low, that two motors will associate directly next to each other. Further, there is some confusion about the difference between the number of motors associated to the bead and the number of motors that can simultaneously interact with the cytoskeletal filament. Since the beads are often rather large (1 μm diameter) compared to the size of the motors (~4 nm diameter), even if many motors bind to the bead, the probability that all motors are bound on the same side is extremely low due to entropy. This can be calculated as follows: The surface area of a 1 μm-diameter bead is ~3 μm^2. If there are 100 motors, spread evenly on the bead, the average distance between two motors will be about 50 nm (Fig. 11.1). Given this distance, and depending on the geometry, 1–3 motors could be in contact with the filament at one time, to transport the bead. The problem is that the specific geometry cannot be known for sure, due to the randomness of the associations. One mechanism to probe the number of engaged motors is to use an optical tweezer to measure the stall force of the moving bead. Since force is additive, the total stall force should be equal to

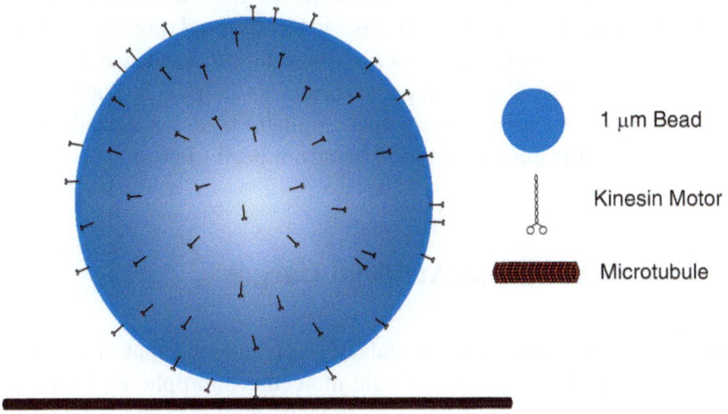

Fig. 11.1 Distribution of motors bound to a bead. This figure illustrates the random distribution of 100 kinesin motors associated with a 1 μm bead. While there are many motors bound to the bead, the probability of more than one motor simultaneously engaging with the microtubule filament is low. Figure is drawn to scale

the sum of the individual motor stall forces, which can be used to determine if there is more than one motor bound to the filament. Unfortunately, most traps have an upper limit on the force they can measure, so in practice, it is only possible to differentiate whether one or more than one motor is engaged in transport.

11.2.3 Attachment of Dissimilar Motors

Once an attachment scheme between a bead and motors is determined, changing the number of actively engaged motors only requires incubating the beads with higher concentrations of motors. If motors of different types are to be employed on the same bead, all motors must attach to the bead using the same mechanism. If the binding ability is similar, they will both populate the bead randomly. Thus, two motors purified from endogenous sources using nonspecific interactions should simultaneously bind to the bead with even distributions. If one motor has a specific binding (such as an antibody) or a higher affinity, that motor type will dominate the binding to the bead. These considerations may need to be examined, if you seem to only be able to bind one type of motor to your bead.

Overall, the use of beads as artificial cargos is an advantageous system because it is relatively simple for motors to attach to beads by simply incubating beads with purified protein or cell extract. Incubation of beads with cell extract allows for the binding of multiple species of endogenous motors, creating cargos that more closely resemble cellular cargos. The beads can be fluorescent and imaged in wide-field epifluorescence, without the need for total internal reflection fluorescence (TIRF) imaging. Many large beads can be observed directly using transmitted light, given their high index of refraction. In addition, beads can be trapped using optical tweezers. One disadvantage of beads as cargos is that while they are relatively simple to execute, the exact number, placement, and orientation of motors on the bead is unknown and varies from bead to bead.

11.3 Quantum Dots as Artificial Cargos

Recently, small fluorescent nanocrystals of semiconductor materials, such as CdSe, have become popular to serve as single-molecule fluorophores for motor protein measurements in cells and in vitro. Due to their small size and inherently quantum nature, these fluorescent particles are called quantum dots (Qdots). The diameter of the Qdot determines the color because it is literally a single quantum well where electrons are excited and emit within that well. When Qdots are in the size range of 2–10 nm, they absorb UV light and fluoresce in the visible range from violet for the smallest to far red for the largest dots.

The main benefits of using Qdots are their brightness and photostability. They can emit 1000× more photons than organic or protein fluorophores, enabling faster acquisition times for super-resolution or FIONA-style spot detection and localization routines [37–40]. Where a GFP motor requires a 0.5–1 s exposure time in order to collect enough photons for FIONA [41], a Qdot can be imaged at frame rates as fast as 30 frames/s and still be fit with 6 nm accuracy using a two-dimensional Gaussian [38, 42]. The advantage of these fast acquisition times is that experiments do not have to be performed at artificially low ATP concentrations to slow motors down, which might alter the stepping dynamics. Further, although TIRF imaging is best for obtaining a higher signal to noise ratio, Qdots are bright enough to be observed with epifluorescence, making them a versatile fluorescent label for single and multi-motor cargo assays. The increased photostability of quantum dots compared with traditional fluorophores enables imaging over long time durations, even with rapid acquisition times. For the study of motors, this property allows for motor dissociation events to be distinct and not to be confused with photobleaching events.

Qdots are made from heavy metals (CdSe) that are inherently insoluble in aqueous solution and poisonous to biology. In an effort to counteract these negatives, so that these bright, stable fluorescent markers can be used for biological applications, soluble and nontoxic capping shells have been developed. Polymeric capping layers, such as polyethylene glycol, can be further modified to enable cross-linking to numerous biological molecules such as antibodies, streptavidin, or carboxyl groups. Although they are marketed as stains for labeling cells for immunofluorescence, the modified Qdots have been successfully used for attachment to motor proteins as a fluorescent tag. The capping and subsequent modifications to the Qdot result in particles that are about 20 nm in diameter, similar to the smallest beads. This small size enables diffusion-limited binding, similar to, but slower than, fluorescently labeled single motors. Thus, unlike large beads that will settle in a chamber, Qdot-labeled motors are free to dissociate and diffuse similar to single motors or small cargos.

Qdots have been used as cargos both in cells and in vitro. In cells, they are endocytosed and encapsulated in an endocytic vesicle. The encapsulated Qdots are associated with various motors through the cell's cargo-docking machinery, many of which are unknown. The brightness of the Qdots allows for the individual fast stepping at high ATP concentrations in the cell to be discerned with high resolution [43–45]. In addition, Qdots have been labeled with motors prior to being introduced into cells via microinjection, allowing for the study of a single motor type in vivo [46–49].

In vitro, Qdots have been used to label individual motors to detect individual steps of the motor at fast time scales using the same attachment schemes detailed for beads [37–40, 50–53]. As with beads, the number of motors can be systematically lowered to achieve single motors associated with the cytoskeletal filaments.

Qdots have been used to investigate the activity of multiple motors associated to the same Qdot to serve as a small cargo. Experiments have been performed examining the transport of Qdot cargos conjugated to multiple copies of the same motor or different types, including kinesin-1 and myosin-V [54]. Recent work has shown that Qdots can associate multiple motors nonspecifically as the Qdot is transported along filaments crowded with motors [42]. Although the nonspecific interaction is of unknown origin [55], it offers a new mechanism by which Qdots can incorporate multiple motors. Importantly, these nonspecific interactions occur only when Qdots are mixed with high local concentrations of kinesin (such as on a crowded microtubule) and are not observed when Qdots and motors are mixed in solution [55]. This suggests that the binding of motors to Qdots in solution through specific linkages does not result in additional nonspecific interactions of motors with Qdots.

Here, we detail some of the issues to consider when using Qdots as a label.

11.3.1 Attachment Scheme

Motors can attach to Qdots through biotin-streptavidin interactions, antibodies, or through nonspecific interactions. Similar to beads, these attachments need to be identical for different motors. For the nonspecific interaction, since it not clear how it is attaching [55], each motor type will need to be specifically tested to determine its ability to bind.

11.3.2 Number and Location of Bound Motors

Much like beads, the number and orientation of motors associated to the Qdot cargos can be difficult to assess. The problem does not have as many free parameters, since the surface area of the Qdot is physically smaller than a large bead, but the attachments are still random and unknown. Much like a bead, many motors can bind to the Qdot that are not active—yet still attached to the Qdot. As with beads, probability dictates when you are in the single motor regime. Unlike beads, the single-molecule regime cannot be confirmed with optical trap measurements of the stall force. Typically the run length is used as a read out. If the run length of Qdot-associated motors is identical to the run length of GFP-labeled motors, it is very likely that there is only a single motor engaged [3, 4, 56–60].

11.4 Scaffolds to Link Motors

One problem with using beads and quantum dots as cargos is that the motors associate to the surface randomly with various distances between the motors and different densities from cargo to cargo. Thus, any two beads or Qdots will have a variable number of motors interacting. Although this may mimic the cellular architecture, it is not systematically controlled. In an effort to control the relative position of motors, several groups have turned to making scaffolds from DNA or proteins to link multiple motors together in a systematic and controlled way. DNA has been particularly effective since the sequence of the DNA can be specified to create one-, two-, and three-dimensional DNA origami structures that can specifically place motors at known locations using the DNA sequence.

Pioneering studies sought to investigate the force–velocity relationship of 2-motor complexes of the same motor type. They designed a DNA- and protein-based scaffold that holds two kinesin motors at a specified distance. A bead, bound to one end, can be used as a handle, and the force can be directly measured using an optical tweezer [61–64]. These experimental measurements resulted in new understandings of cooperative interactions between multiple motors on a single cargo, both in the presence and absence of applied load. Results with myosin-V motors showed that, like kinesin-1, complexes with two myosin-V motors are able to travel over larger distances compared to a single motor. However, in contrast to kinesin-1 motors, the velocity of these two-motor myosin-V complexes were observed to be slower compared to single motors, suggesting that there is negative cooperativity between multiple myosin-V motors [65].

Unlike the motor studies above using identical motors, two recent reports [66, 67] have designed a scaffold to incorporate opposing motors to force a "tug-of-war" (Fig. 11.3). These studies employed very different DNA architectures to measure the activity of like and different motors undergoing translocation. Both groups used DNA to specifically incorporate several motors. The DNA and/or motors had

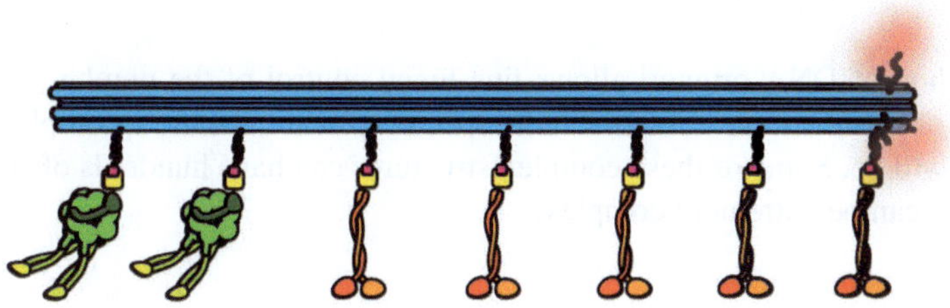

Fig. 11.2 DNA origami scaffolds to link multiple motors. This figure depicts a DNA origami scaffold specifically conjugated to seven motors via SNAP tag linkages. In this case, the conjugated motors are of two different motor species: dynein (*green*) and kinesin (*orange*). The DNA scaffold is labeled with a fluorophore for imaging with TIRF microscopy. Image was provided by S. L. Reck-Peterson

organic fluorescent tags, so they can be directly examined using TIRF microscopy. Unlike beads or Qdots, these small molecule fluorophores are very dim and require TIRF for imaging.

The first group's design was a complex structure of 250+ DNA strands in a multi-helix chassis structure to measure the activities of homogeneous groups of one to seven kinesin motors or dynein motors, as well as mixed populations of kinesin and dynein motors (Fig. 11.2) [66]. Interestingly many of the mixed motor chassis stalled. By engineering photocleavable linkers to either the kinesin or dynein motors, they were able to cut off a specific motor group after photocleaving with 405 nm light. After photocleavage, most of the chassis moved in the direction of the remaining motor type, implying that the motors were stalling each other during a pause.

A second group used a much simpler scheme to achieve multi-motor constructs [67]. This technique used a single piece of DNA that could bind 4–5 smaller pieces with SNAP or Halo-Tag ligands that would incorporate either the kinesin-1 motor or the minus-end-directed, non-processive Ncd motor. With kinesin-only motors coupled, they showed that kinesin motors are not as cooperative with more than two motors using both fluorescence imaging and force measurements. The results for 1 and 2 motors mimicked the results from the Diehl lab, described above. Ncd, on the other hand, was capable of creating cumulative force and processive motion when more than 2 motors were incorporated into the same scaffold. Mixed groups of up to 4 Ncd motors with a single kinesin-1 motor were able to slow the velocity of kinesin-1.

Although scaffolds have been effectively used, they are still cumbersome, and the following aspects should be considered when undertaking this route.

11.4.1 Complexity

The method of DNA origami allows fine-tuned control of the number, type, and position of specific motors, but the more specific you are, the more complicated the origami will be. Some of these complex structures can have hundreds of strands of DNA and can be extremely complex.

11.4.2 Motor Linkages

Some points to consider include how you will link the motor to the DNA. To create a motor construct that will bind to specific regions of the scaffold, some nice attachment schemes include using a SNAP or HaloTag. However, these require the ligand to be incorporated into the scaffold. Another issue with complicated DNA origami is that the motor linkages often dangle off the main chassis, but could also accidentally fold up into the origami [66], or the motor could just not bind at dilute concentrations [67]. Other options are to use protein components that bind tightly, like a split GFP or a biotin–streptavidin linkage.

11.4.3 Checking Your Creation

The ability of your DNA origami to bind motors and walk can be checked using microscopy. If the individual motors are also fluorescent, you can use photobleaching to count the number of motors attached, which will correspond to a steplike loss of intensity in the fluorescence trace over time. If fluorescence cannot be used, electron microscopy or AFM can be used to examine the DNA structures and determine if motors are incorporated [66, 68]. In the case of large motors such as dynein, motor incorporation can also be verified using gel shift assays [66].

11.5 Purified Endogenous Cargos

In the cell, organelle, vesicular, and macromolecular cargos associate motors for transport. Often, there are multiple motors of the same and different types walking along actin or microtubule filaments. Above, we described in vitro methods to simulate such cellular cargos using artificial or man-made constructions, but another method is to simply purify the actual endogenous cargos from cells and examine their motility outside of cells. Examples of purified cargos, sometimes referred to as ex vivo cargos, include vesicles, lipid droplets, and mRNA granules. Purified vesicles with functional motors have been used to study bidirectional transport in vitro using TIRF microscopy. The vesicles are small (~10–50 nm) and were visualized through a fluorescent tag on a motor expressed in transgenic mice [69, 70]. Unlabeled vesicles from brain tissue or squid axoplasm are not fluorescent but can be visualized using DIC microscopy [71, 72]. These provide a robust cargo to mimic axonal transport.

A second recently utilized endogenous cargo that can be used to explore multi-motor activity is the lipid droplet. Lipid droplets are vesicles enclosing fatty lipid particles. The *Drosophila* fly embryo is filled with these cargos, and they were originally used, in flies, to measure the forces of motors during development using an optical tweezer [73]. Recently, these lipid droplets were isolated from yeast, rat livers, and fly embryos and were shown to have active motors that could translocate the lipid droplets [73–75] (Fig. 11.3).

Unlike large lipid droplets that can bind many motors, small cargos, such as mRNAs, have also been purified for quantitative in vitro motility analysis via associated motors [76]. Active transport by motor proteins is used in cells and embryos to transport mRNAs for asymmetric cell division and during neuronal development and maintenance. Both the mRNAs and lipid droplets purified from fly embryos have huge advantages, since flies are a model system that is relatively easy to genetically modify. Purifying cargos from fly lines with motor mutants known to affect development enable the rapid discovery of biophysical mechanisms behind regulation and dysfunction of the motor cargo transport system.

Future work in this area includes purifying mitochondria, endocytic or exocytic vesicles, or melanosomes from tissue with active motors attached. Each will have special considerations and preparations in order to purify the cargo while keeping the motors functional. Another possible method to be explored is to transiently transfect cells with genetically encoded fluorescent proteins as part of the cargo complexes. This method could be very promising because one could also encode tags to enable fast, specific purification, despite the small initial sample volume available from cultured cells. We expect these future in vitro experiments will likely illuminate novel mechanisms of regulation of multi-motor transport.

Using endogenous cargos purified from tissue allows us to explore the actual geometries of the motors as they would exist in the cell. But, unlike the artificial cargos described above, natural cargos cannot be controlled and are more difficult to purify. Here, we describe some specific points to consider when using natural cargos.

11.5.1 Purification and Activity

The largest concern when purifying cargos from tissue is if you can isolate the cargo, intact, with the motors bound and functional. Working out the purification protocol may be the most difficult part of these assays. Using assays from other tissue types and previously determined assays work well, but progress using new

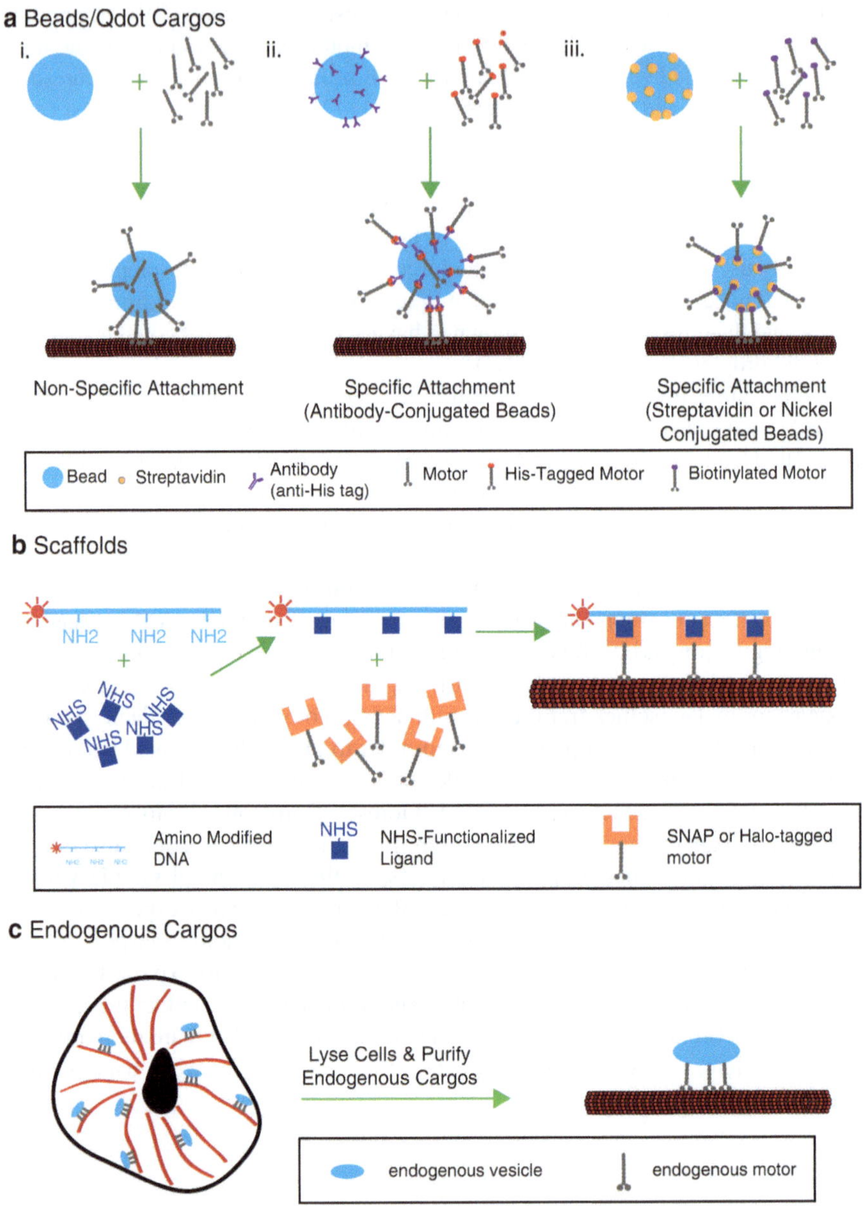

Fig. 11.3 Methods for studying cargo transport. Summary of the methodologies used to study cargo transport by molecular motors. (**a**) Motors can bind to beads through nonspecific or specific interactions. (i) During nonspecific interactions, motors simply bind to beads. Specific interactions can be mediated through (ii) antibodies against the motor tail or against a C-terminal genetically encoded tag (i.e., His or GFP tag) or (iii) molecules conjugated to beads (i.e., streptavidin or Ni-NTA) that bind tags on motors (i.e., biotin or His tag, respectively). (**b**) DNA scaffolds can be used to specifically position motors along a DNA cargo. Amino-modified nucleotides within the DNA scaffold react with NHS-functionalized SNAP or HaloTag ligands. These SNAP or HaloTag ligands mediate the binding of SNAP or Halo-tagged motors, respectively. (**c**) Endogenous cargos

cargos, such as mitochondria, may be hampered until the right mix of biophysicists and biochemists can determine the purification and functional assays.

11.5.2 Number and Types of Motors

As with beads and Qdots, the number of motors on the cargo is unknown, and functional motility assays are able to only hint whether motors are active, whether there is more than one motor, or whether motors of different types are bound. In situ labeling using antibodies to different types of motors can work to determine the types and relative amounts of each type of motor on cargos [74]. A better method would be to create transgenic animals that have fluorescently labeled motors of different colors, so the relative number of each motor type can be measured directly during a fluorescence assay. The fly system for lipid droplets and mRNAs may prove to be the next breakthrough for these experiments given that making transgenic flies is faster and easier than other systems.

11.5.3 Fluorescent Labeling

Fluorescently labeling the motors attached to the cargo is the best method to determine the number and type of each motor associated. Genetic tags can be added to motors expressed in transgenic animals or in transiently transfected cell lines and the cargos can be purified, as described above. Another option is to label the cargo itself. This can be done using dyes that bind to the cargo specifically, such as MitoTracker for mitochondria, or using fluorescent material within cargos, such as endocytosed Qdots to study purified endocytic vesicles [77]. Depending on the amount of fluorescent label, and the size of the cargo, many cargos could be visualized in epifluorescence, without the need for TIRF imaging. Individual motors could be co-localized using TIRF, and quantitative imaging or photobleaching could be used to determine the number of motors engaged at the filament directly [42].

Fig. 11.3 (continued) can be purified from cell lysate. Under the appropriate purification conditions, endogenous motors will remain bound to these cargos

11.6 Conclusions

We have illustrated a number of in vitro methods used to explore the ability of multiple motors to work in concert to enable intracellular cargo transport. Each method has strengths and weaknesses and many open questions still to be probed. Future work with two-dimensional and three-dimensional DNA origami to probe cargo transport among networks of microtubules will take us into increasingly complex and more cell-like reconstitutions. Using artificial cargos that can assemble and disassemble dynamically will illuminate answers as to how slow axonal transport is achieved. Purified cargos with an active assembly of motors will reveal how cellular cargos are organized and transported within the cell.

Acknowledgments Thank you to Michael Gramlich, Ashley Berner-Potts, and Fabian Romano for careful reading of this chapter. L.C. was supported by the Institute for Cellular Engineering, National Science Foundation (NSF) Integrative Graduate Education and Research Traineeship Program Grant DGE-0654128, NSF Grants DMR-1207783 and CMMI-0928540 (to J.L.R.), and a March of Dimes Basil O'Connor Award (to J.L.R.). J.L.R. was also supported by NSF Grant DBI-0923318 and Cottrell Scholars Award #20031 from Research Corporation for Science Advancement.

References

1. Perlson E, Maday S, Fu MM, Moughamian AJ, Holzbaur EL (2010) Retrograde axonal transport: pathways to cell death? Trends Neurosci 33:335–344
2. Vale RD, Schnapp BJ, Mitchison T, Steuer E, Reese TS, Sheetz MP (1985) Different axoplasmic proteins generate movement in opposite directions along microtubules in vitro. Cell 43:623–632
3. Block SM, Goldstein LS, Schnapp BJ (1990) Bead movement by single kinesin molecules studied with optical tweezers. Nature 348:348–352
4. Beeg J, Klumpp S, Dimova R, Gracià RS, Unger E, Lipowsky R (2008) Transport of beads by several kinesin motors. Biophys J 94:532–541
5. Porter ME, Scholey JM, Stemple DL, Vigers GP, Vale RD, Sheetz MP, McIntosh JR (1987) Characterization of the microtubule movement produced by sea urchin egg kinesin. J Biol Chem 262:2794–2802
6. Coy DL, Hancock WO, Wagenbach M, Howard J (1999) Kinesin's tail domain is an inhibitory regulator of the motor domain. Nat Cell Biol 1:288–292
7. Gutiérrez-Medina B, Fehr AN, Block SM (2009) Direct measurements of kinesin torsional properties reveal flexible domains and occasional stalk reversals during stepping. Proc Natl Acad Sci USA 106:17007–17012
8. Wang Z, Sheetz MP (1999) One-dimensional diffusion on microtubules of particles coated with cytoplasmic dynein and immunoglobulins. Cell Struct Funct 24:373–383
9. Schnitzer MJ, Block SM (1997) Kinesin hydrolyses one ATP per 8-nm step. Nature 388:386–390
10. Segal M, Soifer I, Petzold H, Howard J, Elbaum M, Reiner O (2012) Ndel1-derived peptides modulate bidirectional transport of injected beads in the squid giant axon. Biol Open 1:220–231

11. Schroeder HW, Hendricks AG, Ikeda K, Shuman H, Rodionov V, Ikebe M, Goldman YE, Holzbaur EL (2012) Force-dependent detachment of kinesin-2 biases track switching at cytoskeletal filament intersections. Biophys J 103:48–58
12. Wang Z, Khan S, Sheetz MP (1995) Single cytoplasmic dynein molecule movements: characterization and comparison with kinesin. Biophys J 69:2011–2023
13. Schroeder HW, Mitchell C, Shuman H, Holzbaur EL, Goldman YE (2010) Motor number controls cargo switching at actin-microtubule intersections in vitro. Curr Biol 20:687–696
14. Allan VJ, Vale RD (1991) Cell cycle control of microtubule-based membrane transport and tubule formation in vitro. J Cell Biol 113:347–359
15. Ali MY, Homma K, Iwane AH, Adachi K, Itoh H, Kinosita K, Yanagida T, Ikebe M (2004) Unconstrained steps of myosin VI appear longest among known molecular motors. Biophys J 86:3804–3810
16. Mallik R, Gross SP (2004) Molecular motors: strategies to get along. Curr Biol 14:R971–R982
17. Crevel I, Carter N, Schliwa M, Cross R (1999) Coupled chemical and mechanical reaction steps in a processive Neurospora kinesin. EMBO J 18:5863–5872
18. Kojima H, Muto E, Higuchi H, Yanagida T (1997) Mechanics of single kinesin molecules measured by optical trapping nanometry. Biophys J 73:2012–2022
19. Higuchi H, Muto E, Inoue Y, Yanagida T (1997) Kinetics of force generation by single kinesin molecules activated by laser photolysis of caged ATP. Proc Natl Acad Sci USA 94:4395–4400
20. Svoboda K, Schmidt CF, Schnapp BJ, Block SM (1993) Direct observation of kinesin stepping by optical trapping interferometry. Nature 365:721–727
21. Svoboda K, Block SM (1994) Force and velocity measured for single kinesin molecules. Cell 77:773–784
22. Tominaga M, Kojima H, Yokota E, Orii H, Nakamori R, Katayama E, Anson M, Shimmen T, Oiwa K (2003) Higher plant myosin XI moves processively on actin with 35 nm steps at high velocity. EMBO J 22:1263–1272
23. Seamster PE, Loewenberg M, Pascal J, Chauviere A, Gonzales A, Cristini V, Bearer EL (2012) Quantitative measurements and modeling of cargo-motor interactions during fast transport in the living axon. Phys Biol 9:055005
24. Wang Z, Sheetz MP (2000) The C-terminus of tubulin increases cytoplasmic dynein and kinesin processivity. Biophys J 78:1955–1964
25. Muto E, Sakai H, Kaseda K (2005) Long-range cooperative binding of kinesin to a microtubule in the presence of ATP. J Cell Biol 168:691–696
26. Jiang MY, Sheetz MP (1995) Cargo-activated ATPase activity of kinesin. Biophys J 68:283S–284S, discussion 285S
27. Euteneuer U, Koonce MP, Pfister KK, Schliwa M (1988) An ATPase with properties expected for the organelle motor of the giant amoeba, Reticulomyxa. Nature 332:176–178
28. Dreblow K, Kalchishkova N, Böhm KJ (2010) Kinesin passing permanent blockages along its protofilament track. Biochem Biophys Res Commun 395:490–495
29. Ross JL, Shuman H, Holzbaur EL, Goldman YE (2008) Kinesin and dynein-dynactin at intersecting microtubules: motor density affects dynein function. Biophys J 94:3115–3125
30. Richardson RM, Broaddus WC, Holloway KL, Fillmore HL (2005) Grafts of adult subependymal zone neuronal progenitor cells rescue hemiparkinsonian behavioral decline. Brain Res 1032:11–22
31. Gelles J, Schnapp BJ, Sheetz MP (1988) Tracking kinesin-driven movements with nanometrescale precision. Nature 331:450–453
32. Kosik KS, Orecchio LD, Schnapp B, Inouye H, Neve RL (1990) The primary structure and analysis of the squid kinesin heavy chain. J Biol Chem 265:3278–3283
33. Iwaki M, Tanaka H, Iwane AH, Katayama E, Ikebe M, Yanagida T (2006) Cargo-binding makes a wild-type single-headed myosin-VI move processively. Biophys J 90:3643–3652
34. Imanishi M, Endres NF, Gennerich A, Vale RD (2006) Autoinhibition regulates the motility of the C. elegans intraflagellar transport motor OSM-3. J Cell Biol 174:931–937

35. Wolenski JS, Cheney RE, Mooseker MS, Forscher P (1995) In vitro motility of immunoadsorbed brain myosin-V using a Limulus acrosomal process and optical tweezer-based assay. J Cell Sci 108(Pt 4):1489–1496
36. Gelles J, Berliner E, Young EC, Mahtani HK, Perez-Ramirez B, Anderson K (1995) Structural and functional features of one- and two-headed biotinated kinesin derivatives. Biophys J 68:276S–281S, discussion 282S
37. Armstrong JM, Krementsova E, Michalek AJ, Heaslip AT, Nelson SR, Trybus KM, Warshaw DM (2012) Full-length myosin Va exhibits altered gating during processive movement on actin. Proc Natl Acad Sci USA 109:E218–E224
38. Ali MY, Kennedy GG, Safer D, Trybus KM, Sweeney HL, Warshaw DM (2011) Myosin Va and myosin VI coordinate their steps while engaged in an in vitro tug of war during cargo transport. Proc Natl Acad Sci USA 108:E535–E541
39. Zhang D, Grode KD, Stewman SF, Diaz-Valencia JD, Liebling E, Rath U, Riera T, Currie JD, Buster DW, Asenjo AB, Sosa HJ, Ross JL, Ma A, Rogers SL, Sharp DJ (2011) Drosophila katanin is a microtubule depolymerase that regulates cortical-microtubule plus-end interactions and cell migration. Nat Cell Biol 13:361–370
40. Warshaw DM, Kennedy GG, Work SS, Krementsova EB, Beck S, Trybus KM (2005) Differential labeling of myosin V heads with quantum dots allows direct visualization of hand-over-hand processivity. Biophys J 88:L30–L32
41. Yildiz A, Tomishige M, Vale RD, Selvin PR (2004) Kinesin walks hand-over-hand. Science 303:676–678
42. Conway L, Wood D, Tüzel E, Ross JL (2012) Motor transport of self-assembled cargos in crowded environments. Proc Natl Acad Sci USA 109:20814–20819
43. Nan X, Sims PA, Chen P, Xie XS (2005) Observation of individual microtubule motor steps in living cells with endocytosed quantum dots. J Phys Chem B 109:24220–24224
44. Watanabe TM, Higuchi H (2007) Stepwise movements in vesicle transport of HER2 by motor proteins in living cells. Biophys J 92:4109–4120
45. Watanabe TM, Sato T, Gonda K, Higuchi H (2007) Three-dimensional nanometry of vesicle transport in living cells using dual-focus imaging optics. Biochem Biophys Res Commun 359:1–7
46. Nelson SR, Ali MY, Trybus KM, Warshaw DM (2009) Random walk of processive, quantum dot-labeled myosin Va molecules within the actin cortex of COS-7 cells. Biophys J 97:509–518
47. Courty S, Luccardini C, Bellaiche Y, Cappello G, Dahan M (2006) Tracking individual kinesin motors in living cells using single quantum-dot imaging. Nano Lett 6:1491–1495
48. Yoo J, Kambara T, Gonda K, Higuchi H (2008) Intracellular imaging of targeted proteins labeled with quantum dots. Exp Cell Res 314:3563–3569
49. Pierobon P, Achouri S, Courty S, Dunn AR, Spudich JA, Dahan M, Cappello G (2009) Velocity, processivity, and individual steps of single myosin V molecules in live cells. Biophys J 96:4268–4275
50. Reck-Peterson SL, Yildiz A, Carter AP, Gennerich A, Zhang N, Vale RD (2006) Single-molecule analysis of dynein processivity and stepping behavior. Cell 126:335–348
51. Sun Y, Sato O, Ruhnow F, Arsenault ME, Ikebe M, Goldman YE (2010) Single-molecule stepping and structural dynamics of myosin X. Nat Struct Mol Biol 17:485–491
52. Nishikawa S, Arimoto I, Ikezaki K, Sugawa M, Ueno H, Komori T, Iwane AH, Yanagida T (2010) Switch between large hand-over-hand and small inchworm-like steps in myosin VI. Cell 142:879–888
53. Lu H, Kennedy GG, Warshaw DM, Trybus KM (2010) Simultaneous observation of tail and head movements of myosin V during processive motion. J Biol Chem 285:42068–42074
54. Ali MY, Lu H, Bookwalter CS, Warshaw DM, Trybus KM (2008) Myosin V and kinesin act as tethers to enhance each others' processivity. Proc Natl Acad Sci USA 105:4691–4696
55. Conway L, Ross JL (2013) A model system to study transport of self-assembled cargos. Commun Integr Biol 6:e25387

56. Vershinin M, Carter BC, Razafsky DS, King SJ, Gross SP (2007) Multiple-motor based transport and its regulation by Tau. Proc Natl Acad Sci USA 104:87–92
57. Klumpp S, Nieuwenhuizen TM, Lipowsky R (2005) Self-organized density patterns of molecular motors in arrays of cytoskeletal filaments. Biophys J 88:3118–3132
58. Kunwar A, Vershinin M, Xu J, Gross SP (2008) Stepping, strain gating, and an unexpected force-velocity curve for multiple-motor-based transport. Curr Biol 18:1173–1183
59. Korn CB, Klumpp S, Lipowsky R, Schwarz US (2009) Stochastic simulations of cargo transport by processive molecular motors. J Chem Phys 131:245107
60. Müller MJ, Klumpp S, Lipowsky R (2010) Bidirectional transport by molecular motors: enhanced processivity and response to external forces. Biophys J 98:2610–2618
61. Diehl MR, Zhang K, Lee HJ, Tirrell DA (2006) Engineering cooperativity in biomotor-protein assemblies. Science 311:1468–1471
62. Rogers AR, Driver JW, Constantinou PE, Kenneth Jamison D, Diehl MR (2009) Negative interference dominates collective transport of kinesin motors in the absence of load. Phys Chem Chem Phys 11:4882–4889
63. Jamison DK, Driver JW, Rogers AR, Constantinou PE, Diehl MR (2010) Two kinesins transport cargo primarily via the action of one motor: implications for intracellular transport. Biophys J 99:2967–2977
64. Jamison DK, Driver JW, Diehl MR (2012) Cooperative responses of multiple kinesins to variable and constant loads. J Biol Chem 287:3357–3365
65. Lu H, Efremov AK, Bookwalter CS, Krementsova EB, Driver JW, Trybus KM, Diehl MR (2012) Collective dynamics of elastically coupled myosin V motors. J Biol Chem 287:27753–27761
66. Derr ND, Goodman BS, Jungmann R, Leschziner AE, Shih WM, Reck-Peterson SL (2012) Tug-of-war in motor protein ensembles revealed with a programmable DNA origami scaffold. Science 338:662–665
67. Furuta K, Furuta A, Toyoshima YY, Amino M, Oiwa K, Kojima H (2013) Measuring collective transport by defined numbers of processive and nonprocessive kinesin motors. Proc Natl Acad Sci USA 110:501–506
68. Wang P, Ko SH, Tian C, Hao C, Mao C (2013) RNA-DNA hybrid origami: folding of a long RNA single strand into complex nanostructures using short DNA helper strands. Chem Commun (Camb) 49:5462–5464
69. Caviston JP, Ross JL, Antony SM, Tokito M, Holzbaur EL (2007) Huntingtin facilitates dynein/dynactin-mediated vesicle transport. Proc Natl Acad Sci USA 104:10045–10050
70. Hendricks AG, Perlson E, Ross JL, Schroeder HW, Tokito M, Holzbaur EL (2010) Motor coordination via a tug-of-war mechanism drives bidirectional vesicle transport. Curr Biol 20 (8):697–702
71. Brady ST, Lasek RJ, Allen RD (1985) Video microscopy of fast axonal transport in extruded axoplasm: a new model for study of molecular mechanisms. Cell Motil 5:81–101
72. Gill SR, Schroer TA, Szilak I, Steuer ER, Sheetz MP, Cleveland DW (1991) Dynactin, a conserved, ubiquitously expressed component of an activator of vesicle motility mediated by cytoplasmic dynein. J Cell Biol 115:1639–1650
73. Tolić-Nørrelykke IM, Munteanu EL, Thon G, Oddershede L, Berg-Sørensen K (2004) Anomalous diffusion in living yeast cells. Phys Rev Lett 93:078102
74. Bartsch TF, Longoria R, Shubeita GT, Florin E-L (2013) Lipid droplets purified from Drosophila embryos as an endogenous handle for precise single molecule measurements. Biophys J 104:475
75. Barak P, Rai A, Rai P, Mallik R (2013) Quantitative optical trapping on single organelles in cell extract. Nat Methods 10:68–70
76. Amrute-Nayak M, Bullock SL (2012) Single-molecule assays reveal that RNA localization signals regulate dynein-dynactin copy number on individual transcript cargoes. Nat Cell Biol 14:416–423

77. Delehanty JB, Bradburne CE, Boeneman K, Susumu K, Farrell D, Mei BC, Blanco-Canosa JB, Dawson G, Dawson PE, Mattoussi H, Medintz IL (2010) Delivering quantum dot-peptide bioconjugates to the cellular cytosol: escaping from the endolysosomal system. Integr Biol 2:265–277

Chapter 12
Measuring Two at the Same Time: Combining Magnetic Tweezers with Single-Molecule FRET

Marko Swoboda, Maj Svea Grieb, Steffen Hahn, and Michael Schlierf

Abstract Molecular machines are the workhorses of the cell that efficiently convert chemical energy into mechanical motion through conformational changes. They can be considered powerful machines, exerting forces and torque on the molecular level of several piconewtons and piconewton-nanometer, respectively. For studying translocation and conformational changes of these machines, fluorescence methods, like FRET, as well as "mechanical" methods, like optical and magnetic tweezers, have proven well suited over the past decades. One of the current challenges in the field of molecular machines is gaining maximal information from single-molecule experiments by simultaneously measuring translocation, conformational changes, and forces exerted by these machines. In this chapter, we describe the combination of magnetic tweezers with single-molecule FRET for orthogonal simultaneous readout to maximize the information gained in single-molecule experiments.

Keywords Magnetic tweezers and single-molecule FRET • DNA manipulation • Conformational changes • Molecular machines

Abbreviations

CCD	Charge-coupled device
DM	Dichroic mirror
DNA	Deoxyribonucleic acid
dNTP	Deoxyribonucleoside triphosphate
dTTP	Deoxythymidine triphosphate

M. Swoboda • M.S. Grieb • S. Hahn • M. Schlierf (✉)
BCUBE - Center for Molecular Bioengineering, Technische Universität Dresden, Arnoldstraße 18, 01307 Dresden, Germany
e-mail: marko.swoboda@bcube-dresden.de; grieb@bcube-dresden.de; hahn@bcube-dresden.de; schlierf@bcube-dresden.de

C.P. Toseland and N. Fili (eds.), *Fluorescent Methods for Molecular Motors*, Experientia Supplementum 105, DOI 10.1007/978-3-0348-0856-9_12,

dUTP	Deoxyuridine triphosphate
EMCCD	Electron multiplying charge-coupled device
FRET	Förster resonance energy transfer
GPU	Graphics processing unit
HJ	Holliday junction
LUT	Lookup table
PCR	Polymerase chain reaction
RCLED	Resonant-cavity light-emitting diode
RNA	Ribonucleic acid
ROI	Region of interest
TIRF	Total internal reflection fluorescence

12.1 Introduction

Molecular machines are the workhorses of the cell that efficiently convert chemical energy into mechanical motion through conformational changes. They can be considered powerful machines, exerting forces and torque on the molecular level of several piconewtons and piconewton-nanometer, respectively. For studying translocation and conformational changes of these machines, fluorescence methods have proven well suited. Especially powerful is total internal reflection fluorescence (TIRF) microscopy which confines fluorophore excitation to the surface of the sample chamber, thereby minimizing background and allowing the detection of single molecules [1]. Combined with single-molecule Förster resonance energy transfer (smFRET) [2], where single fluorophores provide distance measurements in the molecular frame, these methods have become near ubiquitous in the single-molecule field [3]. Besides fluorescence techniques, force techniques allow us to understand the action of molecular motors, as they are able to characterize the force and torque that cellular machines exert when performing their tasks. The most common of these force techniques are optical tweezers [4], magnetic tweezers [5], and atomic force microscopes (AFM) [6, 7]. They have been applied to many molecular motor systems, such as the bacteriophage Φ29 viral packaging motor [8], kinesins [9], or the RecBCD helicase [10]. Magnetic tweezers are the prime method of choice to study the influence of torque on a given system or the torque exerted by molecular motors, although recent studies have shown that also optical tweezers can be adapted to allow torque spectroscopy employing anisotropic quartz microcylinders [11, 12]. The intrinsic advantage of magnetic tweezers is the anisotropic magnetic field interaction with paramagnetic particles that allows the torque exertion and force application in "routine" experiments [13, 14].

One of the current challenges in the field of molecular machines is gaining maximal information from single-molecule experiments by simultaneously measuring translocation, conformational changes, and forces exerted by these machines. Combining force spectroscopy techniques with fluorescence techniques for orthogonal readouts is therefore becoming increasingly popular [15–18]. This chapter introduces an implementation of a magnetic tweezers instrument combined with smFRET capability. We will introduce the experiments from the bottom-up, first, with a detailed description of the experimental setup, followed by the software principles including a 3D real-time tracking algorithm and the molecular biology approaches for the sample design. We will then present exemplary results and subsections on methods of data analysis.

12.2 Experimental Setup: Concept of Magnetic Tweezers in Combination with TIR Single-Molecule FRET

In analogy to regular tweezers, which are used to grip and manipulate tiny objects, magnetic tweezers allow the manipulation of some of nature's smallest objects, i.e., individual protein or DNA molecules. As the name implies, forces are exerted through magnets, which act on micrometer-sized paramagnetic microspheres or beads by attracting them toward regions of higher magnetic field strength. The beads are commonly manufactured from polystyrene doped with paramagnetic materials. The external magnetic field can be created by permanent magnets, with neodymium–iron–boride magnets being a popular choice [19]. Schemes involving electromagnets have also been developed [20–23], but they are usually limited to lower magnetic fields and require sophisticated cooling of the electromagnets.

In magnetic tweezers experiments, individual molecules are attached at one side on the inner surface of a flow chamber and at the other side to a magnetic bead, with external magnets exerting force and torque, while being observed under a microscope. The magnetic beads can be imaged and tracked quantitatively via computer-based video microscopy using an illumination source, a microscope objective, and a fast camera. From the bead motion, one can deduce the acting forces, molecular conformation, and torsional behavior. In order to track the bead position in 3D, one typically evaluates the diffraction patterns of the bead using a multistep algorithm that will be discussed further below. This diffraction pattern depends on the size of the beads and their position along the optical axis. Already minor changes of the bead height, on an order of magnitude smaller than the illumination wavelength, have a disproportionally great influence on the diffraction pattern ring structure. Here, we use a piezo-actuated XYZ stage to position the flow cell and to calibrate the diffraction patterns along the optical axis and, by extension, deduce the bead height quantitatively. In addition, the rotationally symmetric diffraction pattern from spherical magnetic beads can be analyzed with sub-pixel resolution to obtain the lateral position fluctuations of the bead.

Since magnetic tweezers experiments are surface-tethered experiments, drift and other vibrational sources can severely affect the signal-to-noise level and thus the resolution of the experiment. In order to reduce many effects of drift and other sources of error from flow cell motions, we typically track two beads simultaneously, a so-called reference bead attached to the surface of the flow cell and a tethered bead attached to the molecule of interest. In this type of differential tracking, looking only at the difference of both bead positions can reduce many external sources of error.

Figure 12.1 depicts our experimental setup, which combines a smFRET setup with magnetic tweezers to simultaneously sample two independent reaction coordinates—force and torque versus distance and molecular conformation [16]. In order to prevent crosstalk of the two measurements, we use different wavelength bands of the light spectrum for the detection of the bead position and for FRET measurements. Consequently, the magnetic tweezers operate with illumination in the near infrared (800–920 nm), whereas the remaining parts of the visible spectrum (400–770 nm) remain for smFRET experiments with typical fluorophores. We illuminate our sample chamber with a laser-ignited Xe plasma lamp (Energetiq Inc.) and image the beads in the flow cell with an inverted microscope geometry and a CCD camera (JAI Corp.) with a sample rate of 200 Hz. Each frame of the camera data is read out and analyzed in real time via a PC. Tracking errors resulting from CCD detector shot noise can be consequently reduced with a maximal light flux on each pixel of the camera. While many other magnetic tweezers tend to use resonant-cavity light-emitting diodes (RCLEDs) [24–26] which are available in the visible wavelength range (650 nm), we require an infrared source with sufficient flux to use the visible spectrum for single fluorophore imaging. Infrared laser diodes have the disadvantage of creating unwanted speckle patterns despite their high photon yields and narrow bandwidths, which would in principle be useful for diffraction pattern generation. We therefore chose to filter the emission from the Xe plasma lamp to select a suitable spectral region. The lamp emission is fiber coupled to the setup and collimated before being focused on the sample for illumination.

The dichroic mirrors DM1 to DM4 are used to filter and direct light of various wavelength bands to detectors and to the sample. In this regard, DM1 (Semrock FF750-SDi02-25x36) is used to separate the infrared illumination from the fluorescence and direct it onto the CCD camera used for tracking the magnetic beads. DM2 (Chroma zt532/635/NIR rpc) is used to separate the emitted photons from the excitation light, while DM3 (Chroma 630 DCXR) separates the emission into the two wavelength bands required for smFRET (i.e., donor emission and acceptor emission). The dichroic DM4 (Chroma z620 sprdc) acts as a beam combiner for the excitation light of the two excitation lasers, a frequency-doubled, diode-pumped Nd:YAG at 532 nm (Coherent Sapphire 532) and a diode laser at 640 nm (Coherent OBIS 640LX), which are used for the excitation of many common FRET dye pairs. The fluorescence of the single FRET pairs are recorded with a very sensitive EMCCD camera (Andor DU-897D) and evaluated in a custom written post-experiment analysis.

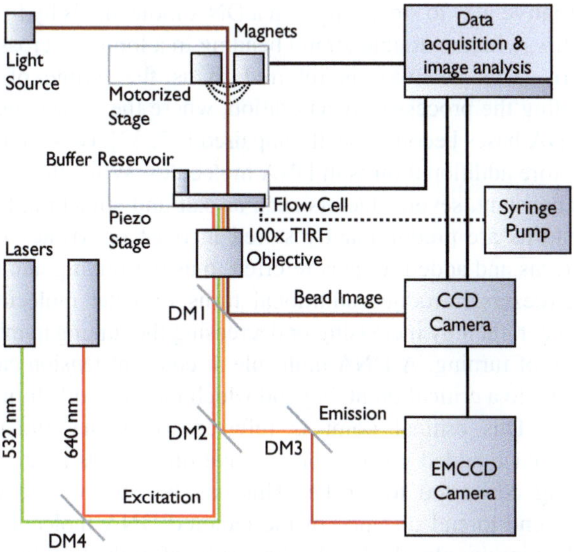

Fig. 12.1 Sketch of the experimental setup. The light collected by the objective is split into an imaging path for magnetic bead tracking and a fluorescence path, detected by a CCD and an EMCCD camera, respectively. Lasers at 532 nm and 640 nm are used for fluorescence excitation; the microscope illumination operates in the near infrared between 800 nm and 920 nm. A piezo stage is used for the calibration of bead diffraction images and a motorized stage for turning and moving the permanent magnets. Bead tracking and data recording is implemented via PC

12.3 Magnetic Tweezers Experiments

The intrinsic coiling of double-stranded B-DNA—the canonical double-helix structure of DNA—is often altered in positive or negative direction during enzymatic processing. This leads, for instance, to so-called supercoiled structures of DNA—the result of adding twist to the DNA beyond its mechanically relaxed state. Supercoiling can also result in the melting or local unpairing of individual bases in the DNA strand. The formation of Z-DNA—a left-handed helix structure of DNA with different base pair spacing—has also been observed with induced negative DNA coiling [14, 27]. The most common observation is probably the formation of plectonemes—colloquially known from telephone cords that form looping coils of the cord with itself with increasing twist. Polymerases, for instance, follow the helical pitch and can thus induce positive supercoiling in DNA double strands ahead of its position [28, 29]. The additional twisting can be mimicked in an experimental setup by adding turns to a torsionally constrained molecule [30]. Additional torque can be stored in a DNA molecule predominantly in two different ways. The individual strands of double-stranded DNA wind around each other with one full turn approximately every 10.5 base pairs. With adding torque, this strand winding density can increase and storing torque in such a way is referred to as twist. An alternative way to store torque in a DNA molecule is by forming loops or coils of the DNA, with the double-strand bending in a loop and crossing itself once per added turn. This behavior is referred to as the writhe of the molecule [31]. Disregarding the process of denaturation, where the two single strands separate and the DNA bases become locally unpaired [32, 33], twist and writhe are the two means to store additional turns in DNA molecules. While the number of added turns remains constant, several factors such as salt and stretching force can influence whether turns are predominantly stored in twist or writhe. The sum of all natural DNA turns and added turns is referred to as the linking number.

Magnetic tweezers introduce additional turns into the molecular system by turning the magnet, thereby increasing or decreasing the linking number, dependent on the direction of turning. A DNA molecule at constant tension can only absorb torque as twist up to a critical point, beyond which the first coils form—an increase of writhe [34]. This critical point is referred to as the buckling transition [35]. Beyond it, each added positive or negative turn results in an additional coil, with twist being converted to writhe. This can be observed in the significant decrease in the end-to-end distance of the tethered DNA molecule (as shown in Fig. 12.2a). Most intuitively, the DNA with writhe after the buckling transition can be imagined like telephone cords that also have a tendency to form such coils with themselves with added turns.

Figure 12.2 shows force-dependent supercoiling measured on a 6-kbp-long DNA molecule for various salt concentrations. The salt concentration was varied from 10-mM sodium chloride, to 50-mM magnesium chloride, to 50 mM each, and finally to 150-mM sodium chloride (Figure 12.2b-e, respectively). The extension of each molecule is plotted relative to its length at zero turns. The four forces were chosen to depict the transition from the lower (0.11 pN) to intermediate (0.65 pN) force regime [30]. At each salt concentration and low forces, we observe the force-dependent buckling transition with negative supercoiling. This transition originates from the plectoneme formation [30, 34], which is accompanied by significant extension decrease with each added turn. The supercoiling curves for positive and negative supercoiling are symmetric with respect to the zero of turns added when relatively low forces (<0.2 pN) are applied. At higher force, negative supercoiling induces denaturation of the DNA [32], which alters the mechanical properties of the DNA molecule with localized formation of other structural forms of DNA, such as Z-DNA [14, 27] or melting and separation of the double-stranded DNA. In this high-force regime, the extension change is no longer observed since no plectonemes are formed. Thus, higher forces lead to asymmetric supercoiling curves, and the buckling transition at positive turns occurs significantly later [35]. At low forces, a plateau extends around the apex (zero turns) before the buckling transition occurs, with the plateau width showing salt dependence. This effect can be attributed to the shielding of repulsive DNA backbone charges at increasing salt concentrations. Increased charge shielding results in a higher mechanical flexibility, i.e., reduced persistence length, and therefore smaller plectonemic loop sizes in relation to the end-to-end distance of the DNA molecule [35, 36].

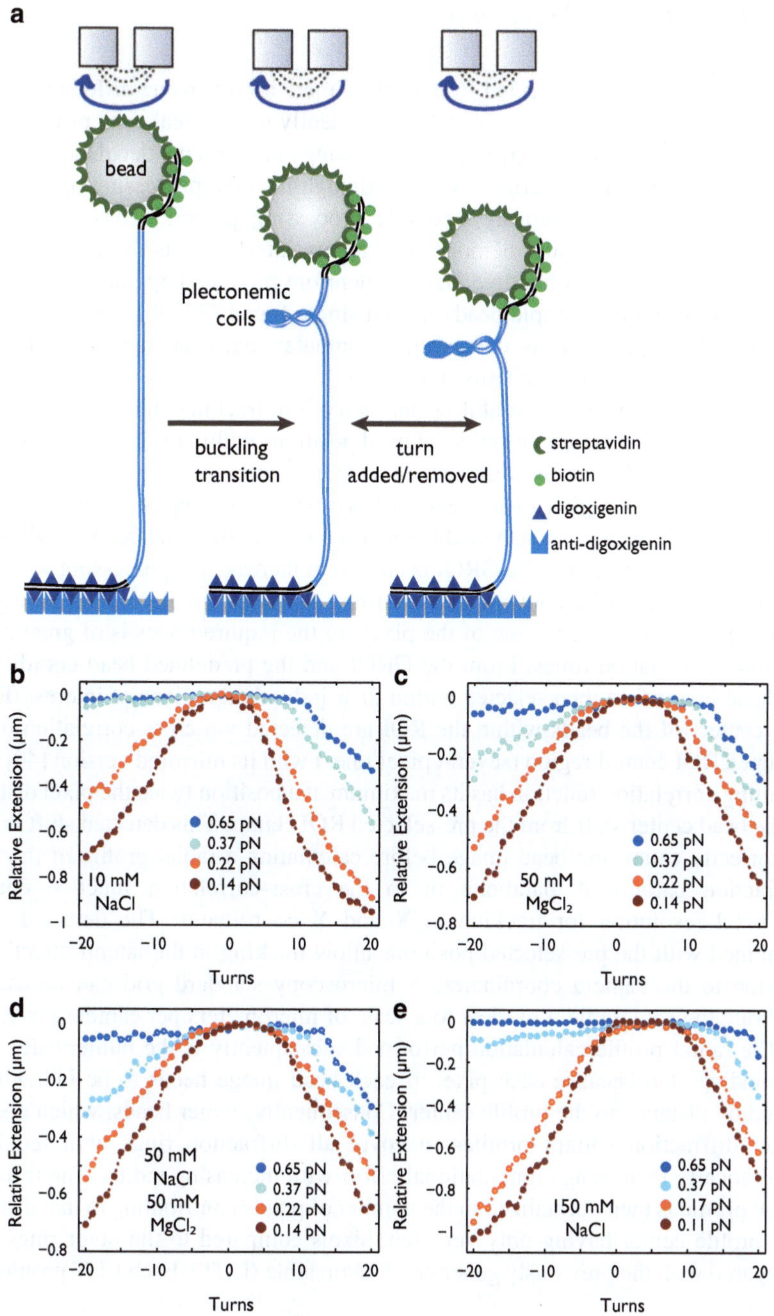

Fig. 12.2 Salt and force dependence of DNA supercoiling behavior. (**a**) Sketch of the sample arrangement. A DNA strand is tethered via biotin–streptavidin and digoxigenin–antidigoxigenin interactions between magnetic bead and flow cell surface. External magnets exert force and torque, leading to supercoiling after critical torque is reached (*left to right*). (**b–e**) Salt- and force-

12.3.1 Tracking Algorithm

Bead positions can be extracted from CCD camera images using different tracking algorithms [5, 13, 22]. These algorithms frequently rely on real-time processing of frames from video microscopy or post-processing of recorded video data, especially for massively parallel experiments [5]. Although the post-processing option is not limited by the instantaneously available processing power, it has its limits in conventional PC hardware memory and camera interfaces, as these cameras can record videos at several hundred Hz and therefore produce large amounts of data. Parallel handling of multiple beads from a single frame [37, 38] is also becoming increasingly popular, as is outsourcing computationally demanding processing steps to graphics processing units (GPUs) [39].

Figure 12.3 depicts a general flow diagram of our tracking algorithm implementation. Inputs (circles) are pre-selected bead positions within the field of view of the tracking camera. Combined with a recent camera frame, these positions are used to evaluate a greater region of interest (GROI), which encompasses all beads with their surroundings and some added pixel space surrounding the beads to allow for fluctuations of the beads. The GROI is chosen to increase the processing speed. As the image needs to be converted to an array of 8-bit numerical pixel values for further processing, a reduction of the pixels to the required ones is of great use to optimize calculation times. From the GROI and the predefined bead coordinates, the bead images are then selected within their individual regions of interest (ROI). The centers of the beads within the ROI are detected via cross-correlation of the summed ROI central region (several pixel lines) with its mirrored version [40]. The resulting correlation function has its maximum at a position twice the pixel distance of the bead center shift from the pre-selected ROI center. This detected shift is used to correctly center the bead image before calculating a radial profile of the bead diffraction pattern. A parabolic fit to the cross-correlation function enables sub-pixel resolution for tracking in **X** and **Y** coordinates. The detected shifts combined with the pre-selected positions allow tracking in the lateral direction in relation to the camera coordinates. A microscopy standard grid can be used to calibrate the camera field of view to a scale of micrometers per camera pixel.

The radial profile calculation performed subsequently is the numerically most demanding step because each pixel of each bead image needs to be individually sorted by distance to the profile center. Consequently, larger beads, which possess larger diffraction pattern profiles to cover all diffraction rings, will require a quadratically increasing computational effort with increasing radius. The resulting radial profile is then normalized to the number of pixels contributing to that distance (the profile center having only very few pixels compared to the outer rings) and compared with the previously generated lookup table (LUT). Each LUT profile s_i of

---

Fig. 12.2 (continued) dependent measurements of the end-to-end distance of tethered 6-kbp DNA double strands depending on the change of the linking number by the addition or removal of turns

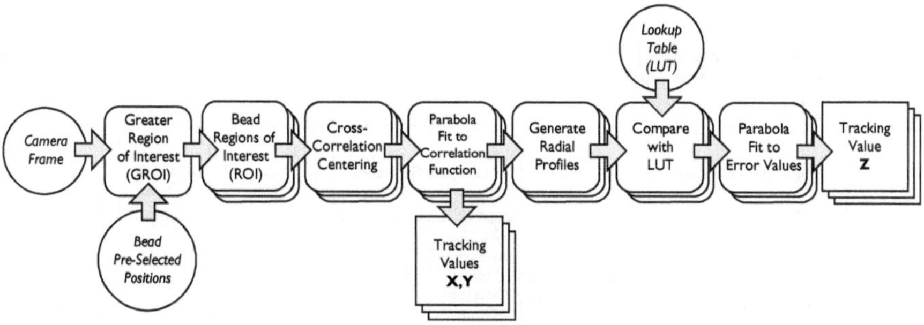

Fig. 12.3 Flow diagram of the tracking algorithm. *Circles* indicate external input parameters and *sharp-cornered squares* the outputs (the tracked coordinates in **X, Y, Z**)

length L (the size of the bead image) is subtracted from the normalized radial profile, and the error values χ_i are calculated as:

$$\chi_i = \sum_{n=1}^{L} (s_n - s_{i,n})^2. \tag{12.1}$$

The resulting distribution of error values χ_i possesses a minimum at the index i most similar to the frame profile, i.e., the corresponding LUT profile associated with a particular height—the measured **Z** value. To reach a higher precision than the spacing of LUT values in the axial direction, a parabola fit to the error values is performed to more precisely determine the actual bead position in the **Z** coordinate from the surrounding error values. With this approach, axial resolutions of 10 nm or better can be achieved at a typical LUT spacing of 150 nm. Alternative fit functions did not show significant improvement, while parabola fits have the benefit of being numerically undemanding.

This tracking procedure can be performed in parallel for multiple beads, with a minimum of two beads (reference and tethered bead). Our current setup allows parallel tracking of up to ten beads without exhausting the available PC resources. Further acceleration of these calculations can be achieved by GPU processing that is highly optimized for parallelization [39].

The typical LUT generation procedure is outlined in Fig. 12.4, with the repetitions for each individual piezo step omitted. It can be summarized as **Z**-scanning of the piezo stage over a range of several microns in steps of a defined spacing and recording a reference profile at each step. Each reference profile is used later for comparison of bead diffraction images with the LUT to determine their height. In a subsequent verification loop, the piezo stage scans through the newly acquired reference positions, and the LUT is used to track the beads at these positions and to compare with the position given by the piezo stage readout. Additionally, this procedure detects drifts in the bead diffraction images, which usually are signs of leaks in the flow cell. The LUT acquisition may suffer from errors introduced by fluctuations in the course of creating the reference profiles. Therefore, for each LUT

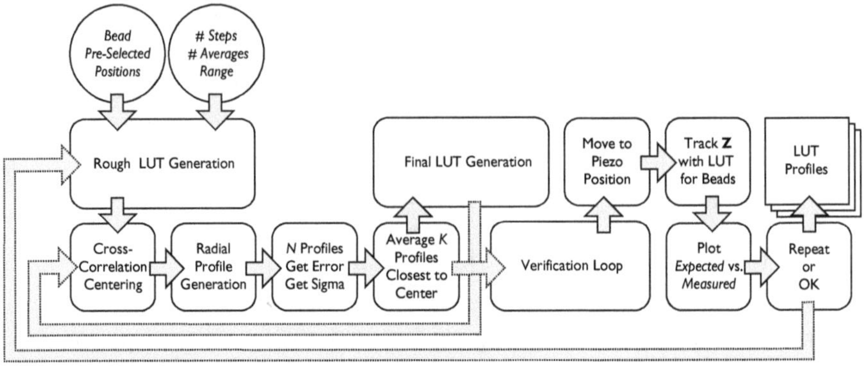

Fig. 12.4 Flow diagram of the lookup table (LUT) generation algorithm. A rough LUT is generated at first to assess the quality of the later acquired final LUT. *Circles* indicate input parameters, *squares* outputs of the algorithm

position, a sample of frames is collected, compared with a previously acquired "rough" LUT, and only the most similar profiles with the lowest error are averaged to serve as the final profile. At a given position, the bead oscillates around the reference position. The resulting profiles closest to the mean will be the ones closest to the actual position, thus improving the LUT profile precision and similarity at a given position.

12.3.2 Force Calibration

Magnetic tweezers require calibration of the forces applied to tethered molecules. The bead magnetization inhomogeneity, the sample mounting precision, and the nonlinearity of the magnetic field exerted by the permanent magnets in the setup [41] all act as sources of error and reduce reproducibility between experiments. Several approaches to calibrate forces have been described elsewhere [42–46].

Figure 12.5a depicts typical data used in the force calibration of a magnetic tweezers instrument. The lateral fluctuations in both **X** and **Y** of a surface-tethered bead are recorded while the magnet distance from the sample is increased stepwise (red curve). The variance of these fluctuations $\langle \delta x^2 \rangle$ can be used in force calibration by applying the equipartition theorem. With $\langle \delta x^2 \rangle$ determined, the equipartition theorem yields the applied force as:

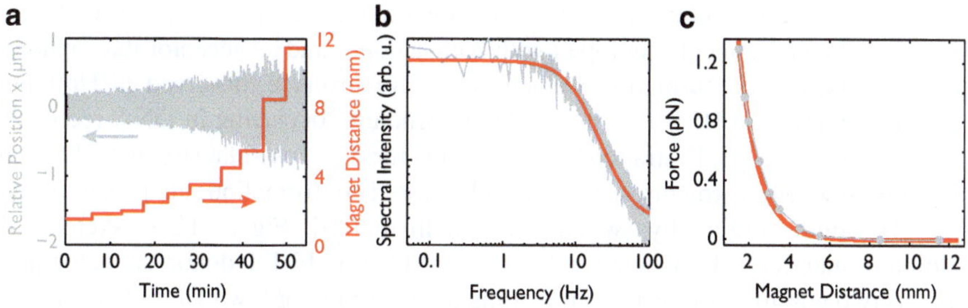

Fig. 12.5 Force calibration data and methods. (**a**) Magnetic bead position in **X**, tracked over time with varying magnet distance from the sample. Note the increasing fluctuation width with decreasing trapping forces, (**b**) power spectrum of bead oscillations with Lorentzian fit (corner frequency $f_c = 11.4$ Hz), (**c**) forces obtained from the measurement in (**a**) using the real-space method, dependent on the magnet distance from the sample, fitted with a single exponential

$$F = \frac{k_B T L_{ext}}{\langle \delta x^2 \rangle} \tag{12.2}$$

with L_{ext} being the extension of the DNA at a given force F and temperature T. This can be done in real space [34] or via the power spectrum of the bead fluctuations [45], as shown in Fig. 12.5b. For the forces obtained from the data in Fig. 12.5, performed with 1-µm magnetic spheres, the choice of the method is not crucial, as demonstrated by Vilfan et al. [43]. Only for forces significantly exceeding 1 pN, the two methods diverge and the forces are systematically overestimated when using the real-space method, due to averaging effects within a camera frame. At low forces, the power spectrum method requires long acquisition times, since the corner frequency becomes very low and, therefore, the equipartition theorem is the preferred method. An alternative method recently put forth to be used in force calibration is the Allan variance [46]. It is designed to converge for most naturally occurring noise and is half the mean of the squared differences of neighboring intervals of a given length, evaluated on time-series data like tracked bead coordinates [47].

12.4 Single-Molecule FRET

FRET measurements are based on the extent of non-radiative energy transfer between two fluorescent dye molecules, which varies with the distance between both molecules [2]. One fluorophore (referred to as the *donor*) is fluorescently excited and transfers its excitation to the second fluorophore, the *acceptor*, through dipole–dipole interaction. The transfer efficiency E depends on the distance R between donor and acceptor fluorophore and a characteristic length, the Förster radius R_0: $E = 1/(1 + (R/R_0)^6)$. By recording the fluorescence of both molecules, their distance can be inferred and used to report on, for example, the conformational changes of a biomolecule equipped with both, a donor and an acceptor fluorophore.

These typical conformational changes are, for instance, observed in Holliday junctions (HJ). The HJ is a four-way DNA structure, occurring in DNA recombination [48]. Four single-stranded DNA strands form a cross-like structure that can fold into stacked-X structures. Two of such stacked conformations are possible, and the HJ would continually switch between them [49]. Figure 12.6a sketches a Holliday junction (HJ) labeled with Cy3 and Cy5 as FRET donor–acceptor pair and its two possible conformations leading to either high or low FRET. This system has been previously described as a "nanometronome" to provide a standard of alternating FRET efficiency for the calibration of FRET instruments [49]. The system can be readily extended to show force-dependent oscillations by attaching a tether on one arm of the HJ to externally modulate the ticking behavior via optical tweezers or any other single-molecule force technique [16]. Furthermore, the oscillation frequency of the HJ between high- and low-FRET efficiency is salt dependent. This allows the tuning of the transition rates to the frame rate of the EMCCD camera used to record the fluorophores, as well as the fluorophore lifetime, i.e., the number of observable oscillations before bleaching.

Figure 12.6b depicts smFRET data obtained from the nanometronome recorded at 30 frames per second. The upper panel shows the fluorescence intensity from the donor Cy3 (green) and the acceptor Cy5 (red) fluorophores, recorded in two channels on our EMCCD camera. When both fluorophores are in close proximity, the observed fluorescence is dominated by the acceptor channel—a high red signal—whereas when the fluorophores are further apart, the fluorescence emission observed is mainly in the donor channel—a high green signal. Both conformers are sketched in Fig. 12.6a. It should be noted that while the transition rates are salt dependent, the individual transition time between the conformer states is beyond the temporal resolution of the camera, as can be seen by the transitions occurring within single recorded frames. The lower panels and black lines in Fig. 12.6 depict the FRET efficiency calculated from the donor and acceptor channel intensities, using the simplified calculation $E = I_A/(I_A + I_D)$. Figure 12.6c emphasizes the alternating FRET efficiency (lower panel) and anticorrelation of donor and acceptor intensity. Photophysical bleaching can be observed in Fig. 12.6b, where the acceptor intensity drops to the background level after ~35 s and the donor bleaches subsequently at ~43 s. Although the acceptor molecule does not emit any further after this transition, it still accepts distance-dependent energy transfer from the donor molecule, as can be seen by the oscillations in the donor intensity. A histogram of the FRET values from all frames in the trace until the acceptor enters its dark state is shown in Fig. 12.6d. Characteristically, for the HJ system, two well-separated FRET distributions are present, representing the two different conformations.

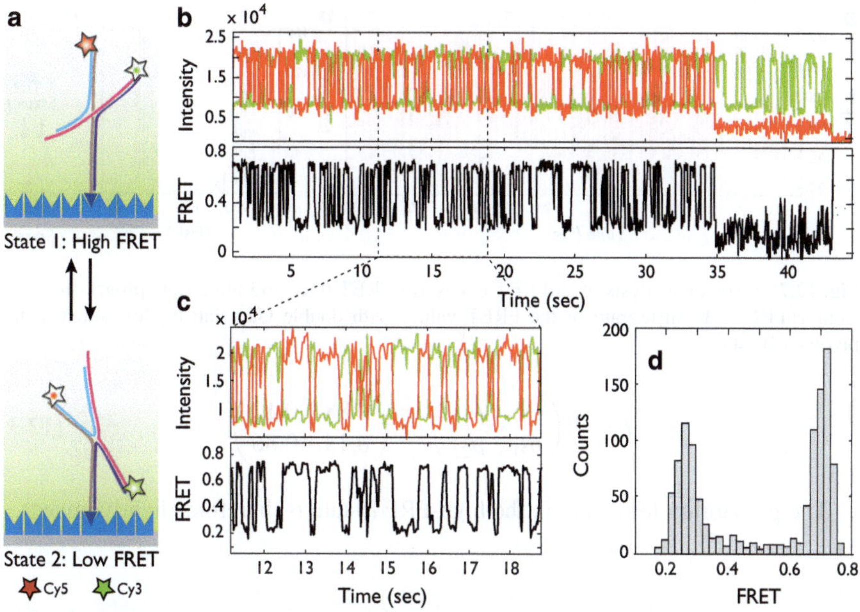

Fig. 12.6 Conformational oscillations of a FRET-labeled Holliday junction. (**a**) Sketch of the experimental arrangement with the nanometronome adopting either high-FRET or low-FRET conformation, (**b**) measured donor (*green*) and acceptor (*red*) fluorescence signal (*upper panel*), with calculated FRET values (*lower panel*), (**c**) zoom of a subsection of (**b**) to highlight the anticorrelated fluorescence between donor and acceptor and the oscillation behavior of the HJ, (**d**) histogram of the FRET values of (**b**), showing two well-separated FRET-value populations

12.4.1 Analysis of smFRET Data

Single-molecule kinetics can be readily extracted from FRET experiments. Various schemes have been developed over the past decade to extract states and transition rates from smFRET data [2, 50, 51]. Here, we chose to use the freely available vbFRET software, provided by [50] and implemented in MATLAB.

Figure 12.7a depicts the FRET efficiency trace from Fig. 12.6, combined with a state-progression path determined by vbFRET. Figure 12.7b shows the histogram from the FRET data, fitted with two Gaussians to obtain the FRET efficiencies of both states of the HJ as $E_1 = 0.71$ and $E_2 = 0.27$.

From the variational Bayesian expectation maximization (VBEM) in vbFRET, we also obtain the transition matrix between the states at $E_{1,\mathrm{vb}} = 0.71$ and $E_{2,\mathrm{vb}} = 0.30$, in excellent agreement with the two histograms. Transition rates between states in the HJ system are both salt and force dependent. Here, we kept the salt concentration at 30-mM $MgCl_2$ and 50-mM NaCl. For the force-free case in Fig. 12.6, the transition probability matrix T_{ij}, giving the probability p_{ij} for a transition from a state i to state j, is found as:

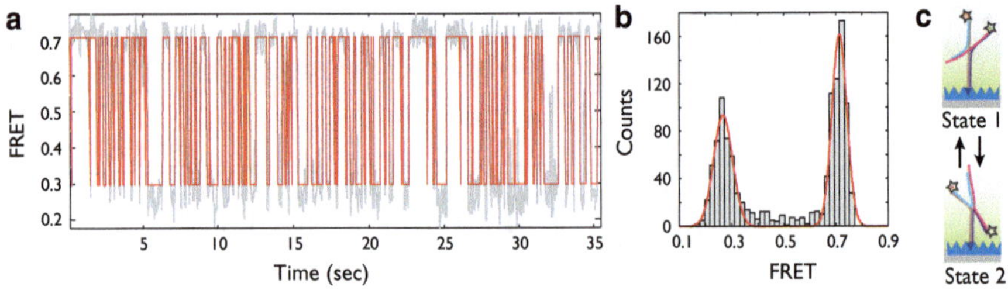

Fig. 12.7 Software analysis of smFRET traces. (**a**) FRET trace and fitted state-progression path from vbFRET, (**b**) histogram of the FRET values with double Gaussian fit, (**c**) sketch of the involved HJ states

$$T_{12} = \begin{pmatrix} p_{11} & p_{12} \\ p_{21} & p_{22} \end{pmatrix} = \begin{pmatrix} 0.88 & 0.12 \\ 0.14 & 0.86 \end{pmatrix}. \tag{12.3}$$

The probability to remain in the high-FRET state is therefore slightly greater.

12.5 Sample Design for Combined Magnetic Tweezers and smFRET

Combined magnetic tweezers and single-molecule FRET experiments require a sophisticated sample design. In this section, we will illustrate a typical workflow for the construction of a sample (Fig. 12.8). The immobilization of the DNA sample in the flow cell on both ends is accomplished by non-covalent interactions between antidigoxigenin and digoxigenin or biotin and streptavidin, respectively. To apply superhelicity to the DNA sample with magnetic tweezers, the molecule must be torsionally constrained. Therefore, both DNA strands of the double-stranded sample have to be attached to the glass slide and the magnetic bead at two or multiple points. The sample of interest is therefore "equipped" with so-called handles on either end. Handles are double-stranded DNA pieces of several hundreds of base pairs in length that carry multiple biotinylated or digoxigenin-modified bases on both strands. These modifications can then interact with proteins on the glass slide or magnetic beads to form attachment points (Fig. 12.8a).

The biotin and digoxigenin handles are obtained by polymerase chain reaction (PCR). The biotin handle PCR product of 1.6 kbp (Fig. 12.8a) is digested by the restriction enzyme KpnI, which results in two 800-bp handles with KpnI specific overhangs. The digoxigenin handle is produced similarly, but with a different restriction enzyme, EagI. Furthermore, a long (5 kbp) spacer sequence is used to prevent interaction of the streptavidin-modified bead with the antidigoxigenin-covered surface of the sample chamber. The spacer is PCR amplified, e.g., from lambda phage DNA, including a single KpnI and NcoI restriction site. The generation of the DNA sequence of interest including restriction sites is discussed in

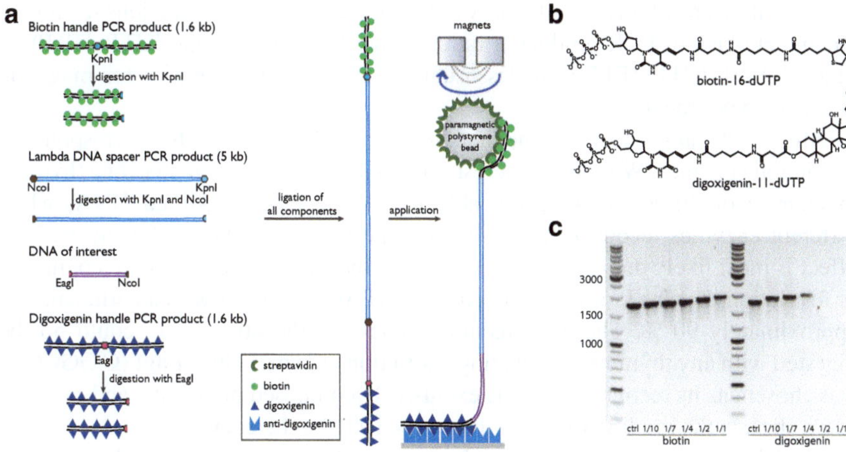

Fig. 12.8 Construction of DNA samples for the application in magnetic tweezers. (**a**) The sample is composed of four parts: a biotin handle, a long spacer sequence, the DNA of interest and a digoxigenin handle, all produced by standard molecular cloning procedures. The four components are joined in a one-step ligation and used in a magnetic tweezers setup as shown *on the right*, (**b**) structures of biotin- and digoxigenin-modified bases used in a PCR reaction to add modifications to amplified DNA, (**c**) agarose gel of the handle PCR using different ratios of modified dUTP to dTTP, reflecting different extents of modification density. Results for biotin-16-dUTP are *on the left*, for digoxigenin-11-dUTP *on the right*

detail below. The handles, the spacer, and the DNA of interest are finally mixed in equimolar amounts and joined together in a one-step ligation reaction that results in the final sample (Fig. 12.8a center). A subsequent purification by gel extraction can increase the purity of the product but is not strictly necessary. For subsequent application in the magnetic tweezers setup, the sample is incubated with streptavidin-coated magnetic beads before flushing them into an antidigoxigenin-coated flow cell (Fig. 12.8a right).

One of the challenges of the sample construction is typically the handle PCR, followed by the restriction digestion of the heavily modified DNA. In order to include several modifications on both strands of the product, modified nucleotides are mixed into the dNTPs for the PCR amplification. The nucleotide deoxyuridine triphosphate (dUTP) coupled to biotin or digoxigenin via a C16 or C14 linker, respectively, is commercially available from various sources (Fig. 12.8b). This nucleotide is incorporated into the PCR product instead of thymine (dTTP). Tuning the ratio of both nucleotides yields a higher or lower incorporation frequency and thus a varying labeling degree of the handles with biotin or digoxigenin, respectively. Figure 12.8c illustrates the effect of increased ratio of modified dUTPs to dTTPs in the range of 1:10 to 1:1. The shift of the bands to greater size with increasing dUTP concentrations is caused by the higher molecular weight of the modified dUTP in comparison to dTTP. Digoxigenin-modified handles are shifted to higher apparent sizes than biotin-modified handles due to the significantly higher molecular weight of digoxigenin (Fig. 12.8b). Furthermore, increasing dUTP concentrations result in a significantly reduced yield in the PCR reactions leading to a complete lack of PCR product at ratios above 1:4 (digoxigenin-dUTP:dTTP). A ratio of 1:10 dUTP:dTTP provides usually a good labeling density for magnetic tweezers experiments.

After PCR amplification, the handles are digested with the corresponding restriction enzymes. We found that restriction enzymes with more than one thymine in their recognition site were not able to cut the biotin-modified DNA, while different enzymes recognizing one thymine perform at varying efficiencies. This effect is most likely due to steric hindrance by biotin during diffusion, recognition, or restriction. For the biotin handle, KpnI (GGTAC|C) digests with an efficiency of approximately 90 %. The digoxigenin handles, on the other hand, could not be digested with any thymine recognizing restriction enzyme. Thus, EagI (C|GGCCG) was chosen, as its recognition site is exclusively composed of guanine and cytosine and it digests the modified DNA with nearly 100 % efficiency.

The DNA sequence of interest can be generated in several ways: (1) the easiest is a PCR reaction followed by enzymatic digestion. This is especially useful when handling long sequences above 200 bp. (2) In cases where shorter DNA stretches are investigated or specific modifications such as fluorophores are required, a different assembly method can be applied using synthesized oligonucleotides. Figure 12.9 depicts such a construction of a short DNA of interest that carries a FRET pair at its boundaries. Here, the synthetic oligos are designed such that the restriction enzyme-specific sticky ends (EagI and NcoI) are exposed on both ends after annealing. Implementing a thymine residue modified with a primary amine allows for the conjugation of a fluorophore modified with a *N*-hydroxysuccinimide ester (NHS ester) [52].

12.6 Applications to Biological Systems

Many biomolecular interactions exert forces on substrates, cofactors, and other cellular structures. Some of the conformational changes that facilitate these interactions, however, occur on length scales far smaller than most typical means to exert forces, e.g., magnetic or silica beads in tweezers experiments. Here, smFRET can work as an orthogonal probe decoupled from force to detect minute distance changes. This orthogonal probe would then allow detecting force-induced structural changes within a biomolecular sample, such as DNA secondary structures. DNA secondary structures stem from at least partially single-stranded DNA pairing with sections of itself to form metastable structures. These structures can exist, for example, in the form of hairpins [53] or the human telomere G-quadruplex [18]. As previously shown, the conformational changes of Holliday junctions [49] and the alternative structural form of DNA, Z-DNA can also be observed through FRET observations [17].

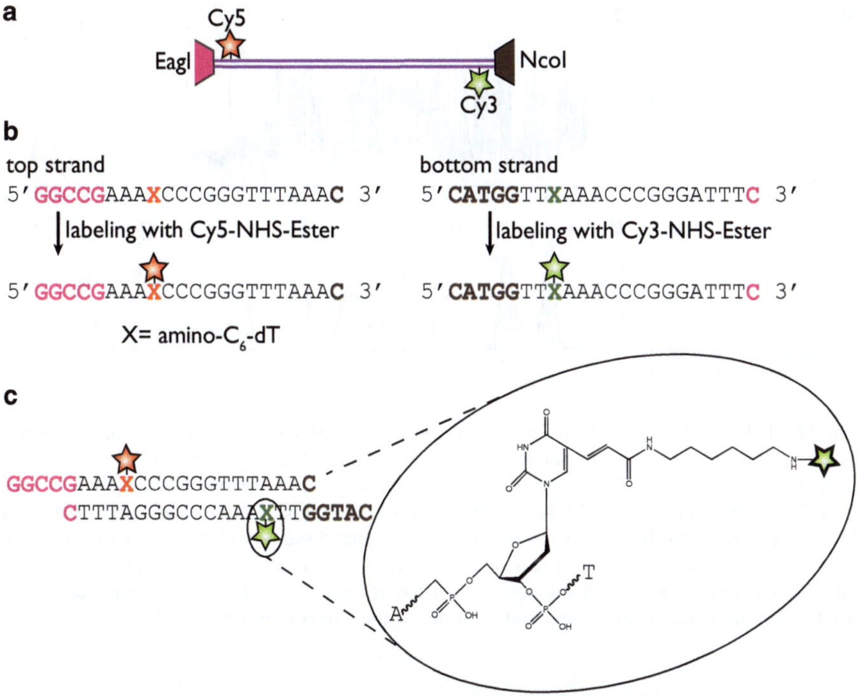

Fig. 12.9 Construction of a FRET-labeled DNA sequence for magnetic tweezers. (**a**) Cartoon of DNA of interest from Fig. 12.8 with a Cy3 and a Cy5 label and an EagI and KpnI restriction site on its end. (**b**) A top and a bottom strand are purchased that are partly complementary except for the 5′-ends that correspond to enzyme-specific sticky ends. The strands contain an amino modified thymine residue (amino-C_6-dT), which can be conjugated to fluorophores modified with an NHS ester. (**c**) Annealing of the two strands results in a FRET-labeled DNA molecule with open restriction sites on both ends. The zoom shows how the fluorophore is attached to the modified thymine via a C_6-linker. Before ligation of this molecule, the 5′ ends need to be phosphorylated by a kinase

12.6.1 Holliday Junctions

The measurements of the unperturbed Holliday junction (HJ) from Sect. 12.4 can be combined with force exerted through attachment of a paramagnetic bead with a DNA spacer (Fig. 12.8). Previous studies showed that the conformation of the HJ is force dependent with an increased probability of being in the low-FRET conformation with increasing force (for our labeling scheme) [16].

Figure 12.10 shows a force-dependent Holliday junction oscillation. The FRET signal is recorded from a molecule with an attached paramagnetic bead. The populations of the two conformations can be manipulated by applying an external force on the paramagnetic bead. Figure 12.10a depicts a sketch of the performed experiment. At greater magnet distance and therefore lower force, the HJ is in a nearly undisturbed environment sampling both its conformations with equal

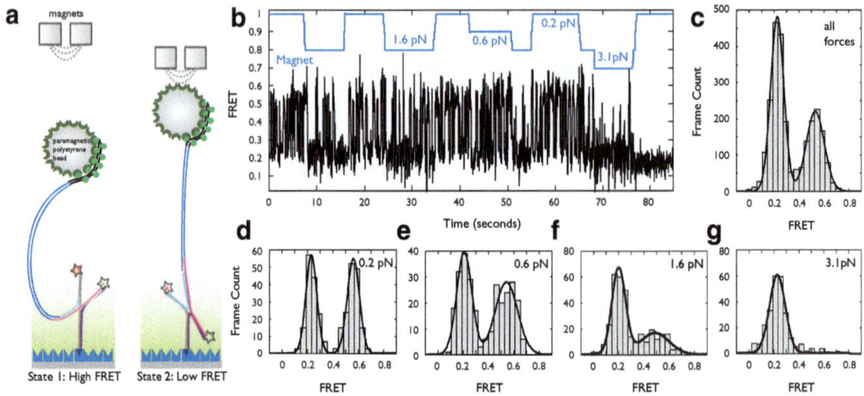

Fig. 12.10 Force-dependent FRET measurements of an oscillating Holliday junction. (**a**) Sketch of the experiment: at larger magnet distance and lower force, the equilibrium of HJ oscillation is shifted toward the high-FRET conformation. At higher force, the second conformation is energetically favored and the equilibrium shifts toward the low-FRET conformation, (**b**) measured FRET values recorded from a single molecule. The magnet distance with the determined forces is shown in the top trace, (**c**) histogram of all FRET values in (**b**) before fluorophore bleaching (at ~78 s), (**d–g**) histograms of subregions at different forces (from **b**). Double Gaussian fits are used to determine the relative areas and positions of the two conformation peaks

probability. When applying force as shown on the right side of Fig. 12.10a, the equilibrium is gradually shifted to the conformation of the low-FRET state making the transition to the second conformation less probable [16]. This behavior can be observed in Fig. 12.10b, where the varying force is depicted by the upper steplike curve (blue), indicating the magnet distance from the sample. A closer magnet position increases the force on the paramagnetic sphere, as indicated by the given force values. When the histograms of the FRET efficiency at different forces are generated, the transition from an almost equal distribution between both conformations at low force (0.2 pN in Fig. 12.10d) to near-exclusive population of the low-FRET conformation at high forces (3.1 pN in Fig. 12.10g) can be directly observed.

In contrast to the data presented by Hohng et al. [16], where each force condition was measured on a different molecule using optical tweezers, all data shown here was obtained from a single molecule. Optical traps most likely contribute to the photobleaching of fluorophores via reactive oxygen species or other secondary effects created by the strong trapping lasers [6, 54], despite the presence of oxygen scavenging systems and triplet state quenchers [2, 3, 55, 56]. The long survival of the FRET pair in Fig. 12.10 is therefore an indicator that magnetic traps, when combined with smFRET, allow for longer average fluorophore lifetimes than optical traps.

Sub-traces of Fig. 12.10 can further be analyzed using again VBEM for the state distribution and kinetics, yielding the corresponding transition probability matrices. The histograms in Fig. 12.11c–e show the progression of the state population with

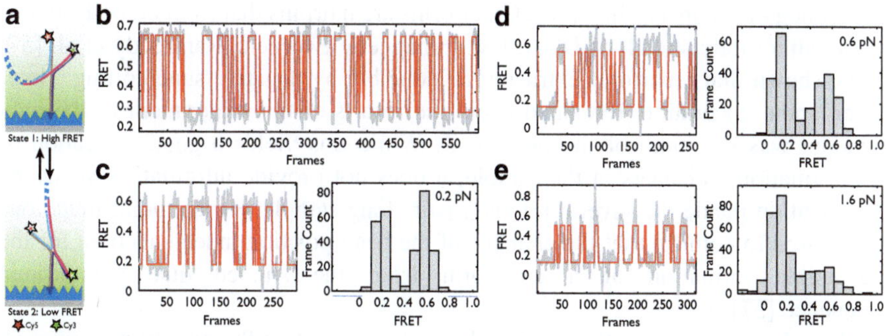

Fig. 12.11 State-progression paths from vbFRET analysis for force-dependent Holliday junction experiments. (**a**) Sketch of the involved HJ states, (**b**) sub-trace from the force-free case of Fig. 12.6, for comparison, (**c–e**) FRET values and state-progression paths with corresponding histograms, for 0.2 pN, 0.6 pN, and 1.6 pN, respectively

increasing force where the change is reflected in the transition probabilities. These can be summarized in the following transition matrices, $T_{0.2}$, $T_{0.6}$, and $T_{1.6}$ for 0.2 pN, 0.6 pN, and 1.6 pN external force, respectively:

$$\begin{aligned} T_{0.2} = \begin{pmatrix} p_{11} & p_{12} \\ p_{21} & p_{22} \end{pmatrix} &= \begin{pmatrix} 0.89 & 0.11 \\ 0.13 & 0.87 \end{pmatrix} \\ T_{0.6} &= \begin{pmatrix} 0.87 & 0.13 \\ 0.11 & 0.89 \end{pmatrix} \\ T_{1.6} &= \begin{pmatrix} 0.86 & 0.14 \\ 0.07 & 0.93 \end{pmatrix}. \end{aligned} \tag{12.4}$$

p_{22} is an example of a force-dependent transition probability, representing the likelihood to remain in the low-FRET state and varying from 87 % at low force to 93 % at high force. The opposite is reflected in the converse probability p_{11} to remain in the high-FRET state, varying from 89 % to 86 %. More striking, perhaps, is the change in the transition probability p_{21} to go from the low-FRET to high-FRET state, which varies from 13 % to 7 % with increasing force, i.e., by a factor of almost two.

12.6.2 Cruciform Formation

Magnetic tweezers are not only capable of changing DNA conformations by pulling forces, but they can also add positive or negative twist on dsDNA. By adding negative twist, it was shown that dsDNA can convert from B-DNA to Z-DNA [14, 27], melt [32], or extract cruciforms [57], all effects also dependent on the applied force. Recently, some biological function was attributed to cruciform formation under negative supercoiling [58]. Cruciforms are hairpins that form on both strands of a palindromic stretch of DNA. Using magnetic tweezers, Ramreddy et al. [57] have shown that cruciform extrusion is inducible by negative supercoiling. However, some uncertainty remained about the results of this experiment as the only readout was the jump of the bead in **Z**. While this yields a good indication of conformational changes in the sample, it does not provide information about the localization inside the several kilobase pairs long sample. Hence, point mutations were used to confirm that the changes of the bead position indeed account for the reversible formation of a cruciform from the studied quasi-palindromic sequence [57].

In order to overcome this drawback and to ensure that the cruciform formation indeed reflects the correct formation of the cruciform of interest, a hybrid single-molecule FRET and magnetic tweezers setup would allow a unique real-time measurement of both, the bead position and additionally a change in FRET efficiency and thus yield direct access of the conformations. As sketched in Fig. 12.12, negative supercoiling at low forces leads to the formation of plectonemes, which reduce the end-to-end distance of the molecule. Magnetic tweezers can detect this process as downward displacement of the magnetic bead attached to the DNA. This supercoiled state presents events of local DNA melting from which cruciform extrusion can arise. Cruciform formation results in the release of plectonemes, due to conservation of the linking number, leading to a detectable upward movement of the bead. Attaching a FRET fluorophore pair on the boundaries of the palindromic sequence would lead to a low-FRET signal in the canonical B-DNA and the plectonemic structure due to spatial separation. Upon hairpin formation, however, the distance between donor and acceptor would decrease, which would lead to a high-FRET signal. This high-FRET signal can, however, only be observed if the hairpin structure extrudes within the DNA sequence region with fluorophore labels. Hairpins forming in other parts of the DNA sample will not change the fluorophore distance and thus not lead to a change in the FRET efficiency. The position of the magnetic bead can be monitored simultaneously. Finally, both sets of data can be correlated to give unambiguous insight into conformational changes undergone by the sample molecule, as well as the localization of these changes. Furthermore, transitional states are more easily identified using FRET as an additional readout [59].

12.7 Conclusions and Outlook

Various other questions from single-molecule experiments remain unsolved due to the limited access to multiple parameters at a time. Indirect control experiments often strengthen the conclusions. However, a direct observation of an enzymatic conformation or oligomerization state during its active cycle while at the same time being manipulated or activated by force would provide better insights into the mechanistic principles of the molecular machines. We anticipate that further efforts

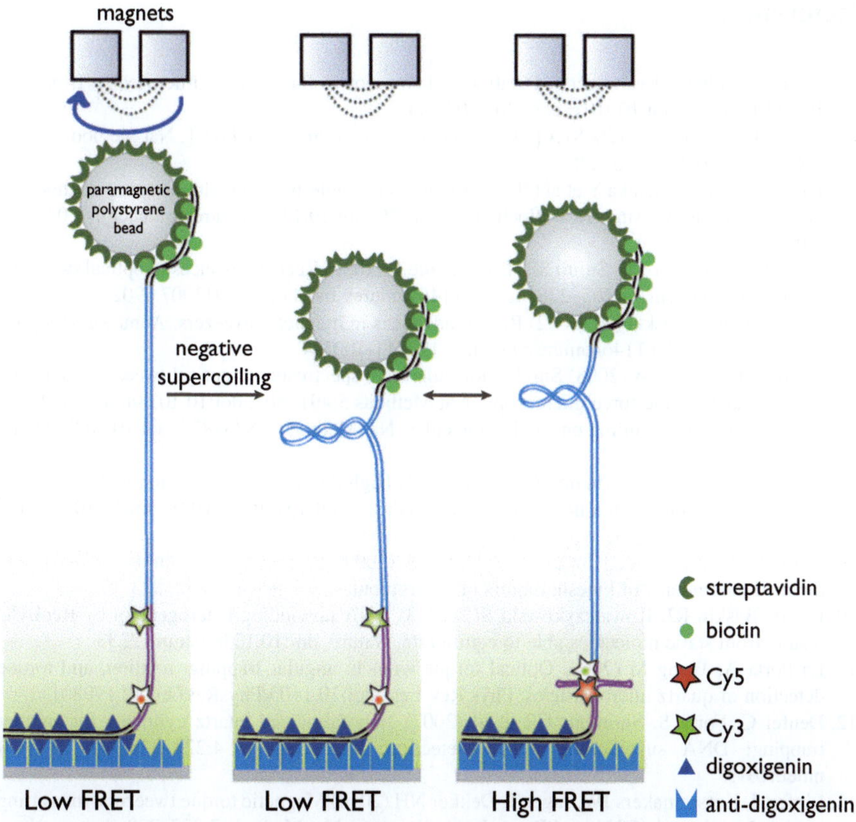

Fig. 12.12 Combining magnetic tweezers and FRET to study cruciform formation induced by supercoiling: The sample contains a palindromic sequence flanked by a FRET pair that gives a low-FRET signal in the relaxed B-DNA conformation. Negative supercoiling of the DNA introduced by magnetic tweezers lead to formation of plectonemes, a shorter end-to-end distance of the sample and a detectable downward movement of the bead. Eventually, the supercoiling can induce cruciform formation by assisting double-strand melting. Cruciform extrusion is the palindromic sequence extruding on either strand as a hairpin. This structural rearrangement releases plectonemic coils, which in turn increases the end-to-end distance. At the same time, the FRET pair distance is reduced which results in a high-FRET signal

in developing a next generation of hybrid instruments to study single molecules in a multidimensional space can open up new avenues of understanding the evolution of multifunctional efficient cellular machines.

Acknowledgements We gratefully acknowledge numerous discussions with the group of Ralf Seidel, in particular Ralf Seidel, Alexander Huhle, and Friedrich Schwarz. Furthermore, discussions and software support by members of the Schlierf lab are highly appreciated. This work was supported by grants from BMBF 03Z2EN11 (to M.S.).

References

1. Axelrod D (2013) Evanescent excitation and emission in fluorescence microscopy. Biophys J 104:1401–1409. doi:10.1016/j.bpj.2013.02.044
2. Roy R, Hohng S, Ha T (2008) A practical guide to single-molecule FRET. Nat Methods 5:507–516. doi:10.1038/nmeth.1208
3. Joo C, Balci H, Ishitsuka Y et al (2008) Advances in single-molecule fluorescence methods for molecular biology. Annu Rev Biochem 77:51–76. doi:10.1146/annurev.biochem.77.070606.101543
4. Moffitt JR, Chemla YR, Smith SB, Bustamante C (2008) Recent advances in optical tweezers. Annu Rev Biochem 77:205–228. doi:10.1146/annurev.biochem.77.043007.090225
5. De Vlaminck I, Dekker C (2012) Recent advances in magnetic tweezers. Annu Rev Biophys 41:453–472. doi:10.1146/annurev-biophys-122311-100544
6. Neuman KC, Nagy A (2008) Single-molecule force spectroscopy: optical tweezers, magnetic tweezers and atomic force microscopy. Nat Methods 5:491–505. doi:10.1038/nmeth.1218
7. de Souza N (2012) Pulling on single molecules. Nat Methods 9:873–877. doi:10.1038/nmeth.2149
8. Chistol G, Liu S, Hetherington CL et al (2012) High degree of coordination and division of labor among subunits in a homomeric ring ATPase. Cell 151:1017–1028. doi:10.1016/j.cell.2012.10.031
9. Bormuth V, Varga V, Howard J, Schaffer E (2009) Protein friction limits diffusive and directed movements of kinesin motors on microtubules. Science 325:870–873
10. Liu B, Baskin RJ, Kowalczykowski SC (2013) DNA unwinding heterogeneity by RecBCD results from static molecules able to equilibrate. Nature. doi:10.1038/nature12333
11. La Porta A, Wang M (2004) Optical torque wrench: angular trapping, rotation, and torque detection of quartz microparticles. Phys Rev Lett. doi:10.1103/PhysRevLett.92.190801
12. Deufel C, Forth S, Simmons CR et al (2007) Nanofabricated quartz cylinders for angular trapping: DNA supercoiling torque detection. Nat Methods 4:223–225. doi:10.1038/nmeth1013
13. Lipfert J, Kerssemakers JW, Jager T, Dekker NH (2010) Magnetic torque tweezers: measuring torsional stiffness in DNA and RecA-DNA filaments. Nat Methods 7:977–980
14. Oberstrass FC, Fernandes LE, Lebel P, Bryant Z (2013) Torque spectroscopy of DNA: base-pair stability, boundary effects, backbending, and breathing dynamics. Phys Rev Lett 110:178103. doi:10.1103/PhysRevLett.110.178103
15. Comstock MJ, Ha T, Chemla YR (2011) Ultrahigh-resolution optical trap with single-fluorophore sensitivity. Nat Methods 8:335–340. doi:10.1038/nmeth.1574
16. Hohng S, Zhou R, Nahas MK et al (2007) Fluorescence-Force spectroscopy maps two-dimensional reaction landscape of the Holliday junction. Science 318:279–283. doi:10.1126/science.1146113
17. Lee M, Kim SH, Hong SC (2010) Minute negative superhelicity is sufficient to induce the B-Z transition in the presence of low tension. Proc Natl Acad Sci USA 107:4985–4990. doi:10.1073/pnas.0911528107
18. Long X, Parks JW, Bagshaw CR, Stone MD (2013) Mechanical unfolding of human telomere G-quadruplex DNA probed by integrated fluorescence and magnetic tweezers spectroscopy. Nucleic Acids Res 41:2746–2755. doi:10.1093/nar/gks1341
19. Strick TR, Dessinges MN, Charvin G et al (2002) Stretching of macromolecules and proteins. Rep Prog Phys 66:1
20. Haber C, Wirtz D (2000) Magnetic tweezers for DNA micromanipulation. Rev Sci Instrum 71:4561. doi:10.1063/1.1326056
21. Leuba SH, Wheeler TB, Cheng C-M et al (2009) Structure and dynamics of single DNA molecules manipulated by magnetic tweezers and or flow. Methods 47:214–222. doi:10.1016/j.ymeth.2008.10.022

22. Gosse C, Croquette V (2002) Magnetic tweezers: micromanipulation and force measurement at the molecular level. Biophys J 82:3314–3329. doi:10.1016/S0006-3495(02)75672-5
23. Fisher JK, Cribb J, Desai KV et al (2006) Thin-foil magnetic force system for high-numerical-aperture microscopy. Rev Sci Instrum 77:023702. doi:10.1063/1.2166509
24. Kauert DJ, Kurth T, Liedl T, Seidel R (2011) Direct mechanical measurements reveal the material properties of three-dimensional DNA origami. Nano Lett 11:5558–5563. doi:10.1021/nl203503s
25. Brutzer H, Schwarz FW, Seidel R (2012) Scanning evanescent fields using a pointlike light source and a nanomechanical DNA gear. Nano Lett 12:473–478. doi:10.1021/nl203876w
26. Schwarz FW, Toth J, van Aelst K et al (2013) The helicase-like domains of type III restriction enzymes trigger long-range diffusion along DNA. Science 340:353–356. doi:10.1126/science.1231122
27. Oberstrass F, Fernandes L (2012) Torque measurements reveal sequence-specific cooperative transitions in supercoiled DNA. Proc Natl Acad Sci USA 109:6106–6111. doi:10.1073/pnas.1113532109
28. Howan K, Smith AJ, Westblade LF et al (2012) Initiation of transcription-coupled repair characterized at single-molecule resolution. Nature 490:431–434. doi:10.1038/nature11430
29. Ma J, Bai L, Wang MD (2013) Transcription under torsion. Science 340:1580–1583. doi:10.1126/science.1235441
30. Strick TR, Allemand JF, Bensimon D, Croquette V (1998) Behavior of supercoiled DNA. Biophys J 74:2016–2028
31. Strick T, Allemand J, Croquette V (2000) Twisting and stretching single DNA molecules. Prog Biophys Mol Biol 74:115–140
32. Salerno D, Tempestini A, Mai I et al (2012) Single-molecule study of the DNA denaturation phase transition in the force-torsion space. Phys Rev Lett 109:118303. doi:10.1103/PhysRevLett.109.118303
33. Tempestini A, Cassina V, Brogioli D et al (2013) Magnetic tweezers measurements of the nanomechanical stability of DNA against denaturation at various conditions of pH and ionic strength. Nucleic Acids Res 41:2009–2019. doi:10.1093/nar/gks1206
34. Strick TR, Allemand JF, Bensimon D et al (1996) The elasticity of a single supercoiled DNA molecule. Science 271:1835–1837
35. Brutzer H, Luzzietti N, Klaue D, Seidel R (2010) Energetics at the DNA supercoiling transition. Biophys J 98:1267–1276. doi:10.1016/j.bpj.2009.12.4292
36. van Loenhout MTJ, de Grunt MV, Dekker C (2012) Dynamics of DNA supercoils. Science. doi:10.1126/science.1225810
37. Ribeck N, Saleh OA (2008) Multiplexed single-molecule measurements with magnetic tweezers. Rev Sci Instrum 79:094301. doi:10.1063/1.2981687
38. De Vlaminck I, Henighan T, van Loenhout MTJ et al (2011) Highly parallel magnetic tweezers by targeted DNA tethering. Nano Lett 11:5489–5493. doi:10.1021/nl203299e
39. Lansdorp BM, Tabrizi SJ, Dittmore A, Saleh OA (2013) A high-speed magnetic tweezer beyond 10,000 frames per second. Rev Sci Instrum 84:044301. doi:10.1063/1.4802678
40. Cheezum MK, Walker WF, Guilford WH (2001) Quantitative comparison of algorithms for tracking single fluorescent particles. Biophys J 81:2378–2388
41. Lipfert J, Hao X, Dekker NH (2009) Quantitative modeling and optimization of magnetic tweezers. Biophys J 96:5040–5049. doi:10.1016/j.bpj.2009.03.055
42. Berg-Sørensen K, Flyvbjerg H (2004) Power spectrum analysis for optical tweezers. Rev Sci Instrum 75:594. doi:10.1063/1.1645654
43. Vilfan ID, Lipfert J, Koster DA et al (2009) Handbook of single-molecule biophysics. Springer, Heidelberg, pp 371–395. doi:10.1007/978-0-387-76497-9_13
44. Norrelykke SF, Flyvbjerg H (2010) Power spectrum analysis with least-squares fitting: amplitude bias and its elimination, with application to optical tweezers and atomic force microscope cantilevers. Rev Sci Instrum 81:075103. doi:10.1063/1.3455217

45. te Velthuis AJW, Kerssemakers JWJ, Lipfert J, Dekker NH (2010) Quantitative guidelines for force calibration through spectral analysis of magnetic tweezers data. Biophys J 99:1292–1302. doi:10.1016/j.bpj.2010.06.008
46. Lansdorp BM, Saleh OA (2012) Power spectrum and Allan variance methods for calibrating single-molecule video-tracking instruments. Rev Sci Instrum 83:025115. doi:10.1063/1.3687431
47. Czerwinski F, Richardson AC, Oddershede LB (2009) Quantifying noise in optical tweezers by allan variance. Opt Express 17:13255–13269
48. Stahl FW (1994) The Holliday junction on its thirtieth anniversary. Genetics 138:241
49. Buranachai C, McKinney SA, Ha T (2006) Single molecule nanometronome. Nano Lett 6:496–500. doi:10.1021/nl052492p
50. Bronson JE, Fei J, Hofman JM et al (2009) Learning rates and states from biophysical time series: a Bayesian approach to model selection and single-molecule FRET data. Biophys J 97:3196–3205. doi:10.1016/j.bpj.2009.09.031
51. McKinney SA, Joo C, Ha T (2006) Analysis of single-molecule FRET trajectories using hidden Markov modeling. Biophys J 91:1941–1951. doi:10.1529/biophysj.106.082487
52. Hermanson GT (2013) Bioconjugate techniques. Academic Press, London
53. Bikard D, Loot C, Baharoglu Z, Mazel D (2010) Folded DNA in action: hairpin formation and biological functions in prokaryotes. Microbiol Mol Biol Rev 74:570–588. doi:10.1128/MMBR.00026-10
54. Landry MP, McCall PM, Qi Z, Chemla YR (2009) Characterization of photoactivated singlet oxygen damage in single-molecule optical trap experiments. Biophys J 97:2128–2136. doi:10.1016/j.bpj.2009.07.048
55. Aitken CE, Marshall RA, Puglisi JD (2008) An oxygen scavenging system for improvement of dye stability in single-molecule fluorescence experiments. Biophys J 94:1826–1835. doi:10.1529/biophysj.107.117689
56. Swoboda M, Henig J, Cheng H-M et al (2012) Enzymatic oxygen scavenging for photostability without pH drop in single-molecule experiments. ACS Nano 6:6364–6369. doi:10.1021/nn301895c
57. Ramreddy T, Sachidanandam R, Strick TR (2011) Real-time detection of cruciform extrusion by single-molecule DNA nanomanipulation. Nucleic Acids Res 39:4275–4283. doi:10.1093/nar/gkr008
58. Loot C, Ducos-Galand M, Escudero JA et al (2012) Replicative resolution of integron cassette insertion. Nucleic Acids Res. doi:10.1093/nar/gks620
59. Swoboda M, Grieb MS, Hahn S, et al. manuscript in preparation

Chapter 13
Using Fluorescence to Study Actomyosin in Yeasts

Daniel P. Mulvihill

Abstract This year marks the 30th anniversary of the first description of the cellular distribution of actin within a yeast cell. Since then advances in both molecular genetics and imaging technologies have ensured research within these simple model organisms has blazed a trail in the field of actomyosin research. Many yeast proteins and their functions are functionally conserved in human cells. This, combined with experimental speed, minimal cost and ease of use make the yeasts extremely attractive model organisms for researching diverse cellular processes, including those involving actomyosin. In this chapter, current state-of-the-art fluorescence methodologies being applied to yeast actomyosin research, together with an honest appraisal of their limitations, such as the pitfalls that should be considered when fluorescently labelling proteins interacting within a dynamic cytoskeleton, will be discussed. Papers describing the established techniques developed for yeast localisation studies will be highlighted. This will provide the reader with an informed overview of the arsenal of imaging techniques available to the yeast actomyosin researcher and encourage them to consider novel ways these simple unicellular eukaryotes could be used to address their own research questions.

Keywords Actin • Myosin • *Schizosaccharomyces pombe* • *Saccharomyces cerevisiae* • Fission yeast • Budding yeast • Microscopy

Abbreviations

ABP	Actin-binding protein
Arp	Actin-related protein

D.P. Mulvihill (✉)
School of Biosciences, University of Kent, Canterbury, Kent CT2 7NJ, UK
e-mail: d.p.mulvihill@kent.ac.uk

C.P. Toseland and N. Fili (eds.), *Fluorescent Methods for Molecular Motors*, Experientia Supplementum 105, DOI 10.1007/978-3-0348-0856-9_13,

CAR	Cytokinetic actomyosin ring
CCD	Charge couple device
CMOS	Complementary metal oxide semiconductor
emCCD	Electron multiplying charge couple device
FP	Fluorescent protein
Tm	Tropomyosin

13.1 Introduction

13.1.1 Background

The mid-1980s marked the birth of research in yeast actin cell biology with the publication of a series of landmark papers describing the cellular distributions of actin within the yeasts *Saccharomyces cerevisiae* [1], *Schizosaccharomyces pombe* [2] and *Candida albicans* [3]. Each study made use of the actin-stabilising agent phalloidin. Originally purified from the mushroom *Amanitas phalloides*, this actin stabiliser binds to the interface between adjacent actin monomers, locking them together. When conjugated to an appropriate fluorescent analogue, phalloidin can be used to label each F-actin structure within any appropriately fixed cell [4]. Once the fixation methodologies were perfected, rhodamine-conjugated phalloidin was used to localise F-actin structures in yeasts. In each yeast, the distribution of actin coincided precisely with sites of cell growth, consistent with actin's role in endocytosis, and facilitating the deposition of new cell wall material (Fig. 13.1).

In yeast, polymerised F-actin is incorporated into actin patches and cables that extend throughout the cytoplasm within interphase cells. Actin patches, the yeast equivalent to cortical actin, are made up of short lengths of actin, nucleated by the Arp2/3 complex. In contrast, actin cables are nucleated from formins and are stabilised by tropomyosin (Tm). During mitosis Tm-stabilised actin cables are incorporated into the cytokinetic actomyosin ring (CAR). The activity of each of these structures is analogous to their different counterparts in mammalian cells, making yeast an attractive system to interrogate their regulation and function in a simple genetically tractable single cell context.

13.1.2 Why Study Actomyosin in Yeast?

So what are the attractions of using yeast for studying actin organisation? Well, as eluded to previously, the actin cytoskeleton shares the same regulatory mechanisms and functions in yeast and mammals, and homologues of the majority of human actin-binding proteins (ABPs) responsible for regulating actin dynamics are found in yeasts [5–7]. The yeast contains the basic cellular actomyosin toolbox, making them an excellent system to study global regulation and function of this

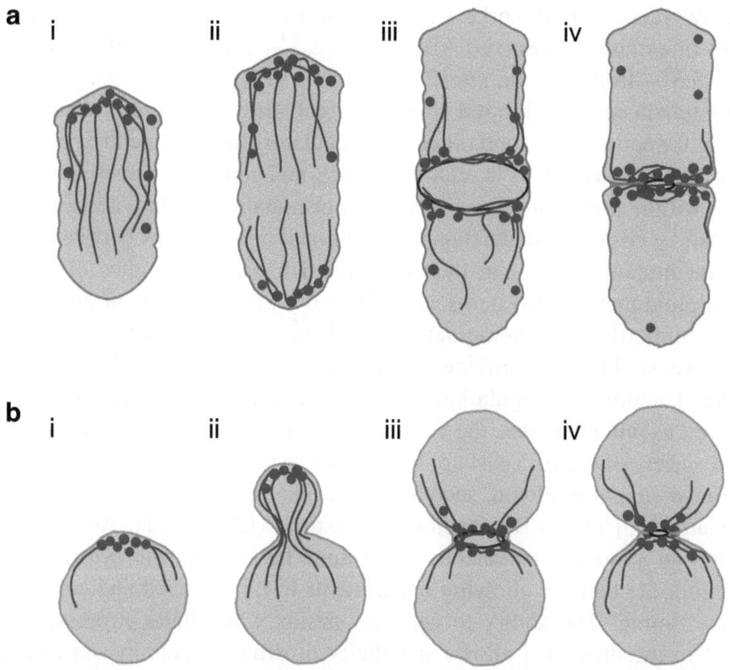

Fig. 13.1 Illustration of actin distribution in the cell cycle of yeasts. Cortical actin patches (*dark dots*) concentrate at growing regions in both fission (**a**) and budding (**b**) yeasts. In fission yeast this coincides with growth from one cell end in early interphase (i) or both ends in late interphase (ii). In budding yeast interphase, actin patches initially concentrate at the site of bud formation during early interphase (i) and then to growth regions of the developing bud (ii). Actin cables (*dark lines*) extend from the regions of cell growth into the cytoplasm and provide a track on which myosin V can deliver cargoes to sites of cell extension. During early mitosis (iii), the actin cables reorganise and are incorporated into the nascent CAR. During cytokinesis (iv) coincidental with CAR constriction, actin patches concentrate to the cell equator to promote formation of the septum, which will eventually split in the middle and become the new ends of the subsequent daughter cells

cytoskeleton. For example, while a mammalian cell can express up to 30 different myosins, yeast can express up to 5 different myosins which represent only 3 different myosin classes. Therefore, they provide an excellent opportunity in which to develop an understanding of how all myosins within a cell contribute to its lifecycle.

The yeasts are relatively easy and fast to grow and allow themselves to rapid molecular genetics, biochemistry, as well as detailed live-cell imaging. The "power of yeast genetics" may seem a well-worn cliché to many, but the ability to undertake specific and random phenotype-based mutant screens has allowed unique insights into the mechanisms behind an array of cellular processes, such as regulation of the actin cytoskeleton and cell cycle. Indeed many of the temperature-sensitive alleles identified in the original fission yeast cell division cycle mutant screen by Nurse and co-workers [8] were subsequently shown to encode mutant versions of essential conserved ABPs (e.g. *cdc3*$^{+}$—profillin, *cdc4*$^{+}$— myosin light chain; *cdc8*$^{+}$—tropomyosin; *cdc12*$^{+}$—formin [9–12]).

Researchers are not limited to the current array of mutant alleles in genes encoding for actin, ABPs, and other actin regulators to facilitate their physiological and biochemical studies. It is relatively quick and simple to screen for conditional mutants within almost any gene (e.g. [13]) or mutants within pathways (e.g. [14]) which can be critical in understanding functional relationships between molecules. Dominant mutant alleles can then be applied to in vivo studies of homologues within diploid metazoan systems and help in our understanding of disease states in humans. Similarly the genome deletion collections generated for both fission and budding yeasts [15, 16] provide a potent resource, allowing the opportunity to understand actomyosin regulation and function on a genome-wide scale.

As well as mutant alleles, the yeast actin cytoskeletons can be challenged by the use of a variety of drugs. Yeast are susceptible to a number of poisons that act upon the actomyosin cytoskeleton, such as the latrunculins [17, 18], cytochalasins [19], blebbistatin [20] or 2,3-Butanedione monoxime (BDM) [21]. Addition of these drugs has an almost instantaneous effect upon the actomyosin cytoskeleton of all cells within the culture, allowing experiments to be designed and controlled with exquisite timing. The ability to generate strains with alleles expressing proteins resistant to the drug [17] allows specificity of drugs to actomyosin targets to be easily tested.

The ability to rapidly fuse cDNA encoding for a variety of affinity tags (e.g. poly-His or FLAG tags) or fluorescent proteins (FPs) (e.g. GFP, dsRED) to the termini of genes at their endogenous chromosomal locus that makes yeast attractive for biochemical and cell biological studies. Tagging at the endogenous locus ensures transcription is under the control of the endogenous promoter, ensures comparable expression levels between all cells within the culture, and allows you to determine whether the addition of the tag has an effect on the protein's function.

Each of these resources and methods for studying and challenging the cytoskeleton makes yeasts extremely attractive model organisms for biochemical and cell biological actomyosin research. In this chapter, current fluorescence techniques for studying the actomyosin cytoskeleton within the yeasts will be described and reviewed, initially focusing on fixed methods using phalloidin and antibodies; moving on to live-cell imaging techniques and the caveats the researcher should be aware of; and finally a discussion of how to set up an imaging system optimised for yeast microscopy.

13.2 Yeast Actomyosin Microscopy

13.2.1 Fixed Cell Techniques

Many may consider that there is little use for imaging fixed samples and immunofluorescence techniques to study protein localisation in the age of multicoloured fluorescent-protein live-cell imaging. However, there can be clear advantages for choosing fixed imaging techniques, especially where tagging components of dynamic complexes (e.g. actin or Tm) have a negative impact on their ability to form polymers. Indeed it has only recently become possible to visualise actin filament dynamics within a living cell using fluorescent labels such as Lifeact [22]. However, due to structural constraints, tagging some molecules such as actin or Tm will have an impact upon its ability to form protein–protein contacts, so antibodies currently remain the only way to confidently examine normal Tm localisation within the cell.

When fixing yeast cells, you can use either aldehyde or alcohol. In contrast to some MT-based structures which are best preserved by fixation with a combination of aldehydes [23], the majority of actin structures and actomyosin antibody epitopes are best preserved using either formaldehyde [2, 24–26] or cold methanol fixations [27]. However, if you are trying to use a new antibody for immunofluorescence in yeast, as well as trying a range of both primary and secondary antibody concentrations, it is always worth testing each fixation technique, for each antibody raised in a different animal.

13.2.1.1 Fixations

Aldehydes When using formaldehyde, it has been the experience within this lab and the labs in which I undertook my microscopy apprenticeship that making fresh formaldehyde just prior to the fixation is critical for obtaining the best results. Paraformaldehyde is dissolved in an appropriate buffer until it is almost all in solution, generating a heterogeneous solution containing aldehyde in a state of flux between polymerised states. This may explain why the fixation appears to preserve structures so well. However, that said, many wonderful images have been published using purchased pre-formulated formaldehyde aliquots. Either way it is important to equilibrate the temperature of the fixative to the temperature of the cell culture.

As with all fixation techniques, it is crucial to ensure the cultured cells are in suspension before the fixative is added, and upon its addition the formaldehyde is rapidly mixed into the culture to obtain a homogenous mix. Different labs have different methods for achieving this, but the simplest method is to rapidly add the formaldehyde directly to the shaking culture flask maintained at the culture temperature to a final 3 % w/v concentration. If glutaraldehyde is also used, it is added a minute or so later [23], to allow the formaldehyde a chance to provide a rapid "soft fix", before glutaraldehyde provides further preservation (analogous to cooking meat at two different temperatures—to ensure the everything is cooked and the flavour is sealed inside). The standard period for fixation is 30 min; however, some epitopes are best preserved with fixations as short as 5 min [28]. You need to test the best conditions for your antibody–epitope combination.

Note: *Remember paraformaldehyde, glutaraldehyde and formaldehyde are potent carcinogens and should be worked with under a flow hood. If you can smell it, it's doing damage!*

Solvents A short word on alcohol fixation. This method provides a simple and rapid method for fixing cells for subsequent processing. Due to secondary effects of the alcohol, the subsequent cell wall digestion and membrane permeabilisation steps can be reduced in length and omitted, respectively. The key for this fixation is to ensure the alcohol is cold upon addition of the cells. Different alcohols and temperatures have been described; however, this lab has had reproducible success using the −80 °C methanol fixation developed by J. Petersen (personal communication). Universal tubes containing 10 ml aliquots of alcohol should be stored at −80 °C for at least 24 h before adding filtered cells. The cells are only left to fix 8 min before rapidly washing three times in the buffer appropriate for digestion.

A word of warning: When employing alcohol fixation, cells need to be imaged within 2 days, as cell shrinkage can occur when using this method.

13.2.1.2 Getting the Antibodies In

Unlike animal cells, yeasts have a cell wall through which antibodies must pass. After fixation, holes are digested out of the yeast cell wall using appropriate sugar-digesting enzymes. The enzyme of choice is normally zymolyase, which specifically digests the α-glucan sugars within the cell wall. While other enzymes are available, zymolyase digestion allows the rate of digestion to be regulated and carefully monitored. Generally, digestion is allowed to proceed until ~50 % of cells burst when an aliquot is added to an equal volume of 1 % SDS.

After washing and quenching unreacted glutaraldehyde in an appropriate reducing agent (e.g. sodium borohydride), cell membranes are permeabilised by incubating with a nonionic detergent such as Triton X100, and cells are incubated in buffer containing blocking agents such as BSA (IgG free) and/or lysine to prevent antibodies interacting with nonspecific charged residues. Within an hour, antibodies are applied to the cells in the same buffer. Antibody concentration and incubation times vary significantly and should be tested.

13.2.1.3 Phalloidin and Antibodies

Fluorescent phalloidin conjugates work extremely well in the yeasts [1–3] (Fig. 13.2). An extensive range of fluorophore conjugates is available now, with a broad range of emission spectra to choose from. When used appropriately, these

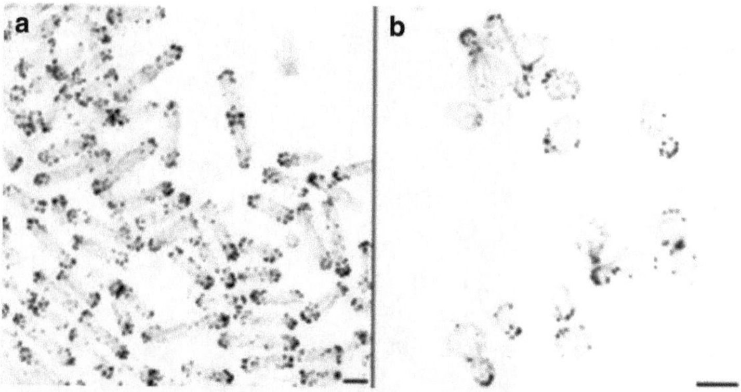

Fig. 13.2 Visualising actin distribution in yeast using rhodamine-conjugated phalloidin. Inverted maximum projection of fixed and rhodamine–phalloidin-stained fission (**a**) and budding (**b**) yeasts. This staining technique allows actin cables, actin patches and CAR to be easily visualised. Scales—5 μm. Thanks to Dr C. Gourlay (University of Kent) for the image (**b**)

have produced images of actin cables, patches and cytokinetic ring of exquisite detail and depth [29–31], and have provided a real insight into the complexity of the yeast actin network. When using phalloidin, it is important to remember to make up the phosphate/magnesium buffer used throughout the fixation fresh on the day of the experiment in order to obtain the best results.

As well as phalloidin, all actin polymers within the cell can also be stained using an appropriate antibody, of which there are a number available. Chicken gizzard actin monoclonal (IgM) antibody works well in both yeasts, and the supernatant from this JLA20 cell line is still available from the Developmental Studies Hybridoma Bank at the time of writing [32].

Staining a specific class of actin structure (e.g. patch) can be achieved using antibodies. These have been generated and used in the past with great success. Examples include anti-Arp2/3 antibodies [27, 33] to label patches and anti-Tm antibodies [24, 34] to label cables (Fig. 13.3). These can either be used in conjunction with appropriately labelled secondary antibodies (indirect immunofluorescence) or can be conjugated directly to a whole spectrum of fluorophores and quantum dots to improve overall signal and signal-to-noise ratio [35].

Antibodies have been raised that specifically recognise a single myosin within yeasts and have been proved to work well for immunofluorescence. In addition using monoclonal antibodies against epitope-tagged versions of the myosins has been shown to work well for a number of cases. Examples of antibodies and studies are listed below in Table 13.1. The majority of these anti-myosin antibodies have been raised against regions within the tail of the myosin interest, as these contain most divergent sequence regions and are less likely to cross-react than those raised against conserved regions within the motor domain. In addition there are a number of published phospho-specific antibodies that can be used to localise specific subpopulations of the myosin within the cell.

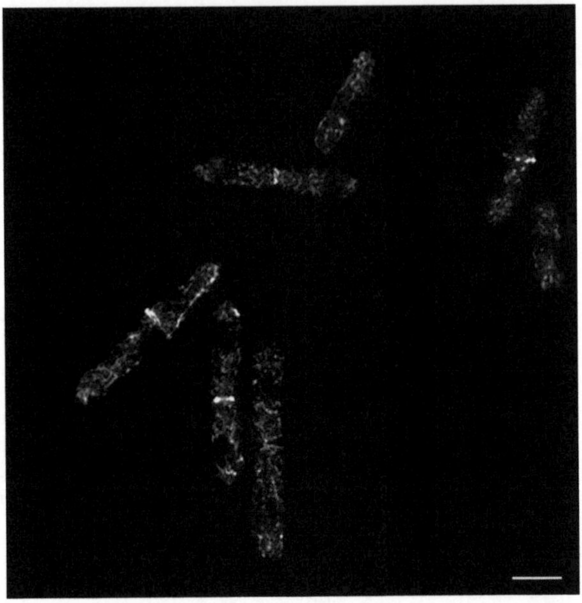

Fig. 13.3 Anti-Tm immunofluorescence. A maximum projection image of fission yeast cells subjected to anti-Cdc8 immunofluorescence. This image illustrates how an anti-Tm antibody is an excellent reagent for specifically labelling formin-nucleated actin cables within the yeast cell. Scale—5 μm

Table 13.1 References of antibodies used for yeast myosin immunofluorescence

	S. cerevisiae		*S. pombe*	
Class	Myosin	Reference	Myosin	Reference
I	MYO3	–	Myo1	[36]
	MYO5[a]	[37]		
II	MYO1	[38]	Myo2	[25, 39]
			Myp2	[39, 40]
V	MYO2	[41, 42]	Myo52	[43]
	MYO4[a]	[44]	Myo51	–

[a]Monoclonal anti-myc antibodies used against tagged protein

If you find there are no appropriate antibodies against the specific ABP you are interested in, generating new antibodies is well worth considering, given the speed and ease by which it can now be done commercially. Similarly if you have a strain expressing a gfp-tagged allele, but the signal is too weak to see (i.e. a low abundance protein), one option you may want to consider is undertaking immunofluorescence using anti-GFP antibodies to enhance the FP signal (e.g. [26, 45]).

13.2.2 Imaging Actomyosin Cytoskeleton Dynamics in the Living Yeast

It is when undertaking live-cell imaging that the yeast model systems come into their own. Not only do their size and shape allow themselves to rapid cell imaging, but it is the ease with which molecular genetics can be undertaken that make yeast such an attractive proposition for investigating the dynamic localisation of endogenous cytoskeleton proteins.

In both budding and fission yeasts, PCR-based approaches allow cDNA encoding for FP tags to be fused to the termini of genes at their genomic locus, e.g. [46–48] in less than a week. This method ensures stable expression equivalent to the wild-type allele, as the single integrated FP-tagged allele remains under the control of the endogenous promoter at the endogenous chromosomal location (important consideration for genes located within heterochromatin rich regions). Yeast allows the opportunity to test functionality, as tags may affect the structure, regulation or protein–protein interactions. In the case of essential genes, such as *S. pombe myo2*, functionality can be tested at the simplest level by the ability of the tagged allele to permit growth. However, the functionality of any tagged gene allele can be tested by examining their ability to maintain normal polarised cell growth in different growth conditions (e.g. media or temperature) or different genetic backgrounds. For example, although the *myo2-gfp* allele described in [49] localises to the correct location in cells with wild-type morphology, it demonstrates synthetic lethality in cells possessing temperature-sensitive mutations in genes encoding for regulators of cytokinesis. It is therefore crucial to consider protein–protein interactions within the cell when considering a tagging strategy. For example, while tagging a yeast myosin I at the carboxyl terminus disrupts their ability to interact with the Arp2/3 complex, fusing a fluorescent protein to the amino terminus of the motor domain has no obvious effect on yeast myosin I function. However, tagging either termini of some proteins, such as actin or Tm, disrupts their ability to function and therefore does not provide a reliable marker for monitoring real actomyosin dynamics. Tagging specific components of the actomyosin cytoskeleton will be discussed later in this chapter.

13.2.2.1 What Colour Suits My Application?

Considering the diverse source, structure and biochemical properties of different FPs, it is not surprising that different FPs can affect the functionality of proteins differently. An obvious example is the differences in functionality between FP alleles of the fission yeast α-tubulin gene, *atb2*. While the gfp-atb2 allele generates microtubules with dynamics equivalent to wild type when under the control of an appropriate promotor [50], red FPs have varying degrees of toxicity to microtubules [51]. This lab has found the same to be true for a number of components of the actomyosin cytoskeleton too (unpublished data).

There is now a smorgasbord of FPs to choose from with which to tag your actomyosin cytoskeleton component, each with different fluorescent properties (i.e. excitation/emission spectra, quantum yield, etc.). This list is continually increasing, with the development of new monomeric high quantum yield FPs, with carefully tuned fluorescence spectra. So which one to choose from? While blue-emitting FPs offer the best spatial resolution (resolution is proportional to the wavelength of the emitted light) on the whole, they are not efficient in transferring the energy from incidence light to emitted fluorescence (i.e. low quantum yield/efficiency) when compared to longer wavelength FPs. In contrast while brighter than cyan variants, red fluorescent proteins are not stable and bleach rapidly, especially when culturing the cells at higher temperatures (e.g. 36 °C).

As well as the most characterised, GFP is still one of the most stable FP used today. There are a number of monomeric versions (important for studying true protein dynamics), and there are a wide range of filter combinations available for a variety of GFP-based fluorescent experiments. Therefore, GFP is still the FP of choice for the majority of yeast-imaging applications.

For long-term analysis of low abundance proteins, many studies have used multiple FP tags to improve the signal (i.e. 3xGFP or 2xTomato tags fused in tandem). Although these proteins are brighter, the consequence the additional mass has upon the dynamics of the protein, protein complex and protein–protein interactions should be carefully considered and tested (e.g. a tdTomato or 3xGFP tag increases the mass of the protein by 50 kDa or 75 kDa, respectively). While on this point, if you feel the mass of the FP is a serious consideration, smaller FPs such as iLOV [52] are now becoming available. This monomeric FP has the same basic fluorescent properties as GFP, but is half its size.

For specialised applications, such as co-localisation, Forster resonance energy transfer (FRET) or fluorescence-lifetime imaging microscopy (FLIM), it is crucial that you carefully match the excitation and emission spectra of the FPs you will use to generate your FP-labelled allele strains. For some applications it is crucial there is minimal overlap (e.g. independent colocalisation), but for others (e.g. FRET) it is desired. While cyan and yellow FPs are the most well characterised for FRET analysis, YFP has been shown to blink and has other properties that make it less suitable for lifetime measurements than other FPs. However, that said Cerulean and YPet [53, 54] are a good combination for FRET analysis in yeast, as both proteins are stable at 36 °C and have optimal wavelength overlap. For yeast FLIM studies GFP and mCherry are currently the best option, as both are stable monomeric FPs in a variety of cellular conditions, with appropriate crosstalk between donor emission and accepter excitation spectra.

13.2.2.2 FP-Labelled Actomyosin Cytoskeleton Components

Adding even a short epitope tag to either end of actin disrupts its ability to polymerise and interact with other proteins. This is also true in yeast. While GFP-actin incorporates into cortical actin patches in wild-type fission yeast [55], it does not polymerise into actin cables in the same cells or complement the function of the wild-type actin protein.

But all is not lost! A number of FP-labelled proteins have been developed that allow actin dynamics to be followed in the cell. In budding yeast, for example, the Pon lab developed FP-labelled versions of the ABPs, Abp1 and Abp140 which elegantly label cortical actin patches and actin cables, respectively [56] (Fig. 13.4).

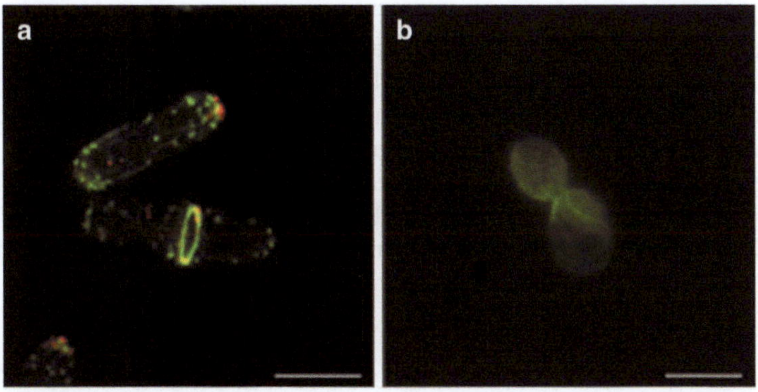

Fig. 13.4 FP-labelled actin polymers in live yeast. (**a**) Maximum projection of fission yeast expressing Myo52-tdTomato (*red*) in which actin polymers have been labelled with GFP-CHD^{Rng2} (*green*). (**b**) Maximum projection of a budding yeast cell expressing GFP-ABP140 which associates with actin cables. Scales—5 μm. Thanks to Dr C. Gourlay (University of Kent) for the image (**b**)

Similarly, actin-associating FP fusions, which allow actin dynamics to be followed in real time have been generated for fission yeast by the Balasubramanian lab. The first was GFP-CHD^{Rng2}, a fusion between GFP and the calponin homology domain of Rng2, the fission yeast IQGAP [57] which associates with all actin polymers within the cell. More recently the Balasubramanian lab generated fission yeast strains expressing red and GFP versions of Lifeact, a short 17 amino acid actin-binding peptide which incorporates into dynamic actin polymer [22] inside the living fission yeast [58] (Fig. 13.4).

As actin is a dynamic polymer that rapidly fluxes between phases of growth and shrinkage, the addition of artificial fluorescent actin-binding molecules has the potential to affect the dynamic properties of the polymer, and therefore actin function within the cell. For example, full expression of GFP-CHD^{Rng2} under the control of the intermediate *nmt41* promoter [59] disrupts cell morphology and reduces cell viability, while even Lifeact can delay cell division. Indeed both actin markers can rescue the growth phenotypes associated with disruption of actin cable integrity [39]. Therefore, expression levels and growth conditions have to be carefully modulated to ensure minimal impact on the cell and its morphology.

If you are looking to follow the dynamics of specific actin structures, there are few options available to the yeast researcher. The best way to visualise actin cables is through the use of one of the actin markers described above. The only realistic other cable-specific ABP option is Tm. Indeed alleles have been described in which the fission yeast Tm, Cdc8, has been FP tagged at either termini (*gfp-cdc8* [24] *cdc8-venus* [60]), and one of the two budding yeast Tms, Tpm2, has been tagged at its carboxyl end (*TPM2-gfp* [61]). While each of these strains are able to at least partially replace the functional copy of the wild-type gene (note—TPM2 is nonessential and the deletion only has subtle effects on actomyosin function [62, 63]), it is hard to imagine how the addition of an FP to either terminus of Tm cannot affect its function, given the nature of the end-to-end contacts between adjacent Tm dimers along the actin polymer [64, 65]. Until biochemical investigation into the consequence of the FP tags on the Tm structure and/or function has been undertaken, these markers should be used with caution.

There is a much larger choice of cortical actin patch markers available, which do not appear to affect actin dynamics. I will not list them as they have been described extensively elsewhere. Cortical actin markers that are used routinely in both yeasts include the myosin I's, Sla2, components of the Arp2/3 complex and cofilin [66–71].

13.2.2.3 Myosins

Budding and fission yeasts strains have been described in which myosins have been tagged with a variety of FPs and are functional according to each assay they have been tested. Imaging these rapidly moving molecules has been extremely challenging in the past, but since the advent of sensitive emCCD with >90 % quantum efficiency, it is possible to visualise even the fastest movements of the myosin V motors as they deliver cargoes throughout the cell (Fig. 13.5). It is even possible to follow the dynamics a myosin I, II and V simultaneously in the same yeast cell (Fig. 13.6). Here I will list considerations for choosing the FP allele for the following myosin dynamics in vivo, followed by a table of published functional FP-labelled myosin allele strains (Table 13.2).

Myosin I The yeast class I myosins all possess long tail domains, each with an acidic Arp2/3 activating domain at their carboxyl termini. Addition of a tag to this end of the protein disrupts its ability to activate Arp2/3-dependent actin nucleation. It is therefore critical that these myosins are labelled with FP at their amino terminus as described in these studies [72, 73, 80].

Myosin II EM studies suggest the long dimerising coiled-coil tail of these proteins is predicted to be unaffected by the addition of tags at the carboxyl terminus [81]. However, interactions with phosphorylation sites towards the end of this tail may be affected by the presence of an GFP [82], but interestingly not an mCherry tag [39]. Tagging the 5′ end of a myosin II gene at its endogenous locus may be the best option for this class of myosin, as described for *S. pombe myp2* [75], as the structure of class II motors indicate addition of an FP tag does not affect functional regions of the protein [83, 84].

Myosin V The majority of fluorescently labelled myosin V alleles encode for motors with an FP tag at their carboxyl terminus, and this does not appear to have any discernable detrimental effect on the cells. These motors can move at speeds of up to 2 μm/s, and it can therefore be challenging to follow their movements in a 3D context. One method to help with this is to generate alleles with multiple FP tags fused to the myosin V gene, but you need to consider the impact this will have on the structure of this dynamic myosin and its ability to

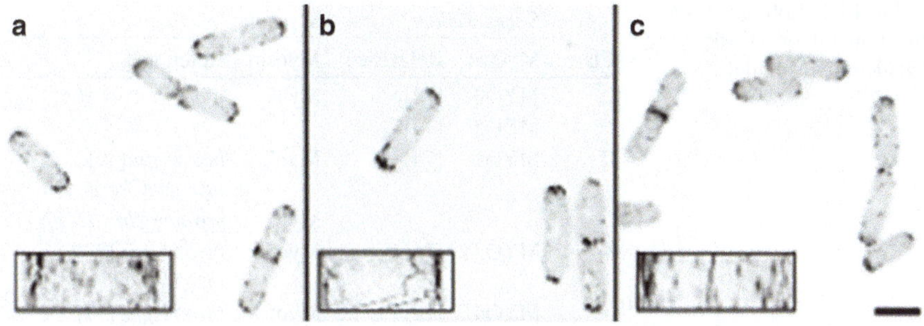

Fig. 13.5 Myosin I, V and calmodulin dynamics. Maximum projections of fission yeast cells expressing (**a**) *cfp-myo1*, (**b**) *myo52-gfp* and (**c**) *cam1-gfp*. *Insets* show 30 time point kymographs of 13 z slice maximum projections from live-cell imaging of each strain (400 ms between frames). Myosin I foci are very dynamic with a lifetime of ~14 s, but have no discernable lateral movement. In contrast myosin V foci move along the length of the cell. The myosin I and V light chain, calmodulin, demonstrates both types of movement. Scale—5 μm

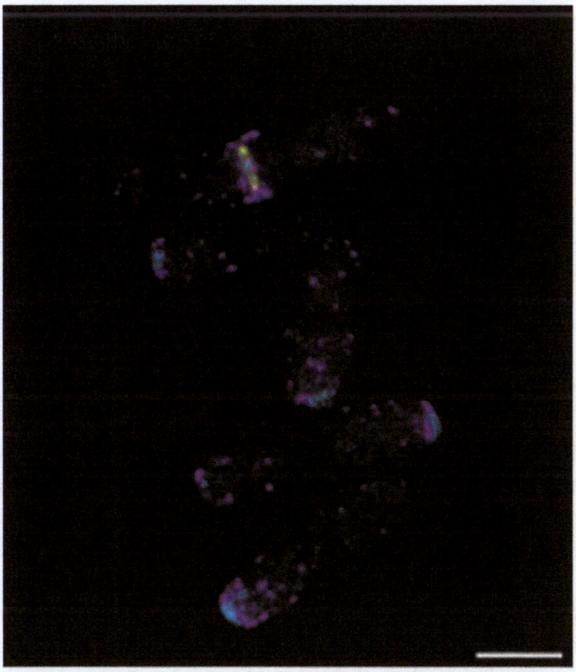

Fig. 13.6 Simultaneous imaging of yeast myosins from three classes. Localisation of myosin I (*cyan*), myosin II (*yellow*) and myosin V (*magenta*) in live yeast cells in which the chromosomal copy of each myosin gene has been fused to cDNA encoding for a different FP. Scale—5 μm

Table 13.2 References of functional fluorescently labelled myosin alleles

	S. cerevisiae		*S. pombe*	
Class	Myosin	Reference	Myosin	Reference
I	MYO3	[72]	Myo1	*cfp/yfp-myo1* [66]
	MYO5	[73]		
II	MYO1	[74]	Myo2	*myo2-gfp* [49] *myo2-mCherry* [39]
			Myp2	*yfp-myp2/myp2-yfp* [75]
V	MYO2	[76]	Myo52	*myo51-gfp* [77] *myo51-mCherry* [78]
	MYO4	[79]	Myo51	*myo52-gfp* [77] *myo52-mCherry* [39]

interact with other proteins (see comments at the beginning of this section) [85, 86]. If your imaging system is not sensitive enough to follow FP-labelled myosin V foci moving throughout the cell or have concerns about adding tags to the myosin, an alternative method you may want to consider is to follow the myosin V cargo, which in some cases can appear brighter (e.g. Budding yeast Sec4 [87]).

Myosin Light Chains The motor activity of each myosin motor is regulated by the association of light chains with IQ motifs within their neck regions. Each myosin class associates with different light chains. Myosin II associates with an essential light chain and a regulatory light chain, while myosin I and V normally associates with calmodulins (Fig. 13.5). However, each light chain does not associate exclusively with a specific myosin. Indeed the calmodulins are predicted to associate within excess of 200 proteins within the yeast cell. Therefore, care should be taken about the conclusions that made when using FP-labelled light chains to follow myosin dynamics.

13.2.2.4 Mounting the Cells and Keeping Them Happy

As in all cell biology experiments, using cells that have been actively growing in a log-phase culture (for a minimum of 48 h) and maintaining them in a non-stressed environment during imaging is critical for studying the function and organisation of active yeast actomyosin cytoskeleton. Otherwise the cells may be starved or have active stress pathways (e.g. TOR signalling/MAP kinase pathways) which can affect cell cycle progression and cell growth. Indeed it is important that the appropriate minimal media is used when culturing the cells (rich yeast extract based media cannot be used as these auto-fluoresce at wavelengths used during live-cell imaging). For example, the nitrogen source used or amino acids added can affect metabolic and signalling pathways. To minimise the effects additional amino acids can have on cells, in some experiments it is advisable to use prototroph strains (i.e. lacks mutations in genes responsible for amino acid synthesis).

It is then critical to minimise stress to the cells during imaging. In addition to minimising heat fluctuations and UV radiation from the imaging system (will be discussed in the next section), the cells need to be mounted onto the coverslip and

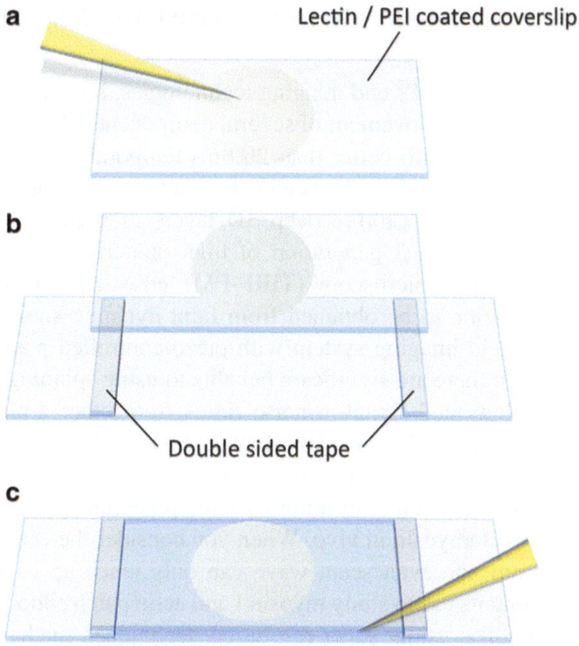

Fig. 13.7 Construction of a simple live-cell imaging chamber for yeast. (**a**) Yeast cells are allowed to adhere to the surface of a lectin/PEI-coated 25 × 50 mm coverslip. Excess cells are washed off with media. (**b**) The coverslip is inverted and stuck down onto a slide with appropriately spaced double-sided sticky tape. (**c**) The space between coverslip and slide is finally filled with preconditioned minimal media before the slide is mounted (coverslip down) onto an inverted microscope

have sufficient nutrients and oxygen available to them for the length of the experiment. The simplest and cheapest way to do this on an inverted microscope is to make a simple flow chamber (Fig. 13.7). Using this method, cells are taken from the liquid culture (without centrifugation) and mounted onto a coverslip that has already been coated with either an appropriate lectin or 0.01 % polyethylenimine (PEI). Once unbound cells have been washed off, the cell-coated coverslip is then fixed with strips of double-sided tape onto a clean slide before filling the chamber with preconditioned media. The slide is then ready to be mounted onto an inverted microscope and can be used for up to 2 h.

For longer experiments, or when media exchange is required (e.g. to either add or remove drugs from the cells during the timelapse experiment), it is advisable to use one of the commercially available chambers. Examples include the Bioptechs FCS2 chamber (Bioptechs, Butler, USA) and the Onix microfluidics system (Cellasics, Hayward, USA). Each system allows media and gas exchange and also the temperature to be carefully controlled for the duration of the experiment, with the latter doing away with the need to "stick" the cells onto the coverslip, as they are held in place using an elastic trapping system.

13.3 Considerations for Yeast-Imaging Systems

With current FPs and imaging technologies, it is possible to simultaneously follow the dynamic movement of several components of the actomyosin cytoskeleton in a 3D context with better than 200 ms temporal resolution. The small size of yeast cells (up to 4 μm diameter) does away with the need for scanning confocal microscopy to capture deep 3D layers as required by larger mammalian cells. Indeed the latest generation of high numerical aperture total internal reflection fluorescence microscopy (TIRF-FM) lenses and CCD cameras allow detailed 3D information to be obtained from faint dynamic signals using a relatively modest wide field imaging system with piezo-controlled positioning of the lens position. Though there are significant benefits to using spinning disc confocal systems (high sensitivity and signal-to-noise ratios from other z-planes but also higher bleach rate), you should consider whether the benefit is worth the additional cost for the experiments you want to undertake.

The thick cell wall minimises the potential applications for TIRF-FM to study yeast actomyosin in vivo. When you consider the cell wall has a thickness of 100–150 nm, the evanescent wave can only reach up to 100 nm into the cytoplasm, limiting its use to study myosin I and actin patch endocytic events at the cell cortex.

Studies using super-resolution techniques such as PALM (photo-activated localisation microscopy) and STORM (stochastic optical reconstruction microscopy) within the yeasts are starting to be published [88–90] and have the potential to generate exquisite images of fixed samples. However, due to the temporal resolution of current systems, it is not possible to follow the highly dynamic actomyosin cytoskeleton.

Here I will go through the major considerations when putting together a microbial imaging system.

Filters Keep your filters as simple as possible for the experiment. There is a wonderful array of high-efficiency magnetron sputter-coated filters available which transmit up to 97 % of the light of the desired wavelength. However, each time additional coatings are added or the emission bandwidth is reduced (necessary for multiwavelength imaging), you are reducing the potential to pick up faint fluorescent signals. Therefore, if you only need to look at one fluorophore, use the simplest filter set with a long-pass emission filter, which will allow you the opportunity to capture every photon that is being emitted from your FP.

Lenses Choose a lens with a high numerical aperture (NA). Current lenses developed for TIRF applications have the highest NAs available (e.g. 1.49 or 1.6 depending on coverslip material). These lenses have a short working distance, which is a benefit for the microbial microscopist. In addition these lenses have the additional benefit in that the higher their magnification, the lower the background signal from other *z*-planes, which has a dramatic positive impact on the signal-to-noise ratio.

Light Sources The options available for wide field systems (in decreasing power) are lasers, halogen bulbs and LEDs. However, LEDs have now become the light source of choice for live-cell imaging. They generate no heat or UV radiation, produce more than enough light for almost all applications, the power output can be controlled precisely and are available in a variety of colours making filter selection easier. In contrast modern halogen systems use 200–300 W bulbs, which radiate significant amounts of heat and UV radiation (this can be minimised using liquid light guides or optical fibres to deliver light to the sample), and light intensity can only be reduced using neutral density filters. While lasers are powerful, controllable, produce light of a precise wavelength and generate minimal heat to the sample, the necessity to have different lasers for each fluorophore or FP being used, together with safety considerations, preclude them from being used in most systems.

Cameras The latest generation of CCD, emCCD and CMOS cameras are extremely sensitive (>90 % quantum efficiency for emCCD) and quick (less than 0.5 ms frame rate for CMOS cameras). Beware—the sensitivity is directly proportional to price. However, by using binning and high gain settings, it is possible to follow the dynamic cytoskeleton at a reasonable frame rate without impacting image size when using the larger chipped CMOS and CCD cameras.

13.4 And Finally a Word on In Vitro Analysis of the Yeast Cytoskeleton

As well as in vivo studies, which have been the focus of this chapter, the yeast actomyosin cytoskeleton allows itself to a variety of in vitro analyses. It is possible to extract and purify actin, myosin and tropomyosin from yeasts [24, 91–96] and baculovirus expression systems [97–99] for in vitro motility assays as well as other biochemical studies using methods described within other chapters of this book. They also allow the purification of the entire functional actomyosin organelles (i.e. the CAR) for in vitro analysis [60].

Thus, yeast provides the perfect opportunity for cross-discipline fluorescence studies to characterise the regulation and function of an entire actomyosin system within a cell.

Acknowledgements Dan Mulvihill is a Royal Society Industry Fellow. Research within his lab is funded by the Royal Society and the BBSRC.

References

1. Adams AE, Pringle JR (1984) Relationship of actin and tubulin distribution to bud growth in wild-type and morphogenetic-mutant Saccharomyces cerevisiae. J Cell Biol 98(3):934–945
2. Marks J, Hyams JS (1985) Localisation of F-actin through the cell-division cycle of *Schizosaccharomyces pombe*. Eur J Cell Biol 39:27–32

3. Anderson JM, Soll DR (1986) Differences in actin localization during bud and hypha formation in the yeast Candida albicans. J Gen Microbiol 132(7):2035–2047
4. Cooper JA (1987) Effects of cytochalasin and phalloidin on actin. J Cell Biol 105(4):1473–1478
5. Mishra M, Huang J, Balasubramanian MK (2014) The yeast actin cytoskeleton. FEMS Microbiol Rev 38(2):213–227
6. Michelot A, Drubin DG (2011) Building distinct actin filament networks in a common cytoplasm. Curr Biol 21(14):R560–R569
7. Kovar DR, Sirotkin V, Lord M (2010) Three's company: the fission yeast actin cytoskeleton. Trends Cell Biol 21(3):177–187
8. Nurse P, Thuriaux P, Nasmyth K (1976) Genetic control of the cell division cycle in the fission yeast Schizosaccharomyces pombe. Mol Gen Genet 146(2):167–178
9. Balasubramanian MK, Helfman DM, Hemmingsen SM (1992) A new tropomyosin essential for cytokinesis in the fission yeast S. pombe. Nature 360(6399):84–87
10. Balasubramanian MK, Hirani BR, Burke JD, Gould KL (1994) The Schizosaccharomyces pombe cdc3+ gene encodes a profilin essential for cytokinesis. J Cell Biol 125(6):1289–1301
11. McCollum D, Balasubramanian MK, Pelcher LE, Hemmingsen SM, Gould KL (1995) Schizosaccharomyces pombe cdc4+ gene encodes a novel EF-hand protein essential for cytokinesis. J Cell Biol 130(3):651–660
12. Chang F, Drubin D, Nurse P (1997) cdc12p, a protein required for cytokinesis in fission yeast, is a component of the cell division ring and interacts with profilin. J Cell Biol 137(1):169–182
13. MacIver FH, Glover DM, Hagan IM (2003) A 'marker switch' approach for targeted mutagenesis of genes in Schizosaccharomyces pombe. Yeast 20(7):587–594
14. Cullen CF, May KM, Hagan IM, Glover DM, Ohkura H (2000) A new genetic method for isolating functionally interacting genes: high plo1(+)-dependent mutants and their suppressors define genes in mitotic and septation pathways in fission yeast. Genetics 155(4):1521–1534
15. Kim DU, Hayles J, Kim D, Wood V, Park HO, Won M, Yoo HS, Duhig T, Nam M, Palmer G, Han S, Jeffery L, Baek ST, Lee H, Shim YS, Lee M, Kim L, Heo KS, Noh EJ, Lee AR, Jang YJ, Chung KS, Choi SJ, Park JY, Park Y, Kim HM, Park SK, Park HJ, Kang EJ, Kim HB, Kang HS, Park HM, Kim K, Song K, Song KB, Nurse P, Hoe KL (2010) Analysis of a genome-wide set of gene deletions in the fission yeast Schizosaccharomyces pombe. Nat Biotechnol 28 (6):617–623
16. Winzeler EA, Shoemaker DD, Astromoff A, Liang H, Anderson K, Andre B, Bangham R, Benito R, Boeke JD, Bussey H, Chu AM, Connelly C, Davis K, Dietrich F, Dow SW, El Bakkoury M, Foury F, Friend SH, Gentalen E, Giaever G, Hegemann JH, Jones T, Laub M, Liao H, Liebundguth N, Lockhart DJ, Lucau-Danila A, Lussier M, M'Rabet N, Menard P, Mittmann M, Pai C, Rebischung C, Revuelta JL, Riles L, Roberts CJ, Ross-MacDonald P, Scherens B, Snyder M, Sookhai-Mahadeo S, Storms RK, Veronneau S, Voet M, Volckaert G, Ward TR, Wysocki R, Yen GS, Yu K, Zimmermann K, Philippsen P, Johnston M, Davis RW (1999) Functional characterization of the S. cerevisiae genome by gene deletion and parallel analysis. Science 285(5429):901–906
17. Ayscough KR, Stryker J, Pokala N, Sanders M, Crews P, Drubin DG (1997) High rates of actin filament turnover in budding yeast and roles for actin in establishment and maintenance of cell polarity revealed using the actin inhibitor latrunculin-A. J Cell Biol 137(2):399–416
18. Petersen J, Nielsen O, Egel R, Hagan IM (1998) F-actin distribution and function during sexual differentiation in Schizosaccharomyces pombe. J Cell Sci 111(Pt 7):867–876
19. Kanbe T, Akashi T, Tanaka K (1993) Effect of cytochalasin A on actin distribution in the fission yeast Schizosaccharomyces pombe studied by fluorescent and electron microscopy. Protoplasma 176(1–2):24–32
20. Riveline D, Nurse P (2009) 'Injecting' yeast. Nat Methods 6(7):513–514
21. May KM, Wheatley SP, Amin V, Hyams JS (1998) The myosin ATPase inhibitor 2,3-butanedione-2-monoxime (BDM) inhibits tip growth and cytokinesis in the fission yeast, Schizosaccharomyces pombe. Cell Motil Cytoskeleton 41(2):117–125

22. Riedl J, Crevenna AH, Kessenbrock K, Yu JH, Neukirchen D, Bista M, Bradke F, Jenne D, Holak TA, Werb Z, Sixt M, Wedlich-Soldner R (2008) Lifeact: a versatile marker to visualize F-actin. Nat Methods 5(7):605–607
23. Hagan IM, Hyams JS (1988) The use of cell division cycle mutants to investigate the control of microtubule distribution in the fission yeast Schizosaccharomyces pombe. J Cell Sci 89 (Pt 3):343–357
24. Skoumpla K, Coulton AT, Lehman W, Geeves MA, Mulvihill DP (2007) Acetylation regulates tropomyosin function in the fission yeast *Schizosaccharomyces pombe*. J Cell Sci 120 (9):1635–1645
25. Naqvi NI, Eng K, Gould KL, Balasubramanian MK (1999) Evidence for F-actin-dependent and -independent mechanisms involved in assembly and stability of the medial actomyosin ring in fission yeast. EMBO J 18(4):854–862
26. Petersen J, Nielsen O, Egel R, Hagan IM (1998) FH3, a domain found in formins, targets the fission yeast formin Fus1 to the projection tip during conjugation. J Cell Biol 141(5):1217–1228
27. McCollum D, Feoktistova A, Morphew M, Balasubramanian M, Gould KL (1996) The Schizosaccharomyces pombe actin-related protein, Arp3, is a component of the cortical actin cytoskeleton and interacts with profilin. EMBO J 15(23):6438–6446
28. Grallert A, Hagan IM (2002) Schizosaccharomyces pombe NIMA-related kinase, Fin1, regulates spindle formation and an affinity of Polo for the SPB. EMBO J 21(12):3096–3107
29. Motegi F, Nakano K, Mabuchi I (2000) Molecular mechanism of myosin-II assembly at the division site in Schizosaccharomyces pombe. J Cell Sci 113(Pt 10):1813–1825
30. Pelham RJ, Chang F (2002) Actin dynamics in the contractile ring during cytokinesis in fission yeast. Nature 419(6902):82–86
31. Arai R, Mabuchi I (2002) F-actin ring formation and the role of F-actin cables in the fission yeast Schizosaccharomyces pombe. J Cell Sci 115(Pt 5):887–898
32. Lin JJ (1981) Monoclonal antibodies against myofibrillar components of rat skeletal muscle decorate the intermediate filaments of cultured cells. Proc Natl Acad Sci USA 78(4):2335–2339
33. Winter D, Lechler T, Li R (1999) Activation of the yeast Arp2/3 complex by Bee1p, a WASP-family protein. Curr Biol 9(9):501–504
34. Pruyne DW, Schott DH, Bretscher A (1998) Tropomyosin-containing actin cables direct the Myo2p-dependent polarized delivery of secretory vesicles in budding yeast. J Cell Biol 143 (7):1931–1945
35. East DA, Mulvihill DP, Todd M, Bruce IJ (2011) QD-antibody conjugates via carbodiimide-mediated coupling: a detailed study of the variables involved and a possible new mechanism for the coupling reaction under basic aqueous conditions. Langmuir 27(22):13888–13896
36. Attanapola SL (2009) Analysis of the regulation and function of the fission yeast class I myosin, Myo1. University of Kent, Canterbury
37. Anderson BL, Boldogh I, Evangelista M, Boone C, Greene LA, Pon LA (1998) The Src homology domain 3 (SH3) of a yeast type I myosin, Myo5p, binds to verprolin and is required for targeting to sites of actin polarization. J Cell Biol 141(6):1357–1370
38. Watts FZ, Miller DM, Orr E (1985) Identification of myosin heavy chain in Saccharomyces cerevisiae. Nature 316(6023):83–85
39. Coulton AT, East DA, Galinska-Rakoczy A, Lehman W, Mulvihill DP (2010) The recruitment of acetylated and unacetylated tropomyosin to distinct actin polymers permits the discrete regulation of specific myosins in fission yeast. J Cell Sci 123:3235–3243
40. Bezanilla M, Wilson JM, Pollard TD (2000) Fission yeast myosin-II isoforms assemble into contractile rings at distinct times during mitosis. Curr Biol 10(7):397–400
41. Schott D, Ho J, Pruyne D, Bretscher A (1999) The COOH-terminal domain of Myo2p, a yeast myosin V, has a direct role in secretory vesicle targeting. J Cell Biol 147(4):791–808
42. Catlett NL, Duex JE, Tang F, Weisman LS (2000) Two distinct regions in a yeast myosin-V tail domain are required for the movement of different cargoes. J Cell Biol 150(3):513–526

43. Motegi F, Arai R, Mabuchi I (2001) Identification of two type V myosins in fission yeast, one of which functions in polarized cell growth and moves rapidly in the cell. Mol Biol Cell 12 (5):1367–1380
44. Munchow S, Sauter C, Jansen RP (1999) Association of the class V myosin Myo4p with a localised messenger RNA in budding yeast depends on She proteins. J Cell Sci 112 (Pt 10):1511–1518
45. Mulvihill DP, Petersen J, Ohkura H, Glover DM, Hagan IM (1999) Plo1 kinase recruitment to the spindle pole body and its role in cell division in Schizosaccharomyces pombe. Mol Biol Cell 10(8):2771–2785
46. Bahler J, Wu JQ, Longtine MS, Shah NG, McKenzie A 3rd, Steever AB, Wach A, Philippsen P, Pringle JR (1998) Heterologous modules for efficient and versatile PCR-based gene targeting in Schizosaccharomyces pombe. Yeast 14(10):943–951
47. Wach A, Brachat A, Alberti-Segui C, Rebischung C, Philippsen P (1997) Heterologous HIS3 marker and GFP reporter modules for PCR-targeting in Saccharomyces cerevisiae. Yeast 13 (11):1065–1075
48. Janke C, Magiera MM, Rathfelder N, Taxis C, Reber S, Maekawa H, Moreno-Borchart A, Doenges G, Schwob E, Schiebel E, Knop M (2004) A versatile toolbox for PCR-based tagging of yeast genes: new fluorescent proteins, more markers and promoter substitution cassettes. Yeast 21(11):947–962
49. Mulvihill DP, Barretto C, Hyams JS (2001) Localization of fission yeast type II myosin, Myo2, to the cytokinetic actin ring is regulated by phosphorylation of a C-terminal coiled-coil domain and requires a functional septation initiation network. Mol Biol Cell 12(12):4044–4053
50. Grallert A, Beuter C, Craven RA, Bagley S, Wilks D, Fleig U, Hagan IM (2006) S. pombe CLASP needs dynein, not EB1 or CLIP170, to induce microtubule instability and slows polymerization rates at cell tips in a dynein-dependent manner. Genes Dev 20(17):2421–2436
51. Snaith HA, Anders A, Samejima I, Sawin KE (2010) New and old reagents for fluorescent protein tagging of microtubules in fission yeast; experimental and critical evaluation. Methods Cell Biol 97:147–172
52. Chapman S, Faulkner C, Kaiserli E, Garcia-Mata C, Savenkov EI, Roberts AG, Oparka KJ, Christie JM (2008) The photoreversible fluorescent protein iLOV outperforms GFP as a reporter of plant virus infection. Proc Natl Acad Sci USA 105(50):20038–20043
53. Nguyen AW, Daugherty PS (2005) Evolutionary optimization of fluorescent proteins for intracellular FRET. Nat Biotechnol 23(3):355–360
54. Rizzo MA, Springer GH, Granada B, Piston DW (2004) An improved cyan fluorescent protein variant useful for FRET. Nat Biotechnol 22(4):445–449
55. Wu JQ, Pollard TD (2005) Counting cytokinesis proteins globally and locally in fission yeast. Science 310(5746):310–314
56. Huckaba TM, Gay AC, Pantalena LF, Yang HC, Pon LA (2004) Live cell imaging of the assembly, disassembly, and actin cable-dependent movement of endosomes and actin patches in the budding yeast, Saccharomyces cerevisiae. J Cell Biol 167(3):519–530
57. Karagiannis J, Bimbo A, Rajagopalan S, Liu J, Balasubramanian MK (2005) The nuclear kinase Lsk1p positively regulates the septation initiation network and promotes the successful completion of cytokinesis in response to perturbation of the actomyosin ring in Schizosaccharomyces pombe. Mol Biol Cell 16(1):358–371
58. Huang J, Huang Y, Yu H, Subramanian D, Padmanabhan A, Thadani R, Tao Y, Tang X, Wedlich-Soldner R, Balasubramanian MK (2012) Nonmedially assembled F-actin cables incorporate into the actomyosin ring in fission yeast. J Cell Biol 199(5):831–847
59. Maundrell K (1993) Thiamine-repressible expression vectors pREP and pRIP for fission yeast. Gene 123(1):127–130
60. Mishra M, Kashiwazaki J, Takagi T, Srinivasan R, Huang Y, Balasubramanian MK, Mabuchi I (2013) In vitro contraction of cytokinetic ring depends on myosin II but not on actin dynamics. Nat Cell Biol 15(7):853–859

61. Yoshida S, Kono K, Lowery DM, Bartolini S, Yaffe MB, Ohya Y, Pellman D (2006) Polo-like kinase Cdc5 controls the local activation of Rho1 to promote cytokinesis. Science 313 (5783):108–111
62. Drees B, Brown C, Barrell BG, Bretscher A (1995) Tropomyosin is essential in yeast, yet the TPM1 and TPM2 products perform distinct functions. J Cell Biol 128(3):383–392
63. Huckaba TM, Lipkin T, Pon LA (2006) Roles of type II myosin and a tropomyosin isoform in retrograde actin flow in budding yeast. J Cell Biol 175(6):957–969
64. Frye J, Klenchin VA, Rayment I (2010) Structure of the tropomyosin overlap complex from chicken smooth muscle: insight into the diversity of N-terminal recognition. Biochemistry 49 (23):4908–4920
65. Greenfield NJ, Huang YJ, Swapna GV, Bhattacharya A, Rapp B, Singh A, Montelione GT, Hitchcock-DeGregori SE (2006) Solution NMR structure of the junction between tropomyosin molecules: implications for actin binding and regulation. J Mol Biol 364(1):80–96
66. Sirotkin V, Beltzner CC, Marchand JB, Pollard TD (2005) Interactions of WASp, myosin-I, and verprolin with Arp2/3 complex during actin patch assembly in fission yeast. J Cell Biol 170(4):637–648
67. Nakano K, Mabuchi I (2006) Actin-depolymerizing protein Adf1 is required for formation and maintenance of the contractile ring during cytokinesis in fission yeast. Mol Biol Cell 17 (4):1933–1945
68. Chen Q, Pollard TD (2013) Actin filament severing by cofilin dismantles actin patches and produces mother filaments for new patches. Curr Biol 23(13):1154–1162
69. Galletta BJ, Chuang DY, Cooper JA (2008) Distinct roles for Arp2/3 regulators in actin assembly and endocytosis. PLoS Biol 6(1):e1
70. Kaksonen M, Toret CP, Drubin DG (2005) A modular design for the clathrin- and actin-mediated endocytosis machinery. Cell 123(2):305–320
71. Alvarez-Tabares I, Grallert A, Ortiz JM, Hagan IM (2007) Schizosaccharomyces pombe protein phosphatase 1 in mitosis, endocytosis and a partnership with Wsh3/Tea4 to control polarised growth. J Cell Sci 120(Pt 20):3589–3601
72. Evangelista M, Klebl BM, Tong AH, Webb BA, Leeuw T, Leberer E, Whiteway M, Thomas DY, Boone C (2000) A role for myosin-I in actin assembly through interactions with Vrp1p, Bee1p, and the Arp2/3 complex. J Cell Biol 148(2):353–362
73. Jonsdottir GA, Li R (2004) Dynamics of yeast Myosin I: evidence for a possible role in scission of endocytic vesicles. Curr Biol 14(17):1604–1609
74. Bi E, Maddox P, Lew DJ, Salmon ED, McMillan JN, Yeh E, Pringle JR (1998) Involvement of an actomyosin contractile ring in Saccharomyces cerevisiae cytokinesis. J Cell Biol 142 (5):1301–1312
75. Wu JQ, Kuhn JR, Kovar DR, Pollard TD (2003) Spatial and temporal pathway for assembly and constriction of the contractile ring in fission yeast cytokinesis. Dev Cell 5(5):723–734
76. Karpova TS, Reck-Peterson SL, Elkind NB, Mooseker MS, Novick PJ, Cooper JA (2000) Role of actin and Myo2p in polarized secretion and growth of Saccharomyces cerevisiae. Mol Biol Cell 11(5):1727–1737
77. Win TZ, Gachet Y, Mulvihill DP, May KM, Hyams JS (2001) Two type V myosins with non-overlapping functions in the fission yeast Schizosaccharomyces pombe: Myo52 is concerned with growth polarity and cytokinesis, Myo51 is a component of the cytokinetic actin ring. J Cell Sci 114(Pt 1):69–79
78. Doyle A, Martin-Garcia R, Coulton AT, Bagley S, Mulvihill DP (2009) The fission yeast type V myosin Myo51, is a meiotic spindle pole body component, with discrete roles during cell fusion and spore formation. J Cell Sci 122(24):4330–4340
79. Kruse C, Jaedicke A, Beaudouin J, Bohl F, Ferring D, Guttler T, Ellenberg J, Jansen RP (2002) Ribonucleoprotein-dependent localization of the yeast class V myosin Myo4p. J Cell Biol 159 (6):971–982
80. Lee WL, Bezanilla M, Pollard TD (2000) Fission yeast myosin-I, Myo1p, stimulates actin assembly by Arp2/3 complex and shares functions with WASp. J Cell Biol 151(4):789–800

81. Bezanilla M, Pollard TD (2000) Myosin-II tails confer unique functions in Schizosaccharomyces pombe: characterization of a novel myosin-II tail. Mol Biol Cell 11(1):79–91
82. Mulvihill DP, Win TZ, Pack TP, Hyams JS (2000) Cytokinesis in fission yeast: a myosin pas de deux. Microsc Res Tech 49(2):152–160
83. Kliche W, Fujita-Becker S, Kollmar M, Manstein DJ, Kull FJ (2001) Structure of a genetically engineered molecular motor. EMBO J 20(1–2):40–46
84. Rayment I, Rypniewski WR, Schmidt-Base K, Smith R, Tomchick DR, Benning MM, Winkelmann DA, Wesenberg G, Holden HM (1993) Three-dimensional structure of myosin subfragment-1: a molecular motor. Science 261(5117):50–58
85. Bendezu FO, Martin SG (2011) Actin cables and the exocyst form two independent morphogenesis pathways in the fission yeast. Mol Biol Cell 22(1):44–53
86. Lo Presti L, Chang F, Martin SG (2012) Myosin Vs organize actin cables in fission yeast. Mol Biol Cell 23(23):4579–4591
87. Schott DH, Collins RN, Bretscher A (2002) Secretory vesicle transport velocity in living cells depends on the myosin-V lever arm length. J Cell Biol 156(1):35–39
88. Lando D, Endesfelder U, Berger H, Subramanian L, Dunne PD, McColl J, Klenerman D, Carr AM, Sauer M, Allshire RC, Heilemann M, Laue ED (2012) Quantitative single-molecule microscopy reveals that CENP-A(Cnp1) deposition occurs during G2 in fission yeast. Open Biol 2(7):120078
89. Dodgson J, Chessel A, Yamamoto M, Vaggi F, Cox S, Rosten E, Albrecht D, Geymonat M, Csikasz-Nagy A, Sato M, Carazo-Salas RE (2013) Spatial segregation of polarity factors into distinct cortical clusters is required for cell polarity control. Nat Commun 4:1834
90. Saner N, Karschau J, Natsume T, Gierlinski M, Retkute R, Hawkins M, Nieduszynski CA, Blow JJ, de Moura AP, Tanaka TU (2013) Stochastic association of neighboring replicons creates replication factories in budding yeast. J Cell Biol 202(7):1001–1012
91. Lord M, Pollard TD (2004) UCS protein Rng3p activates actin filament gliding by fission yeast myosin-II. J Cell Biol 167(2):315–325
92. Reck-Peterson SL, Tyska MJ, Novick PJ, Mooseker MS (2001) The yeast class V myosins, Myo2p and Myo4p, are nonprocessive actin-based motors. J Cell Biol 153(5):1121–1126
93. Takaine M, Mabuchi I (2007) Properties of actin from the fission yeast Schizosaccharomyces pombe and interaction with fission yeast profilin. J Biol Chem 282(30):21683–21694
94. Ti SC, Pollard TD (2011) Purification of actin from fission yeast Schizosaccharomyces pombe and characterization of functional differences from muscle actin. J Biol Chem 286(7):5784–5792
95. Liu HP, Bretscher A (1989) Purification of tropomyosin from Saccharomyces cerevisiae and identification of related proteins in Schizosaccharomyces and Physarum. Proc Natl Acad Sci USA 86(1):90–93
96. Chen X, Rubenstein PA (1995) A mutation in an ATP-binding loop of Saccharomyces cerevisiae actin (S14A) causes a temperature-sensitive phenotype in vivo and in vitro. J Biol Chem 270(19):11406–11414
97. Hodges AR, Bookwalter CS, Krementsova EB, Trybus KM (2009) A nonprocessive class V myosin drives cargo processively when a kinesin- related protein is a passenger. Curr Biol 19 (24):2121–2125
98. Hodges AR, Krementsova EB, Bookwalter CS, Fagnant PM, Sladewski TE, Trybus KM (2012) Tropomyosin is essential for processive movement of a class V myosin from budding yeast. Curr Biol 22(15):1410–1416
99. Clayton JE, Pollard LW, Sckolnick M, Bookwalter CS, Hodges AR, Trybus KM, Lord M (2014) Fission yeast tropomyosin specifies directed transport of myosin-V along actin cables. Mol Biol Cell 25(1):66–75

MIX
Papier aus verantwortungsvollen Quellen
Paper from responsible sources
FSC® C105338

If you have any concerns about our products,
you can contact us on
ProductSafety@springernature.com

In case Publisher is established outside the EU,
the EU authorized representative is:
Springer Nature Customer Service Center GmbH
Europaplatz 3, 69115 Heidelberg, Germany

Printed by Libri Plureos GmbH
in Hamburg, Germany